Human Biology

Evolution, Genetics, Behavior, & Impact on the Environment

Klaus Kalthoff
University of Texas - Austin

Kendall Hunt
publishing company

Cover images:
Center image of 3 Skulls © Kevin O'Farrell CONCEPTS.
Background image (c) 2010, Shutterstock, Inc.

Kendall Hunt
publishing company

www.kendallhunt.com
Send all inquiries to:
4050 Westmark Drive
Dubuque, IA 52004-1840

Copyright © 2010 by Klaus Kalthoff

ISBN 978-0-7575-7623-2

Kendall Hunt Publishing Company has the exclusive rights to reproduce this work,
to prepare derivative works from this work, to publicly distribute this work,
to publicly perform this work and to publicly display this work.

All rights reserved. No part of this publication may be reproduced,
stored in a retrieval system, or transmitted, in any form or by any
means, electronic, mechanical, photocopying, recording, or otherwise,
without the prior written permission of the copyright owner.

Printed in the United States of America
10 9 8 7 6 5 4 3 2

CONTENTS

ACKNOWLEDGEMENTS V

Part I: About this Book
CHAPTER 1: Why Human Biology?... 3

Part II: Human Evolution
CHAPTER 2: Who Is Who Among the Primates?...................... 7
CHAPTER 3: Humans Evolved in Response
to a Challenge 21
CHAPTER 4: Milestones of Hominin Evolution 35
CHAPTER 5: Habitual Bipedalism and Hand
Precision Grip 51
CHAPTER 6: Toolmaking of Monkeys, Apes,
and Hominins 63
CHAPTER 7: Language and Brain Evolution.......................... 77
CHAPTER 8: Germs, Colonialism, and
Darwinian Medicine... 91

Part III: Human Genetics and Genomics
CHAPTER 9: The Human Genome Project............................ 111
CHAPTER 10: Stem Cells and Cloning................................... 127
CHAPTER 11: Gene Therapy 145
CHAPTER 12: Aging and Senescence 163
CHAPTER 13: Nature and Nurture... 177
CHAPTER 14: Estimating the Heritability
of Continuously Variable Traits... 191
CHAPTER 15: Human Behavioral Genetics 203

Part IV: Biological Aspects of Human Behavior
CHAPTER 16: Development of Reproductive Organs........ 219
CHAPTER 17: Hormonal Control of Brain
Development 235
CHAPTER 18: Introduction to Sociobiology 249
CHAPTER 19: Sociobiology of Human Behavior................. 261
CHAPTER 20: Sex and Gender 275
CHAPTER 21: Aggression, Cooperation,
and Kindness 287

Part V: Human Population Growth and Associated Problems

CHAPTER 22: Human Population Growth 301
CHAPTER 23: Earth's Carrying Capacity 313
CHAPTER 24: Environmental Degradation 327
CHAPTER 25: Endocrine-Disrupting Chemicals 341
CHAPTER 26: Energy Production and Conservation 355
CHAPTER 27: Species Extinction and Synopsis 371

ACKNOWLEDGEMENTS

I want to thank Kendall Hunt Publishing Company for lending me their professional skills in producing this book. My project coordinator (Michelle Bahr), permissions editor (Caroline Kieler), and acquisitions editor (William England) were a pleasure to work with. Gwen Gage of the School of Biological Sciences at UT Austin cheerfully prepared the many drawings that I have used for teaching and for illustrating this book. I am grateful to my colleagues and students, who helped me to clarify my thoughts about the wide range of topics that I cover. Last but not least, I wish to acknowledge the support of my family and friends, who read draft chapters and encouraged me along the way.

Klaus Kalthoff

PART I

ABOUT THIS BOOK

Humans have always wondered about their place in nature. We are deeply embedded in it, and yet we feel we are different. This puzzle began to occupy my mind me when I was finishing high school, and originally I thought I would write a textbook on human biology for high school students. I had the privilege of attending a traditional German gymnasium, where several of my teachers had Ph.D. degrees. They thought of themselves as scholars first and then as educators. I wanted to be one of them, and so I started studying mathematics, physics, and biology, because those had been my favorite subjects. My degree plan requirements in philosophy and education I satisfied as much as possible with courses in cultural anthropology. I loved to read about the myths and customs of aboriginal peoples, which were strange and yet so true.

As part of my scientific training to become a German high school teacher, I had to do the equivalent of a Master's thesis. I opted for an experimental thesis under the mentorship of Klaus Sander, Professor of Zoology at the University of Freiburg. He gave me a project that led me to characterize regulatory RNA molecules, which females of certain insect species deposit in the anterior pole region of their eggs. They regulate gene activity at the front end of the embryo so as the make a head.

As my thesis work unfolded, Klaus Sander persuaded me to become a university professor. A group of congenial colleagues led me to accept a job offer from the University of Texas at Austin, where I am now a professor in the Section of Molecular Cell and Developmental biology. Here I have continued my research on anterior determinants, taught developmental biology courses, wrote a textbook on the subject, and served as department chairman. While doing the things that university professors do, the idea of writing a human biology text kept stirring in me. Indeed, it became ever more attractive as new and fundamental advances were made, such as the discovery of key hominin fossils, the human genome project, the discipline of sociobiology, and the detection of global warming.

So I began to develop a course in human evolution, genomics, behavior, and impact on the environment, which is now listed as BIO 346 Human Biology ay UT Austin. It is aimed at upper division undergraduates. Preparing for this course was a stretch for me because many of the topics were outside of my expertise. Inviting external speakers for an associated seminar course helped me along. Having taught BIO 346 for more than ten years, I felt it was time to write up the textbook that had always been on my mind.

The publisher and I have tried to keep this text modest in size and price. Readers who are interested in more and colored illustrations, as well as in references to the literature, are welcome to log on to my website for BIO 346, http://www.bio.utexas.edu/courses/kalthoff/bio346/. It contains colored versions of the images used in this text, as well supplementary illustrations. To access these illustrations from the home page, push the button "PowerPoint Presentations for Bio346 in PDF Format" and select the appropriate chapter. The callouts for supplementary illustrations in the printed text have the format "Figure S", followed by the chapter number and a running letter. For example, the callout for the second supplementary image in Chapter 3 is "(Figure S3.b)". The figure legends in the PowerPoint presentations on the website begin with the same callouts. The website also contains references to original literature, exam questions, and other materials. I will try to keep this site up to date, and I welcome any suggestions for improvement that users may have.

CHAPTER ONE

WHY HUMAN BIOLOGY?

The biology of a species is about its salient features and how these are used to make a living. The biological examination of humans differs from the study of other organisms in several respects. On the one hand, it would be unethical to do potentially harmful experiments with humans. On the other hand, there is a treasure trove of medical literature on humans. Also, the human genome project is completed and is now driving many lines of investigation on humans in comparison to other organisms. However, there is concern that applications of genomics and cloning are racing ahead, while the political process of creating an economical and legal framework for these applications is lagging behind. In addition, there is a sense of urgency among ecologists and also in the general public about human population growth and environmental degradation. Biologists have a special responsibility in fostering a broad discussion of these issues.

After this short introduction (Part I), we will explore four related topics in human biology: evolution (part II), genetics and genomics (part III), brain and behavior (part IV), and impact on the environment (part V).

Part II: We humans are unique animals. We have several features that set us apart from all other primates, including habitual bipedalism, hand precision grip, extensive reliance on toolmaking, language, a very large brain, and an exceedingly complex behavior that is shaped by both nature and nurture. Yet our genetic makeup is very similar to that of chimpanzees. How can we reconcile this apparent contradiction? We have to assume that small genetic changes were positively selected during the evolution of our ancestors. Indeed, we are beginning to identify some of the genetic alleles that have made the human species so unique, and we are exploring the environmental conditions that seem to have caused a positive selection for these alleles.

Part III: Cloning and molecular genetics are revolutionizing agriculture, medicine, and law. The use of human genomic data, gene therapy, reproductive and therapeutic cloning, and genetically modified organisms will shape our lives—more so than building space stations. Whether modern biology will be used for better or worse will depend on the legal and administrative frameworks in which the new techniques are used. For the political process of crafting these frameworks, biologists do not have all the answers but they can certainly help to raise the level of discourse.

Part IV: The biological underpinnings of human behavior raise issues that are at the same time biological, political, and personal. Are the brains of males and females different? Are behavioral traits like intelligence, sexual orientation, and aggressiveness hereditary? Exploring these issues should help greatly in developing realistic views of ourselves, and in shaping laws and institutions accordingly. To some people, even discussing genetic and hormonal effects on human behavior smacks of "biological determinism." However, it needs to be kept in mind that statistical data do not negate individual freedom of choice, and that a behavior's adaptive value during the Stone Age does not dignify the same behavior today.

Part V: The number of humans living on Earth has increased enormously, and humans are exploiting and modifying many ecosystems to satisfy their needs and wants. However, the current growth of the human population, and many of our ways of exploiting nature are not sustainable. Of critical importance is how we can improve the standard of living in the underdeveloped parts of the world while containing global warming, which is caused by burning fossil fuels. Solving these problems will be the greatest challenge of this century.

The topics discussed in Parts II through V are connected. The same traits that have sustained us through millions of years of evolution are now bringing us to the brink of disaster. Our superior skills as toolmakers, which originally compensated for our lack of natural weapons, have led to arsenals of weapons that can destroy all of us. The same technical skills have promoted, in the developed countries, an energy-burning lifestyle that wreaks havoc on our environment. This lifestyle is stoked by traditional behaviors through which males acquire status and mating opportunities. An innate aversion to discrimination, which we share with other social apes and monkeys, is now leading developing countries to emulate the environmentally destructive lifestyles of developed countries. Biologists should be well prepared to understand these connections and to promote their appreciation by the general public.

To do my part in this discussion, I have developed a course on the above topics for biology undergraduates. The course, designated BIO 346 at The University of Texas, Austin, has met with great interest. Indeed, it has become the gateway course for an entire degree plan, a B.S. degree in Human Biology. It combines the rigor of a traditional B.S. degree with options for students to take courses in anthropology and sociology, which are focused on the human and reflect on the social implications of modern biology.

PART II

HUMAN EVOLUTION

Three scientific discoveries have caused major upheavals in European culture. The first occurred in the sixteenth century when Nicolaus Copernicus replaced the geocentric model of the universe with a heliocentric model, removing humans from the center of the world to a small planet orbiting in a vast space. Charles Darwin caused the second upheaval with his two books on evolution, the first one on *The Origin of Species by Means of Natural Selection,* published in 1859, and twelve years later its sequel on *The Descent of Man and Selection in Relation to Sex.* The third upheaval came half a century later when Sigmund Freud found that our rational mind is but a thin layer of ice covering deep waters of the unconscious into which even educated people can fall and embarrass themselves, or drown. Each of the three great discoverers—Copernicus, Darwin, and Freud—met with intense criticism from his contemporaries. People have always liked to see themselves center stage, as the pinnacle of creation, and being the masters of their fate.

Most of us have gotten over the Copernican shock. Astronauts have become our heroes, and space exploration is a major item in both popular culture and national budgets. Likewise, the existence of an unconscious mind has come to be widely accepted. However, the majority of the U.S. population still does not believe that humans and apes share common ancestors. These people ignore the work of scientists called paleoanthropologists (Greek: *palaios,* ancient; *anthropos,* the human), who have made great progress analyzing fossil bones, stone tools, and other clues to prehistoric human life. Recently, they have joined forces with biologists who study the behavior and ecology of animals, for instance, how food availability shapes the size and structure of animal groups. Other biologists are sleuthing in laboratories, trying to reconstruct our evolutionary past from subtle differences in the DNA sequences that modern humans of various ancestries are carrying in the cells of their bodies.

From the work of different types of biologists over the past decades, a picture of human ancestors is emerging that can actually fill us with pride, or at least with respect. This was not so in Darwin's time. In 1856, the first human fossil was found in the Neanderthal, meaning the valley of the Neander, a small river near Düsseldorf in Germany. The Neanderthals were stocky people with short, muscular limbs, well adapted to living in Ice Age Europe. With their prominent brow ridges, low forehead, and large nose, Neanderthals became the caricatures of brutish, dim-witted cavemen—not the kind of ancestor that cultured people wanted to be associated with. Since then, we have learned that Neanderthals were the first people to care for their sick and aged and to bury their dead (thus, leaving many complete skeletons). It also turned out that Neanderthals are not the ancestors of modern humans but a line of distant cousins who became extinct.

Digging deeper into our prehistoric past, we have found that all our ancestors have been astounding people. They braved harsh climates, both hot and cold. With sticks and stones, they confronted saber-toothed tigers and giant hyenas. Horrible circumstances at times killed all but a few thousand of them. Eventually, they colonized the entire planet, in the process hunting many large animal species to extinction but also leaving breathtaking cave paintings and sculptures.

Part II of this text, encompassing Chapters 2 through 8, will be essentially about human evolution. Chapter 2 will compare us currently living humans to the great apes: chimpanzee, bonobo, gorilla, and orangutan. The comparisons will reveal which traits we share with these species and which set us apart as a truly unique species. Chapter 3 will explore how the first human ancestors, or hominins, answered the challenge of a cooling climate by forging a new way of life: Tapping into a new source of food while avoiding the associated dangers, they quickly evolved the suite of traits that makes us human. Chapter 4 will sketch out the time and places where this drama unfolded. Chapters 5 through 7 will feature some of its key events, including the evolution of bipedalism, hand precision grip, toolmaking, spoken language, and our exceptionally large and complex brain. Finally, Chapter 8 will explore the coevolution of humans and their parasites as an "arms race" of measures, countermeasures, and counter-countermeasures. It will introduce "Darwinian medicine" as an approach that considers some symptoms of disease as our natural defenses, and will point out the dangers of pathogens that have become resistant to medication.

CHAPTER TWO

WHO IS WHO AMONG THE PRIMATES?

We humans have always been intrigued by apes and monkeys because of their uncanny similarity to us. What do we share with our closest nonhuman relatives, the orangutan, gorilla, chimpanzee, and bonobo, and what sets us apart? Even in this close company, humans stand out by anatomical traits including bipedalism, opposable thumbs, and a much larger brain. Humans also have characteristic traits of behavior, including sophisticated cultures with extensive toolmaking, spoken languages, body decorations, and art. The state of helplessness in which human babies are born, and the long period of juvenile dependency, make it adaptive for men and women to raise their children together. An entire set of reproductive traits that distinguishes humans from other primates can be interpreted as stabilizing a sexual bond between committed partners.

A. How Humans View Their Place in Nature

Comparative anatomy of humans and other mammals took hold in the Greek and Roman civilizations. When a Greek doctor named Galen was appointed chief physician to the gladiators by the city of Pergamum in AD 158, he began an extensive study of human muscles. He also dissected apes and pigs for detailed information on various organs of the human body.

There is no doubt that we humans are **mammals**. Our muscles and bones, our inner organs, and the human female's way of bearing and nursing children are mammalian. Mammals began to evolve during the Mesozoic Era, which began about 250 million years ago (mya). This was the age of the dinosaurs, and the first mammals escaped their fearsome predators by keeping out of their way. Then the dinosaurs became extinct around 65 mya, when a giant meteor hit Earth on the Yucatan peninsula. The impact apparently kicked up firestorms that darkened the skies and interrupted the food web sustaining the dinosaurs, causing their demise. However, some small, shrewlike mammals survived, and they became the new rulers of the planet. Within a few million years, various mammalian classes evolved, including hoofed herbivores, their carnivorous predators, whales, bats, anteaters, and—last but not least—the primates (Figure S2.a).

It is also obvious that humans belong to the order of the **primates**; this name is derived from the Latin word for "first" or "foremost," expressing our view of being at the top of the animal kingdom. Humans share all major primate characteristics (Figure 2.1), including:

- flexible hands and feet with free digits
- flat fingernails and toenails
- teeth with rounded cusps that can chew plant food as well as meat
- parallel eyes for stereoscopic vision
- large brains and an exceptional learning ability

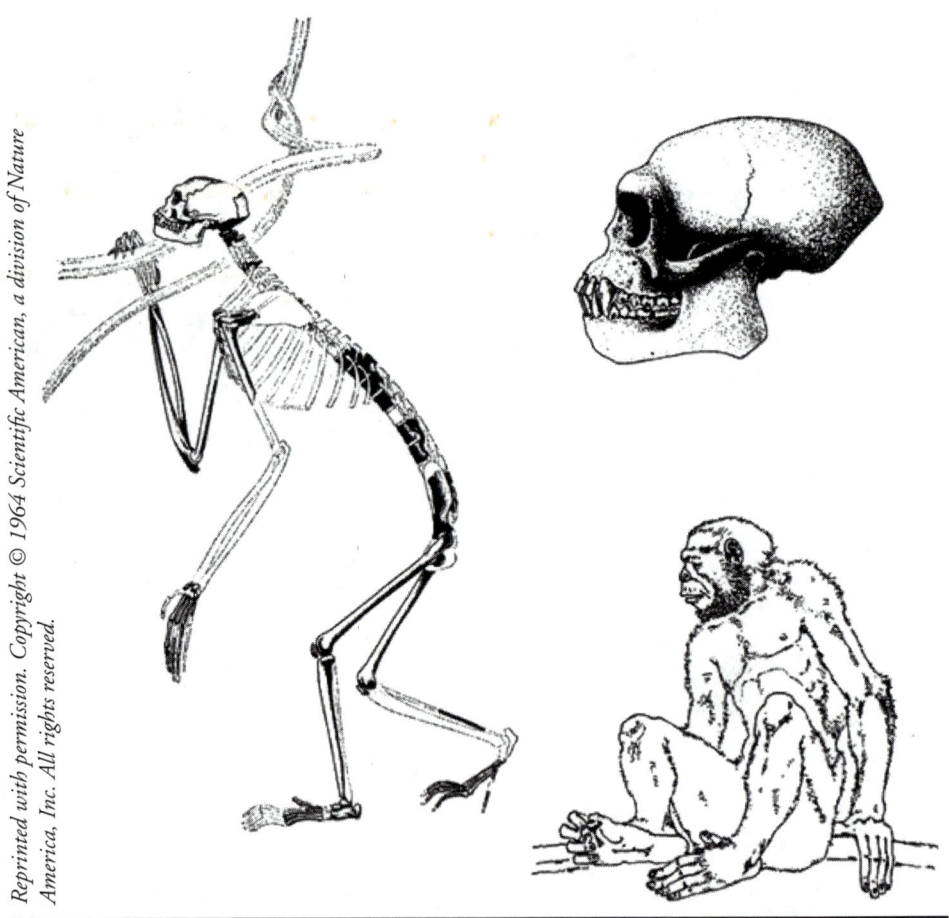

FIGURE 2.1:
Primate characteristics shown in *Pliopithecus*, a fossil primate found at Neudorf (Slovakia) and dated about 20 million years ago. The preserved bone fragments (black) reveal basic primate features.

Many primate traits seem to have evolved as adaptations to the primates' original habitat: the canopy of rain forests. This is a complex three-dimensional environment with uncertain footholds, requiring depth perception and the ability to grasp branches for swift locomotion. Here primates lived on a mixed diet of fruits, leaves, insects, and maybe an occasional small mammal. The fossil record indicates that animals with this lifestyle appeared some 50 mya.

B. The Taxonomic Tree of the Primates

The currently living species of primates are only a small sampling of many more forms that once populated the tropical forests but are now extinct. Today we count about 290 primate species, most of them endangered. Figure 2.2 shows a simplified **taxonomic system** of the primates, which is based on "classical" criteria (morphology, behavior, geographical distribution) as well as DNA sequencing (see Chapter 3). A taxonomic group called the *Prosimii* have preserved many traits of the original mammals, including a small body size and a relatively long snout with a nose pad. They include the lemurs, which have survived only on Madagaskar.

The more evolved primates, called *Anthropoidea,* have shorter snouts without nose pads. Their brains are larger, and their periods of gestation and maturation are

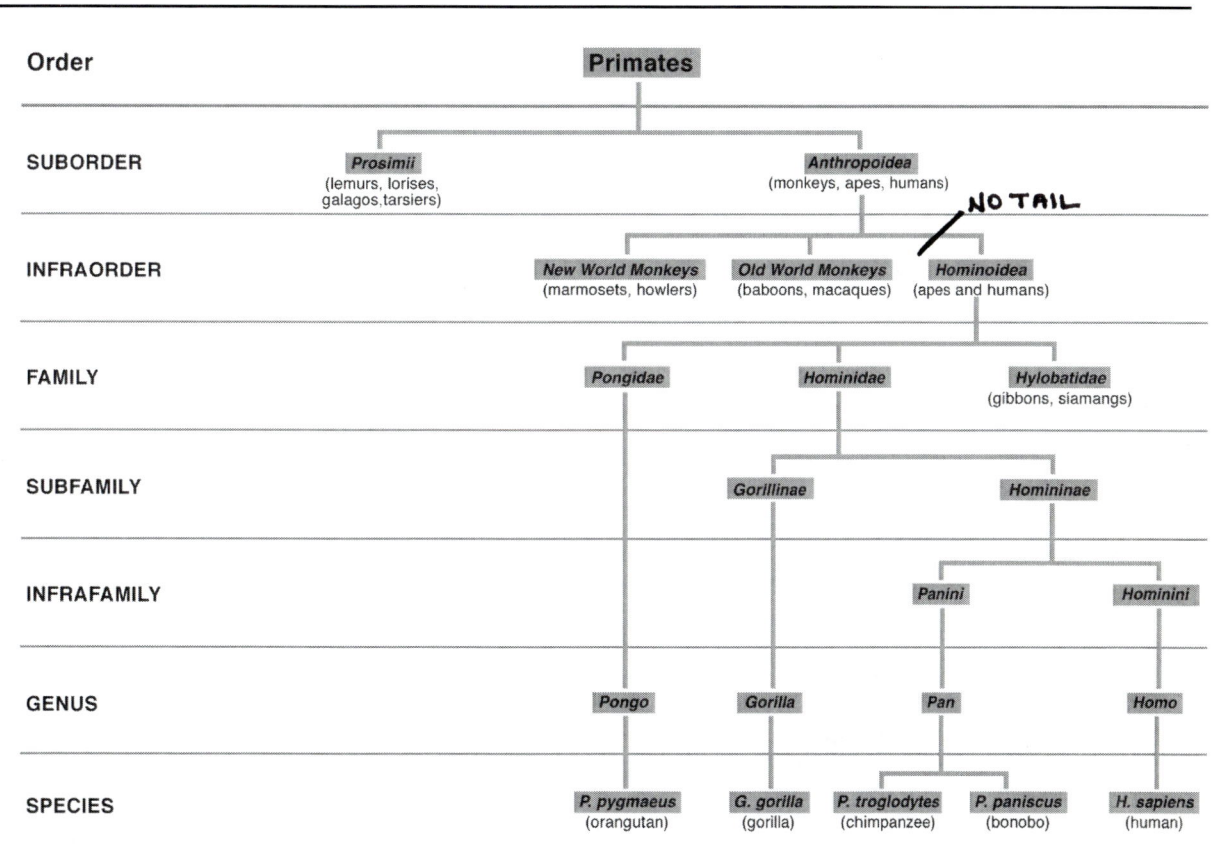

FIGURE 2.2:
Simplified primate taxonomy.

longer. They are subdivided into three major groups: New World monkeys, Old World monkeys, and *Hominoidea*. The last group is the one into which we place ourselves, and the name "*Hominoidea*" is Latin for "humanlike."

The **New World monkeys** are at home in the rain forests of Central and South America, moving skillfully in the tree canopy and often using their long tail as a fifth hand. They range in size from 12-ounce marmosets to 20-pound howler monkeys (Figure S2.b).

The **Old World monkeys** live in various habitats, from the tropical jungles to the deserts of Africa and Asia, and even in seasonally snow-covered areas of northern Japan. Like their New World cousins, they have long tails, which they use for balance and communication. While many species of Old World monkey stay in the trees, others dwell on the ground. During feeding and grooming, they spend much time sitting with their upper bodies held erect (Figure S2.c).

The ***Hominoidea*** are distinguished from other primates by their complex brain and behavior, by very long periods of infant development and dependency, and by the absence of a tail. In taxonomic parlance, the *Hominoidea* are an infraorder. They can be divided into three families: *Hylobatidae*, *Pongidae*, and *Hominidae* (see Figure 2.2).

The ***Hylobatidae***, with body weights from 12 to 25 pounds, are the smallest apes. Living in the forests of southeast Asia and China, they feed on fruits and leaves (Figure S2.d). They are great acrobats: Moving in the tree canopy by arm swinging, they reach speeds of 35 mph and cover branch-to-branch distances of up to 50 feet. Their basic social unit is a monogamous pair with up to three dependent offspring. Males and females are approximately the same size, and both contribute equally to raising their young.

The family of the ***Pongidae*** has only one extant (currently living) species, *Pongo pygmaeus*, the orangutan. Orangutans are the only great apes living outside of Africa today.

The family of the ***Hominidae*** is subdivided into two subfamilies, *Gorillinae* and *Homininae*. The *Gorillinae* comprise only one extant species, scientifically known as *Gorilla gorilla* and commonly called simply the gorilla. Molecular evidence indicates that *Gorillinae* and *Homininae* began to evolve separately some 10 mya, and that the *Homininae* split again about 7 mya into two infrafamilies, *Panini* and *Hominini*. (see Figure 2.2).

The ***Hominini*,** or hominins, comprise us modern humans and our fossil ancestors back to the separation from the *Panini*. In this sense, the earliest hominins mark the birth of our species, *Homo sapiens*. It was a birth out of crisis. Some 7 mya, a climate change in Africa led to the replacement of much forest by open grasslands and swamps. The new habitats attracted herds of antelopes and other plant-eaters, with saber-toothed tigers and other predators following on their heels. While the *Panini* stayed in their familiar, albeit shrinking, tree canopy, the *Hominini* took on a more adventurous way of life: On the fringes of the forest, they scavenged from the carcasses that the large predators had left behind. We will trace these breathtaking events in Chapter 3.

The ***Panini*** comprise our closest nonhuman relatives, *Pan troglodytes* (chimpanzee) and *Pan paniscus* (bonobo). In terms of both anatomy and DNA sequences, they are more similar to each other than to any other great ape or the human. They began to go their separate evolutionary ways about 3 mya. Today, the chimps live north of the Congo river, while the bonobos are limited to a smaller range south of the river.

C. The Great Apes: Our Closest Nonhuman Relatives

Following are thumbnail sketches of the four great apes: orangutan, gorilla, chimpanzee, and bonobo.

ORANGUTAN

The orangutan (*Pongo pygmaeus*) is the only great ape species living outside of Africa. Its ancestors moved to Asia about 17 mya, when the previously isolated African continent became connected to Eurasia by a new land bridge, now the Arabian peninsula and Sinai. Today, the orangutan is restricted to the heavily forested parts of two Indonesian islands, Borneo and Sumatra. Even there, this species faces imminent extinction by humans through habitat destruction and poaching.

Female orangutans weigh around 75 pounds and usually move in the tree canopy, using all four limbs for grasping and support. Males, at 150 pounds or so, are too heavy for the local trees and travel on all fours on the ground. Orangutans eat mostly fruit supplemented with bark, leaves, insects, and occasionally meat. Because they are choosy about their diet, foraging in large groups is not practical except in areas of abundant food supply. For the most part, the lives of orangutans are rather solitary, with basic social units consisting of one adult female accompanied by one or two dependent offspring. The females are visited by roving males who keep track of the females' reproductive cycles. The time of ovulation, and the associated sexual receptivity of mammalian females, is called **estrus**. Estrous orangutan females prefer to mate with large, boisterous males while trying to evade smaller males. However, small adult males are still larger than females and sometimes force them into copulation. Big males try to monopolize mating by defending territories covering the home ranges of several females.

FIGURE 2.3:
Adult orangutan male with cheek pads
Image © 2010 frog-traveler. Used under license from Shutterstock, Inc.

Male orangutans are not only much heavier than females, they also have larger canine (cuspid) teeth. Fully adult males also have conspicuous cheek pads (Figure 2.3). Such obvious differences in body size, shape, color, or weaponry between males and females is generally known as **sexual dimorphism** (Figure S2.e). In many animal species, males are bigger and better armed than females, indicating that males fight over access to females. As a result, dominant males tend to monopolize reproduction. This type of mating system, where one male copulates with several females but not vice versa, is known as **polygyny**. In animals with a **monogamous** mating system, as in the lesser apes mentioned earlier, males and females form long-lasting pair bonds, and both parents contribute to raising the young. Males do not fight excessively over females, and there is little sexual size difference.

GORILLA

The gorilla (*Gorilla gorilla*) is the largest of all living primates: Females weigh around 200 pounds and males up to 400 pounds. Best known, due to the pioneering observations of Dian Fossey, are the mountain gorillas of Rwanda (Figure 2.4). Being too heavy for climbing, gorillas knuckle-walk on the ground, where they feed on low growing forest plants. Habitat destruction, poaching, civil war, and the Ebola virus have driven them close to extinction.

Gorillas live in polygynous family units consisting of a large silverback male, several females, and their young. A **silverback** is a large, dominant male with a saddle-shaped area of silvery hair growing across his back (Figure S2.f). The silverback mates with all females of the family, and he does not tolerate other adult males, except perhaps one or two of his sons. A typical silverback is a benevolent dictator who determines the movements of the family, breaks up fights, plays with the young, and defends the family against large cats and roving gorilla males who have been driven away from their natal group. Young adult females sometimes leave their natal group, too, apparently by their own choice. They will either transfer to another existing family or join bachelor males in forming a new family.

When challenged by another male, a silverback will first display his strength, typically by chest beating and charging. If the intruder is undeterred, the silverback will fight him ferociously. At stake for the resident silverback is his own life as well as the survival of his offspring, because a victorious challenger will take over the resident silverback's harem and proceed to kill his young. Such infanticide is adaptive for the perpetrator. Once the acquired females are no longer nursing, they will soon be in estrus and will bear and raise the victorious male's offspring. Indeed, females seem to accept raids and infanticide by males as a sign of the males' prowess and reproductive value. The death toll from infanticide among mountain gorillas is high: About one out of seven newborns will be killed by raiding males.

FIGURE 2.4:
Gorilla
Image © Digital.Vision Photo Safari.

CHIMPANZEE

The chimpanzee (*Pan troglodytes*) is the best-known great ape (Figure 2.5). Chimps live in a broad belt of equatorial Africa, mostly in forests but also in open woodlands and savannas. With body weights around 90 pounds for females and 120 pounds for males, they are adept at tree climbing, using both hands and feet for grasping branches. When traveling on the ground, they assume a four-legged posture, with feet flat and hands curled, knuckles down. The diet of chimps consists of fruits and other plant parts, supplemented with insects. Groups of males also hunt smaller monkeys and gazelles (Figure S2.g). If they make a kill, they share the highly prized meat with other males to forge alliances, or with females to gain sexual favors. Loosely knit communities of 20 to 200 individuals live in territories that are patrolled by males and occupied by their community for many years. Within the community, chimps form fluctuating parties of 2 to 70 individuals. Party size seems to depend mostly on food distribution, with abundant food sources supporting larger parties.

Male chimps stay in their native community, which means they tend to be genetically related, and they have known one another for many years. They cooperate in hunting and defense, even though they constantly compete for rank. Rank-oriented interactions among males include physical fights, and aggressive displays (fluffing up hair, charging, uprooting small trees, and throwing rocks). There are also gestures of appeasement (bows and grunts), reconciliation (hugging and kissing), and reassurance (dominant male touches or softly bites subordinate male). Male chimps have elaborate ways of forging alliances to move up in rank (see Chapter 21). One way is the sharing of prized food items, especially meat. Another way is mutual grooming to remove fleas, ticks, and other parasites. A grooming session is usually initiated by the lower-ranking male. The ultimate goal of these interactions is to attain **alpha status;** the alpha male gets all the meat and all the copulations he wants.

Males take advantage of their larger size in bullying females. An exception is the alpha male, who tends to protect females from other males and breaks up fights between females. Encounters between males of different communities lead to physical fights and killings. Groups of males sometimes raid a neighboring community with the intent to kill males and bring home females. Jane Goodall has described as "chimp wars," the planned and sustained raids she observed at the Gombe National Park in Tanzania.

FIGURE 2.5:
Chimpanzee
Image © jupiterimages.com

Female chimps are less overtly social except that they raise their infants, and mother–offspring relations last for many years beyond weaning. The females of a community tend to be genetically unrelated since they leave their native community when they reach sexual maturity. Within their adopted community, they establish a rank order, and high-ranking females raise more offspring. Adult females have menstrual cycles of about 35 days. During estrus, their genitalia become pink and puffy, and males compete intensely over estrous females.

Mating occurs in different patterns. In large parties, copulations are liberal and promiscuous. In small parties, dominant males use aggression to monopolize access to females. Lower ranking males sometimes subvert this pattern by striking up temporary but exclusive consortships with females away from the rest of the group. Conversely, females sometimes wander off to copulate with males from neighboring communities. Male chimps protect all young from within the community and contribute to raising them by sharing food. However, they kill infants they suspect have been sired by males from another community.

BONOBO

Closely related to the chimp, and not recognized as a separate species until 1933, is the bonobo (*Pan paniscus*). Being more slender than chimps, bonobo males weigh around 95 pounds, females 75 pounds (Figure 2.6). Their geographic range is limited

FIGURE 2.6:
Bonobo
Image ©jupiterimages.com

to an area of rain forest south of the Congo river. Their diet consists mostly of fruit, supplemented with stem pith from plants and protein from small animals. The species is endangered due to rain forest destruction, poaching, and civil war.

Like chimpanzees, bonobos live in large communities in which males are genetically related, whereas females leave their natal community when they reach sexual maturity. Nevertheless, bonobo males are not as strongly bonded as chimpanzee males, and they are less occupied with competing for rank. While chimp males like to show off their strength in spectacular ways, carrying on for many minutes, bonobo males usually limit displays to brief runs while dragging a few branches behind them. Bonobo males also spend less time jockeying for rank by forging alliances with other males. Instead, even adult males rely on their mothers if they need help. Generally, bonobos are very playful, especially as juveniles (Figure S2.h). They like to tickle one another and to make funny faces, sometimes in long solitary pantomimes.

A major difference in the adult lives of chimpanzees and bonobos is the relative importance of sex and aggression. In chimpanzees, males from different communities will often skirmish and engage in raids or even in sustained war. In contrast, the lifestyle of bonobos can be epitomized by the hippie motto: "Make love, not war!" Serious aggression seems to be rare within communities and between them. In fact, there are reports of peaceable mingling, including sex and mutual grooming, between members of different communities.

Within bonobo communities, casual sex is a way of life. Adults become aroused very easily, and the females are willing to mate throughout their monthly reproductive cycle. Copulations are frequent and promiscuous, and they occur in two positions, face-to-face and male mounting from behind. Partners exchange long mutual gazes as well as facial and vocal signals about the desired position. Despite their heightened sexual activity, the bonobos' rate of reproduction is about the same as that of the chimpanzees. A female gives birth to a single infant every five or six years.

In addition to heterosexual intercourse, bonobo females have what behavioral scientists call genito-genital rubbing, or **GG rubbing.** One female, facing her partner, clings to her with arms and legs. The two then rub their vulvas together, with apparent pleasure. Similarly, two male bonobos may stand back-to-back, one rubbing his scrotum against the buttocks of the other. Males may also engage in "fencing" with their erect penises. All these sexual activities are short, casual, and relaxed. They often serve to resolve the same communal problems that chimps settle by aggression. For example, when a female has hit a juvenile, the latter's mother may lunge at the aggressor but immediately thereafter soothe her by GG rubbing.

GG rubbing is most frequent between adolescent females and in the context of females transferring into a new community. A newly arriving female will typically choose senior resident females for special attention, using GG rubbing and grooming to establish a relationship. After producing her first offspring, the young female gains additional acceptance into the group. Eventually the cycle repeats itself when new immigrants, in turn, seek a good relation with the now established female.

The stronger bonds of bonobo females have important consequences for their relation to males. While a chimpanzee female may be bullied by any adult male in her community, this does not happen to bonobo females who are bonded. In a contest over food, for example, two bonobo females will prevail against a male who could drive off either of the females if she were alone.

D. No Model Primate

The differences in behavior and social structure between chimpanzees and bonobos are remarkable. They are the most closely related ape species, differing only slightly in size, anatomy, and habitat. Yet their societies differ greatly in male versus female dominance, and in the relative importance of aggression and sex. We have to conclude that behaviors are complex adaptations to different needs that may evolve quickly. It is therefore not surprising that even greater changes in behavior and societal structure have occurred during human evolution, because chimps and bonobos have evolved separately for only 3 million years, whereas the human line of evolution split from the common chimp/bonobo line about 7 mya. For the same reason, the hope of some biologists and psychologists to find a primate species that could serve as a model for human dietary needs, behaviors, societal structures, etc., is not realistic. Each species has its own way of making a living, and behaviors evolve faster than anatomical traits.

E. Humans Are Set Apart by Several Features of Anatomy and Behavior

The family of the *Hominidae* is named after our own species, *Homo sapiens*. Our scientific name, Latin for "wise man," still resonates with the optimism of past centuries that the extraordinary capacity of humans for rational thinking would lead to a continuous improvement of society. Although one may wonder whether we are always using our rational capacity wisely, it is clear that our mental abilities are unique in the animal kingdom. The anatomical basis of these abilities is the human **brain**, which is unparalleled in its complexity and relative size (see Chapter 7). However, the brain costs much energy to build and maintain, and this extraordinary expense only pays off in conjunction with other human traits such as toolmaking and language.

Another obvious characteristic of humans is our **habitual bipedalism**. While many mammals are able to walk on their hind legs for short distances, only for humans is this the normal way of getting around. Habitual bipedalism has come with multiple adaptations in our bones and muscles, including long legs for running and a short pelvis that brings the center of gravity close to the hip joints (Fig. S2.i). As we will discuss more fully later, habitual bipedalism began very early in hominin evolution.

Following habitual bipedalism, and presumably promoted by it, was a subtle but functionally important modification of our hands: the **opposable thumb**. While all primates can move the thumb sideways from the palm of the hand, only humans can fully rotate the thumb so that its tip becomes opposed to the tips of the other fingers. Thus, we are better prepared than monkeys and apes to grasp objects forcefully in a power grip or manipulate them delicately in a precision grip (Figure 2.7).

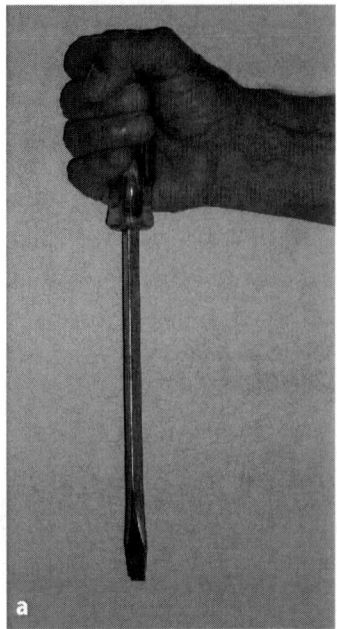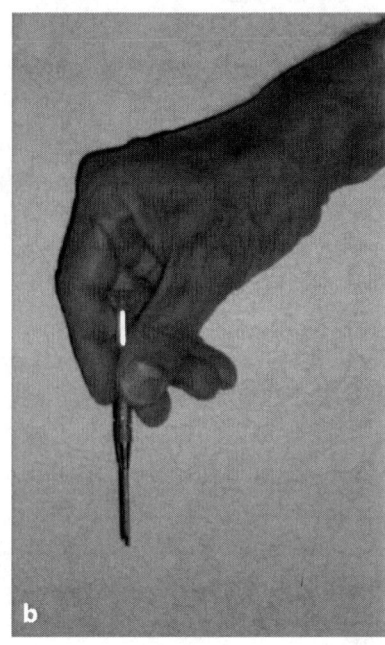

FIGURE 2.7:
(a) Power grip: Object is grasped forcefully between curled fingers and palm, while thumb provides counter pressure. **(b) Precision grip:** Object is carefully manipulated between tips of first two digits and fully opposed thumb

It seems no accident that the primate with the largest brain and the finest hand precision grip came to excel in **toolmaking**. Chimps also make simple tools such as "fishing sticks," twigs stripped of leaves and bitten to appropriate length, which they poke into insect nests to retrieve those insects that cling to the twigs. By comparison, even the simplest human stone tools, some 2.5 million years old, were more demanding to make and more effective to use. Tools were of tantamount importance in human evolution, as we will discuss in Chapters 3 and 6. And of course, the material culture of developed societies today is beyond comparison with the tools of any animal.

Also setting us apart from all nonhuman primates is the **loss of body hair**. This fairly unique trait is best explained as an adaptation to warm habitats, where sweating is needed to keep the body from overheating. An added advantage of shedding body hair was getting rid of most of those lice, flees, and ticks that other primates have to remove from their fur in long mutual grooming sessions.

Another human distinction is our **elaborate spoken languages** (see Chapter 7). Primates, as well as other social animals, use alarm calls and mating calls, to which group members respond by taking appropriate action. Still, human language is by far more complex and powerful than any animal language. A sophisticated language in turn is a prerequisite for other uniquely human traits such as myths, religions, and cultural norms of behavior, which are hallmarks of cooperation in human societies. Thus, language is an integral part of the human way of life, and the drive to speak seems to be inborn.

Yet another characteristically human trait is **art**. Chimps and bonobos, when given paper, paint, and a brush, will put down strokes with a semblance of composition, and they will quit when they feel they are done, but there is no apparent intention of wanting to represent something, or of trying to express a mood. In contrast, human ancestors

FIGURE 2.8:
Cave paintings in Lascaux, France
Image © Corel, Inc.

who lived about 15,000 years ago have created spectacular cave paintings (Figure 2.8). In stark lines and vivid colors, these paintings show the large mammals that were our ancestors' fearsome foes and prized game. There are also marvelous pieces of portable art, such as figurines carved from ivory, which were created 25,000 years ago and would easily blend in with the exhibits of a modern art museum (see Figure 4.7).

F. Several Uniquely Human Traits Revolve around Parenting and Sexuality

A major set of traits that distinguishes us humans from nonhuman primates revolves around **sexuality** and **parenting**. The large human brain and the limited size of the birth canal have made it necessary for human babies to be born in an undeveloped and helpless state. It takes an extremely long time for human infants to become self-sufficient, a task that calls not only for the female but also for the male parent to contribute. Among animals, we find marriage-like family units only in those species that require more effort to raise the young than an average single parent can provide. The best-known examples are bird species, some of which mate not only for a brood period but for life.

In birds and other animals with family units, males gage their parental investment according to their perceived likelihood of paternity. For a man, one of the greatest concerns is that he may be cuckolded. Thus, it is not surprising that virtually all human societies have as a central institution marriage, an exclusive bond between a man and a woman who pledge to raise their children together and to forego extramarital affairs. The latter part is crucial. It provides the husband with an assurance that the children he helps to raise are indeed his. Likewise, marital fidelity assures the wife that her husband's efforts will not be diverted to other women and their children.

For humans, marital faithfulness is more difficult than for other primates. Lesser apes and gorillas have family units, but because these families live in isolation, there are few opportunities for extramarital sex. Our most closely related species, the chimps and bonobos, live in large, promiscuous communities without families, where infants are raised only by their mothers. No other primate has the uniquely human situation of nuclear families embedded in larger communities, which provide opportunities for extramarital sex. Real or perceived adultery disrupts not only familial bonds but also communal cooperation. Homicide is a major drain on primitive societies, and the most common cause is jealousy.

The need to ensure both marital faithfulness and communal peace can help to explain some unique traits of human sexuality, which include the following.

- Concealed estrus, female sexual receptivity throughout reproductive cycle
- Long-lasting intercourse
- Permanent breasts as sexual attractants
- Modesty
- Marriage as a universal societal institution

One human sexual characteristic has been termed the **concealed estrus**. Ovulation is inconspicuous in human females. However, women can be sexually receptive throughout their reproductive cycle, a situation that allows humans (as well as bonobos) to have sex for pleasure, independently of reproduction. Correspondingly, humans take more time for sexual intercourse than nonhuman primates, whose copulations only last for seconds. Moreover, adult human females have **permanent breasts** that are present independently of nursing and act as sexual attractants for males. Of course, the joy of sex by itself does not promote marital fidelity, but it does in conjunction with other human traits such as **modesty** and a preference for having sex in private. In most cultures, humans cover their genitalia and women tend to hide their breasts. Thus, it appears that the unique traits of human sexuality, when acting together, serve to stabilize the bond between married partners while reducing conflict within the community. The biological underpinnings of marital faithfulness are of course reinforced by the societal institution of **marriage**, which is almost universal across cultures.

It is important to realize that behaviors can be adaptive without being conscious. We do not have sex in private in a deliberate effort to stabilize our marriage or to keep the peace in our community, we simply prefer it that way. The evolutionary link is that humans whose brains were wired for sex in private have formed better communities and raised more children. These children developed the same preference for sex in private and in turn raised many children, until eventually they displaced communities of people whose brains were wired for sex in public. Likewise, we do not fall in love in order to start a family; it simply happens to us. And those who fall in love have outcompeted the ones who don't.

Given so many traits that set humans apart from all great apes, it seemed natural for "classical" taxonomic trees to reserve the family of the *Hominidae* just for the human, while lumping together all the great apes as the *Pongidae*. The "modern" taxonomic tree shown in Figure 2.2 takes into account data from DNA sequencing and other molecular investigations. To reconcile the classical and modern taxonomic trees, one has to assume that the *Hominini* have evolved *uncommonly fast*. This means that an unusually large fraction of the random DNA changes that occurred during human evolution were positively selected for. We will explore this proposition in the following chapter.

EXERCISE PAGE FOR CHAPTER 2

Student Last Name _____

Student First Name _____

Discussion Section _____

On the course web site, use the link on the **syllabus** to find a pdf file of your **assigned reading** for this chapter. The same reading is referenced below. Use the bulleted list of questions to test your knowledge of this reading.

De Waal, F. B. M. (1995). Bonobo sex and society. *Scient. Amer.* **March 1995:** 82–88.

- How do chimp and bonobo societies differ in the relative importance of sex and aggression?
- How does heterosexual behavior of bonobos and chimps differ?
- Does the higher level of heterosexual activity in bonobos translate into more offspring?
- What is GG rubbing between females, and what are some of the social functions of this behavior? Are there similar behaviors in males?
- Describe some of the social (nonreproductive) functions of sex in bonobo communities.
- How do males acquire status in chimp and in bonobo societies?
- How do chimps and bonobos differ in regard to male dominance?

Feedback

On a scale from 1 to 10, the being the best, please rank the above reading for

1. Interest and Relevance to the topic (10 being most interesting/relevant) _____
2. Readability (10 being most clearly written, easy to understand) _____

In the remaining space, enter any comments that you may have on this reading.

CHAPTER THREE

HUMANS EVOLVED IN RESPONSE TO A CHALLENGE

Traditional taxonomic trees, based mainly on morphological criteria, have placed humans in their own family, the *Hominidae*, while lumping all great apes together in a separate family, the *Pongidae*. This taxonomy had to be revised with the advent of molecular data, according to which humans, chimps, and bonobos are closely related and equally distant to gorillas and orangutans. This modern taxonomic tree places us humans amidst the great apes even though many of our morphological and behavior traits are uniquely human. A hypothesis that reconciles these seemingly contradictory trees proposes that humans have evolved unusually fast, spurred on by a climate change that left them in a state of poor adaptation to their new habitat. Because of this mismatch, the chances of random mutations to be positively selected were exceptionally high. Indeed, the search for genes that show the molecular footprint of positive selection reveal some of the alleles that seem to have made us human.

A. Taxonomy, Phylogeny, and the Problem of Convergence

In the scientific names of species, the first name is for the *genus* and the second name is for the *species*. In *Homo sapiens*, the designation for the human, *Homo* refers to the genus and *sapiens* refers to the species. This **binomial nomenclature** was introduced by the Swedish naturalist *Carl von Linné* (1707–

1778), also known as *Carolus Linnaeus* (Figure S3.a). His most famous book, entitled *Systema Naturae*, was written in Latin as was customary for the scientific literature of his day. The book listed the names and short description of all known minerals, plants, and animals. Since the work was popular, Linnaeus kept publishing new and expanded editions, which grew from eleven pages in the first edition (1735) to three thousand pages in the final and thirteenth edition (1767).

Linnaeus still believed that species were constant, rather than evolving, and he grouped them into a **taxonomic system** with families, orders, etc. using convenient characteristics. He saw his work as a useful catalog that would allow naturalist to see whether a new organism they encountered had already been described before. Given the compelling similarities between apes and humans, Linnaeus listed the human among the primates, and the Lutheran archbishop of Uppsala accused him of "impiety" for doing so.

When the theory of **evolution**, or descent with modification, became widely accepted in the twentieth century, taxonomic catalogs were expected to be "natural," reflecting the course of evolution. Large taxonomic groups should then look like trees, with the trunk representing a major taxonomic unit, say, the *class* of the mammals (Figure 3.1). The major limbs of the tree would than represent different *orders*, such as the primates. Each order in turn would comprise several families, such as the *Hominidae*. Each branch of a family would carry as twigs one or more *genera* (plural of genus), and each genus would carry leaves as species. The further apart any two parts of the tree were, the less related they would be, and the earlier they would have been separated in evolution.

In the absence of complete fossil records, taxonomic trees have traditionally been based on biogeographic, morphological, and behavioral characteristics of *extant* (currently living) species. The degree of similarity in these features has been used to estimate relative distance of species on such trees.

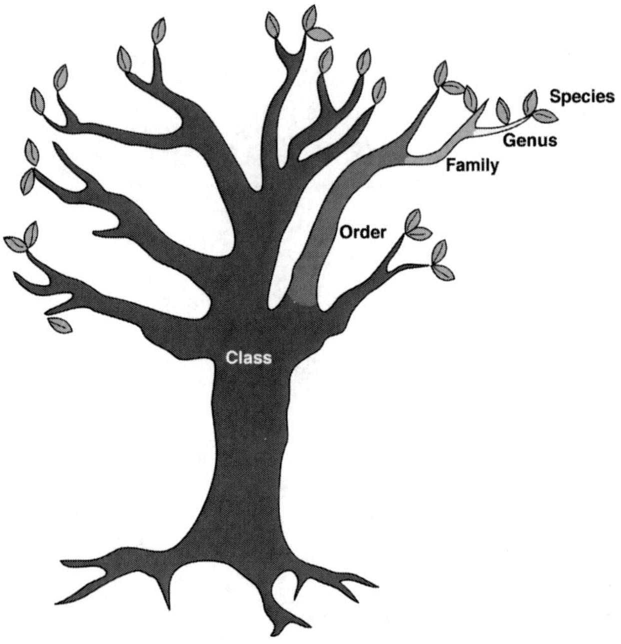

A potential problem with establishing taxonomic trees from extant species is the fact that species may be similar, not only because of common descent, but also because of parallel selection over long periods of time, a phenomenon known as **convergence**. As an example, consider the spindle-shaped bodies and fins of fishes and dolphins. These traits have evolved independently because they allow their carriers to hunt or escape effectively by swimming fast. In the case of the dolphins, their lungs and the anatomy of other internal organs show clearly that they are mammals rather than fishes. However, for taxonomic systems groups such as birds, whose way of life places stringent requirements on *all* of their organs, convergence has made it notoriously hard to establish natural taxonomic trees.

A problem opposite to convergence may arise when evolution occurs unusually fast, making species appear to be more distantly related than they are. This situation may arise when a species colonizes a previously uninhabited area, such as a remote island. Rapid evolution may also

FIGURE 3.1:
The concept of a "natural" taxonomic tree

occur in response to environmental change, as we will discuss for hominin evolution later in this chapter.

Even though retarding and accelerating effects on evolution have been the bane of taxonomists, they demonstrate a driving role of the environment. If an **ecological niche** (a set of environmental conditions under which a species can make a living) is there, then chances are that a species fitting the niche will arise by random mutations and positive selection.

B. Phylogenetic Trees Based on Molecular Data

The **modern synthesis** explains evolution as the combined result of random DNA changes and subsequent selection or "genetic drift." Changes in DNA occur randomly as a result of replication errors as well as exposure to chemical mutagens and cosmic radiation. Any resulting changes in morphology or behavior may be beneficial or detrimental to its carrier, or neutral, depending on the environment.

In an interbreeding population, mutant alleles will spread in the gene pool. However, if a population splits into reproductively isolated subpopulations, then each will begin to accumulate its own mutant alleles. The number of mutations by which the two subpopulations *differ* will then increase with the time. Eventually, individuals from different subpopulations will no longer mate and have fertile offspring, which means the original species has split in two.

As an analogy, consider the differences in vocabulary that evolve when a human subpopulation splits off. For example, U.S. citizens use the terms "gas" and "eraser" for the same objects that the British call "petrol" and "indiarubber," respectively.

When DNA sequencing became available, it turned out that the coding regions of genes were most conserved, reflecting the structural and functional constraints of their protein products. Regulatory regions of genes were generally less conserved, while introns and the DNA between genes were least conserved. The latter types of genomic DNA have been dubbed "junk DNA," because much of it has no known function. Mutations in this DNA seem to be unchecked by selective pressures, representing mostly genetic drift.

C. The Concept of Molecular Clocks

Genetic drift can be used for quantitative estimates of the "genetic distance" between species, and hence, as a way of constructing taxonomic trees that reflect the course of evolution.

Mutations accumulating at a steady rate represent a **molecular clock**, indicating how long any two populations have evolved independently. The evolutionary separation between emerging species can be expressed in years if the "clock" is calibrated. Such a calibration can be achieved if the evolutionary separation of one or more species in the group of interest is known independently, for example, from paleontological evidence.

Extending the language analogy used earlier, consider the accumulation of new words in the language of people who became separated by historic events. For example, there are more differences between the English languages spoken in England and the United States (independent since 1776) than between England and Kenya (independent since 1955).

Potential problems with the molecular clock concept include the following:

- The accumulation of mutations may be distorted by major changes in population size, including "bottleneck" and "founder" effects.
- The rate at which mutations accumulate depends on the efficiency of DNA repair mechanisms, which may change over long periods of time.
- Mutant alleles may be subject to positive or negative selection.

The proportion of "junk" DNA in the entire genomic DNA is about 95 percent for humans and similarly high for other mammals. Under these circumstances, selection pressures on non-junk DNA have only small effects on measurements of genetic distance based on total genomic DNA.

D. DNA Hybridization as an Overall Method for Counting Mutations

Sequencing large segments of DNA is still an expensive effort (see Chapter 9). A simpler method for quantifying the difference between two genomes is **DNA hybridization**. This method has been used to reinvestigate the phylogenetic trees of birds and primates.

The two strands of a DNA double helix are held together by numerous weak hydrogen bonds. The strands come apart when a DNA solution is heated to near the boiling point of water. This process is known as **DNA melting** because single-stranded DNA is less gelatinous, or more liquid, than double-stranded DNA. The process of DNA melting can be monitored by measuring the DNA's absorption of ultraviolet light (UV), which increases as the strands separate (Figure 3.2). The resulting "melting curve," which indicates how much DNA is already single-stranded, rises sharply between 85 and 90°C. The **melting temperature** (T_{50}), at which 50 percent of the DNA is single-stranded, can therefore be measured exactly.

The melting temperature of DNA is lowered by any mismatched base pairs (Figure 3.3). The drop in melting temperature (ΔT) is nearly 1°C for each percent of mismatched bases. In hybrids of DNA strands from two closely related species, a few percent of the base pairs are mismatched, causing ΔT values of a few °C (Figure 3.4).

By mixing small amounts of radiolabeled DNA from one species with excess amounts of unlabeled DNA from a related species, one can create a situation in which nearly all radioactivity is in present in *hybrid DNA*, that is, double stranded DNA consisting of one strand from each species. Thus, by tracing this radioactivity, one can determine exactly the melting curve of the hybrid DNA (see Appendix). The melting temperature of such hybrid DNA is lower, by a differential designated ΔT, as compared to "pure" DNA from either species. Thus, the quantitative value of ΔT is proportional to the fraction of mismatched base pairs in the hybrid DNA, which in turn is a measure of the genetic distance between the two species. This method was first

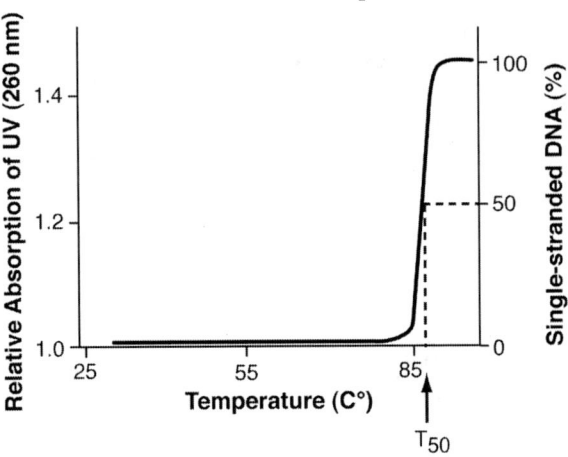

FIGURE 3.2:

Melting of double-standed DNA by heat, monitored by absorption of ultraviolet light at 260 nm (left ordinate). The melting curve also indicates how much of the total DNA that is already single-stranded (right ordinate). The temperature at which 50% of the DNA has melted (T_{50}) is called the DNA's melting temperature.

used to establish a DNA-based taxonomic system for birds, to much acclaim of traditional taxonomists who had been plagued by the convergence problems arising from the use of morphological comparisons between bird species.

When the same method was applied to the *Hominoidea*, the smallest ΔT values were found between chimps and bonobos, and increasing differentials between these two species and human, gorilla, orangutan, and representatives of the *Hylobatidae* and the *Cercopithecoidea* (Figure 3.5). These data were combined with a with a "clock" calibration based on fossil evidence dating the divergence of orangutans from the other great apes to around 17 mya. The resulting evolutionary tree separates chimpanzees and bonobos about 3 mya, humans from chimps and bonobos around 7 mya, gorillas from chimps, bonobos, and humans about 10 mya.

This DNA hybridization tree is also known as the **molecular tree** because it is compatible with DNA sequencing data on corresponding (orthologous) DNA segments from other primates. Immunological and protein sequencing data also argue for the same taxonomic system of the infraorder *Hominoidea* (see Figure 2.2).

E. Molecular versus Classical Data for Hominoid Phylogeny

The molecular taxonomic tree has revolutionized our thinking about primate evolution, especially with regard to the time period between 4 and 25 mya, for which hominoid fossils are scarce. The molecular tree shows some major departures from the **classical tree**, which was based on the morphology, behavior, and biogeographical distribution of primate species (Figure 3.6).

- In the molecular tree, the family of the *Pongidae* has the orangutan (*Pongo pygmeus*) as its only extant species, whereas the family of the *Hominidae* now includes all African great apes as well as the human.
- Humans, chimpanzees, and bonobos share a subfamily, the *Homininae*. This relegates us humans, along with our fossil ancestors and relatives (those who would be placed on the double line in Figure 3.6), to an infrafamily called the *Hominini*, or hominins.

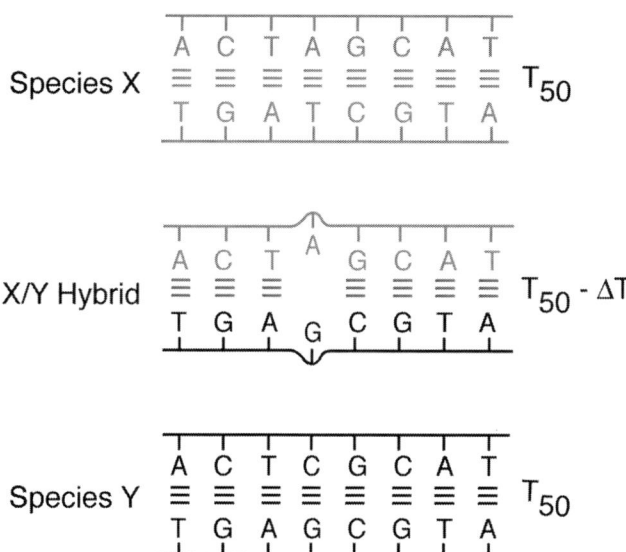

FIGURE 3.3:
Decrease in DNA melting temperature by base-pari mismatches

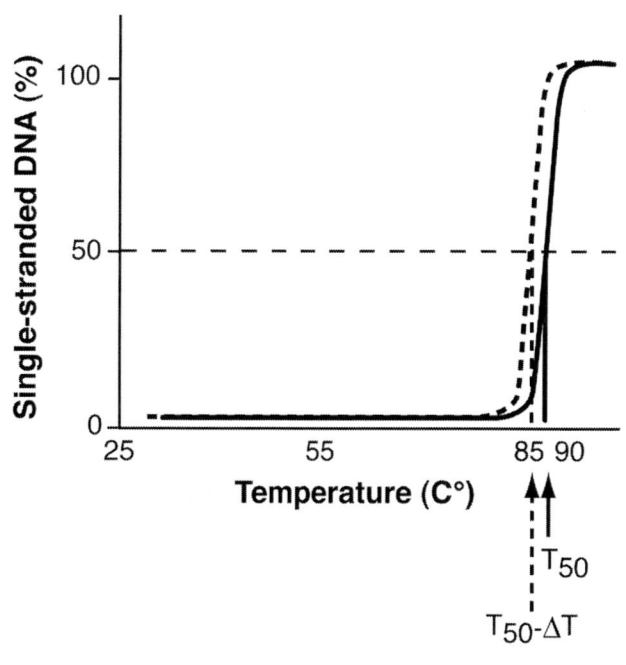

FIGURE 3.4:
Melting curve of hybrid DNA with complementary strands from different species (dotted curve) as compared to pure DNA (solid curve)

- Our "sister" infrafamily, the *Panini,* consists of bonobos and chimpanzees, which separated from each other about 3 mya.
- The phylogenetic separation (~7 mya) of the *Hominini* from the *Panini* in the molecular tree is more recent than the separation (~17 mya) of the *Hominidae* from the *Pongidae* in the traditional tree.

How can we reconcile the two taxonomic trees? On the one hand, the molecular tree almost certainly represents the actual course of evolution because it is much less affected by confounding phenomena such as convergence as discussed earlier. On the other hand, the classical tree seems to reflect much better the fact that there are many uniquely human traits, such as bipedalism, large brain, and complex spoken languages, which set us apart from *all* nonhuman primates. It seems as if relatively small changes in human DNA have caused major differences in anatomy and lifestyle.

Based on DNA hybridization data and on more recent comparisons of thousands of DNA sequences, humans and chimps differ in 1.24 percent of their DNA base pairs. Such a difference could have large or small effects, depending on where in the genome the mutations occur and on the way the mutations play out in gene expression and development.

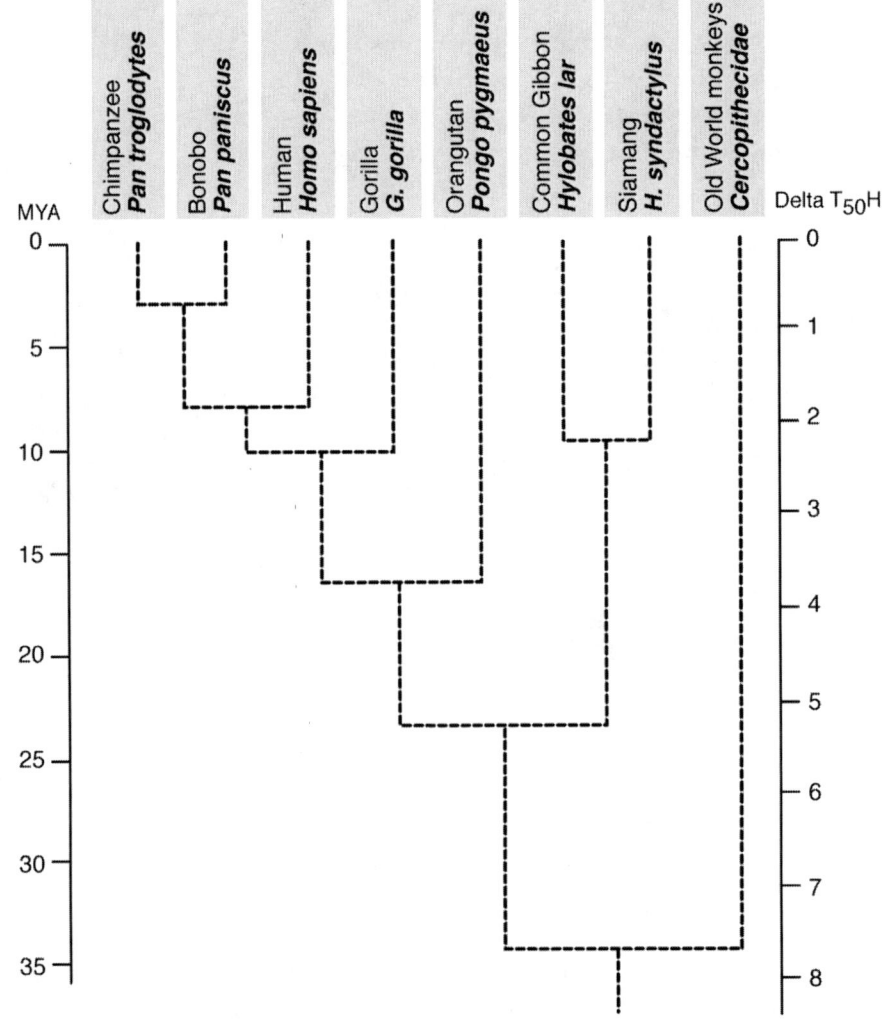

FIGURE 3.5:
Genetic distance based on DNA hybridization data. **Right ordinate:** shifts in melting temperature (ΔT) of hybrid DNA from pair-wise combinations of species. **Left ordinate** is calibrated in million years ago (MYA) based on fossil data about orangutan. Redrawn after data of Sibley C. G. and Ahlquist J. E. (1984) The phylogeny of the hominid primates, as indicated by DNA-DNA hybridization. *J. Molec. Evol.* **20**: 2–15.

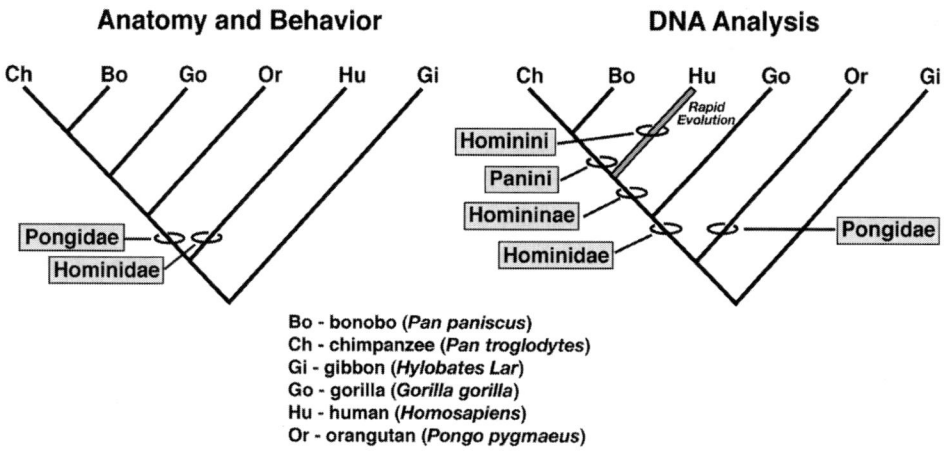

FIGURE 3.6:
Classical and molecular taxonomic trees of the *Hominoidea*

Synonymous mutations (those that do not cause amino acid substitutions) are likely to be inconsequential. Likewise, mutations in introns or other "junk" DNA will probably have no effects. **Nonsynonymous mutations** in exons are more likely to count, especially if the resulting amino acid substitutions occur in active domains of the encoded proteins.

Dramatic effects can be expected from nonsynonymous mutations that change the binding of transcription factors to their target genes. Such effects are best known in the fruit fly *Drosophila melanogaster*, where point mutations in genes encoding certain transcription factors can replace antennae with legs, or turn the normally two-winged fly into a four-winged mutant (Figure S3.b). Similar mutations have been generated in mice. In most mammals including humans, mutations in the *SRY*⁺ gene may turn a male into a sterile female (see Chapter 16). In proteins that self-assemble into larger units, single amino acid exchanges can have major effects. For example, the substitution of valine for glutamine in position 6 of ß-globin makes hemoglobin "sticky," causing sickle-cell anemia in homozygotes but making heterozygotes resistant to malaria (Figure S3.c; see also Chapter 8).

In order to reconcile the classical and molecular trees for higher primates, we have to assume that *an unusually large fraction of the mutations occurring in hominins had phenotypic effects that were positively selected for*. This assumption would explain why humans differ significantly from *all* great apes in morphology and behavior while being similar in their DNA sequences, especially to chimps and bonobos. In other words, hominins must have evolved *unusually fast*, while chimps and bonobos have stayed close to our common ancestor and to their more distant relatives, gorillas and orangutans.

Positive selection can be expected if the environment changes in a way that creates a new ecological niche, to which the existing species are not well adapted. Under these circumstances, any new allele that improves the adaptation of its carrier to the changed environment will have a greater chance of being passed on to following generations. As an analogy, think about the "evolution" of passenger automobiles. Four-door family cars running on cheap, readily available gasoline have changed little in their basic design between 1950 and 2000. Only since gasoline has jumped in price do we see smaller and lighter car bodies, more fuel-efficient engines, electric batteries that weigh less and hold more charge, and so on. These trends also help to answer the challenge of global warming (see Chapters 24, 26). So just as it took an economic change to build a new kind of automobile, it took a climate change to build a new primate.

F. Hominins Evolved Rapidly in Response to a Climate Change

Our closest nonhuman relatives, the chimpazees, bonobos and gorillas, are now living in Africa. It is therefore likely that our common ancestors were also residents of Africa.

The earliest known hominin fossil, named *Sahelanthropus tchadensis,* was found near Toros-Menala in Chad, Central Africa and is dated 6 to 7 mya. The material includes a nearly complete braincase (Figure 3.7) and fragmentary lower jaws and teeth. *S. tchadensis* has many apelike features, such as a small brain, whereas other traits, including small canine teeth, are reminiscent of later hominins. The protrusion of the jaws and the thickness of tooth enamel are intermediate between chimpanzees and more recent hominins. This mixture of traits, along with the fact that the fossils are

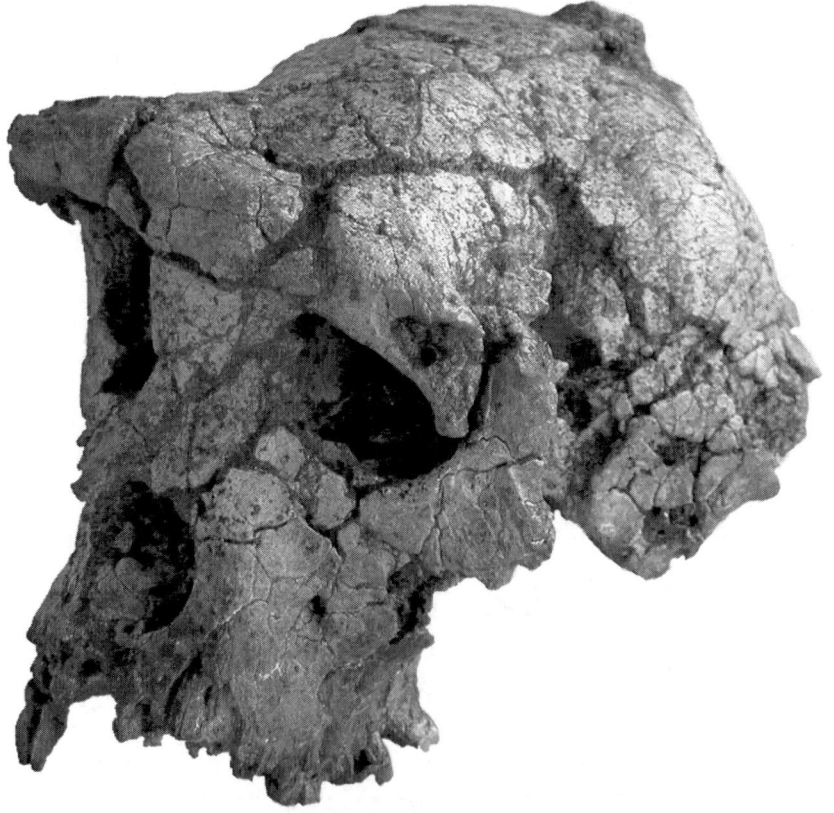

FIGURE 3.7:

Sahelanthropus tchadensis. Reconstructed skull of the oldest known hominin
Photograph courtesy of Dr. Michel Brunet, leader of the paleoanthropological team that discovered *Sahelanthropus*

dated close to the time when the human and chimp/bonobo evolutionary lines have diverged, suggests that *S. tchadensis* may be close to the common ancestor.

Around the same time, between 10 and 5 million mya, the climate in Africa became *cooler, drier, and more seasonal*. This change in climate caused the *replacement of much tropical rain forest with open habitats,* including savannahs, swamps, and wooded riverbeds. This **habitat fragmentation** is seen in the fossil record, including pollen analysis and the fossilized bones of animal species living in such open habitats.

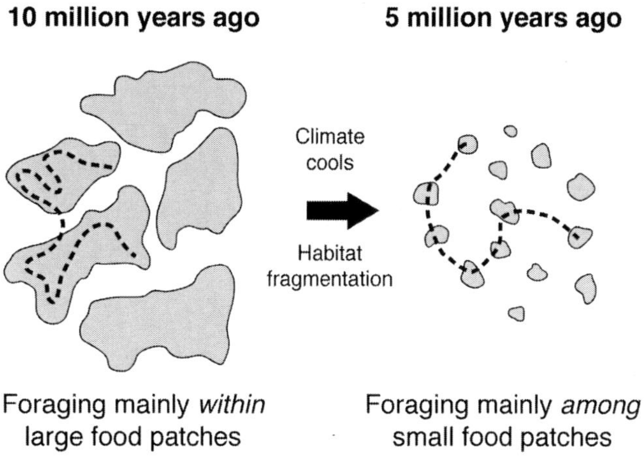

The loss of much rain forest was a major challenge for primates who had relied on trees for a diet of fruit and insects, for shade, and for safe places to sleep. Having to travel from one tree cluster to the next on the ground cost them energy and exposed them to the heat of the sun (Figure 3.8). It also left them vulnerable to large predators, including saber-toothed tigers and giant hyenas. These major changes in climate, flora, and fauna played out just when the evolutionary lines of the *Hominini* and the *Panini* diverged. The *Panini* simply stayed in their familiar tree canopy, their numbers shrinking as the forests receded. Meanwhile the *Hominini* "experimented" with new ways of making a living in a changed environment to which they were poorly adapted.

Those hominins who survived and eventually gave rise to us modern humans underwent a rapid series of evolutionary steps creating a new primate that was habitually bipedal, and had opposable thumbs, a large brain, and the ability to make highly effective stone tools. These bold changes allowed hominins to supplement the inherited plant diet by scavenging meat, brains, and bone marrow from the carcasses left by large predators.

The anatomical and behavioral changes that supported the new way of life would have been selected *against* if our ancestors had continued to live in a state of nearly perfect adaptation to their old habitat. Walking legs and opposable thumbs are no use if you live in a tree canopy. Indeed, the chimps and bonobos, which have maintained their tree-dwelling lifestyle, have evolved very slowly. By contrast, some of the newly identified genetic alleles that support critical human traits (*FOXP2* gene required for speech and *ASPM* gene required for brain growth) show the telltale signs of rapid evolution by positive selection.

FIGURE 3.8:
Fragmentation of forest habitat in Africa by a cooler climate
Redrawn after Lewin R. (1993) *Human Evolution: An Illustrated Introduction.* Cambridge: Blackwell Scienctific Publications.

G. Exact Nature of the Genetic Differences between Humans and Other Primates

Since the genomic DNA sequences of humans, chimpanzees, and other primates are now available for comparison, efforts are under way to identify exactly some of the genomic changes that have contributed to uniquely human traits.

Some studies focus on *nucleotide substitutions* in the *coding regions* of human *genes* and their counterparts in nonhuman primates. Are there traces of positive selection in the DNA of *Hominini* versus *Panini*, or apes versus monkeys?

The study described in the assigned reading (below) is focused on a gene that is critical to human speech. Mutations in this gene cause severe articulation difficulties accompanied by linguistic and grammatical impairment. Comparisons of human, chimpanzee, gorilla, orangutan, rhesus macaque, and mouse DNA indicate that the affected gene has changed to an unusual degree in human evolution.

Other studies look for nucleotide substitutions in *any DNA sequences* rather than selected genes. One such study identified a DNA sequence, *HAR1*, that encodes a nontranslated RNA. It has evolved rapidly in humans, and is expressed specifically in the developing neocortex (see Chapter 7). On a larger scale, some of the genomic changes that occurred during hominin evolution and seem to have been positively selected include gene duplications resulting in more than 200 copies of the same gene.

Still other studies compare the *rates of gene expression*, which may respond strongly, and dependent on body region or developmental stage, to small changes in gene regulatory sequences. In a pioneering study of this kind, the authors compared the rates of expression for thousands of genes in white blood cells, liver, and brain among humans, chimpanzees, orangutans, and macaques. It turned out that rates of transcription have changed particularly in the human brain.

An example of recent hominin evolution in gene expression is the gene encoding lactase, the enzyme that breaks down the milk sugar lactose. In most people, this gene is switched off after weaning, and these people are lactose-intolerant as adults. However, after the beginning of agriculture, mutations keeping the lactase gene on permanently were positively selected for in populations using the milk of domestic animals in their diet.

The effort to identify specific genomic changes that have occurred during hominin evolution will rarely identify one specific allele that completely determines a uniquely human trait such as bipedalism or language. The relationship between genes and phenotypic traits in humans is often complex (Figure 3.9).

- Most genes are **pleiotropic**, that is, affecting more than one phenotypic trait.
- Most traits are **polygenic**, that is, controlled by more than one gene.
- Environmental factors often modulate the effects of gene activity.

Even if a mutant phenotype is available for a gene of interest, it only shows that the gene has a significant effect on the observed trait. It doesn't show how many other genes contribute to the same trait. Conversely, if a genetic allele is pleiotropic, its currently most prominent phenotypic effect is not necessarily the one for which

FIGURE 3.9:
Genes and traits. A given human gene may affect multiple traits (pleiotropy), while a given trait may be affected by several genes (polygeny). In addition to genes, environmental factors contribute to human traits and may influence natural selection.

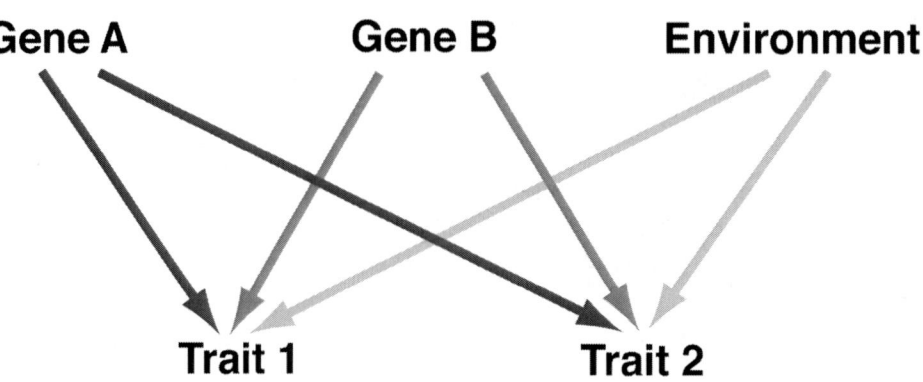

the allele has been selected earlier during evolution. For instance, while the current human *FOXP2* allele is critical to language, its original positive selection in hominins may have been driven by its effects on facial muscles.

Appendix: DNA Hybridization Protocol

The following protocol is based on work of Britten and Kohne (1968) and has been developed by Sibley and Ahlquist (1983, 1984) to establish phylogenetic trees for various groups of organisms.

- Prepare total nuclear DNA from species A
- Sonicate DNA to fragments of about 0.5kb
- Prepare "single-copy" fraction—denature by heat, allow repetitive DNA to hybridize, separate single-stranded from double-stranded DNA by hydroxyapatite (HAP) chromatography
- Radiolabel this "tracer DNA" from species A
- Allow labeled, single-stranded tracer DNA to hybridize with excess amounts of single-stranded, unlabeled "driver DNA" from species A as well as other species (B, C, etc.)
- Let hybrids bind to HAP column at 55°C, elute nonhybridized DNA
- Raise column temperature in small increments and elute any single-stranded DNA from molten hybrids after each step
- Measure radioactivity in each eluate in order to draw a melting curve
- Determine melting temperature (T_{50}) with driver DNA from species A, B, C, and so on. In heterospecific combinations, T_{50} is lower because mismatched double-stranded DNA melts at lower temperature. The temperature differential (ΔT) is a measure for the genetic distance between tracer and driver DNA
- Tabulate ΔT values for all pairs of species
- Convert tabulated values into cladogram, beginning with the species showing the lowest differentials and proceeding to species with higher differentials
- Calibrate "DNA clock" assuming a *uniform average rate (UAR)* of DNA evolution and using related fossils or geological events that have been dated independently and accurately. (For both primates and birds, a ΔT of 1°C seems equivalent to about 4.5 million years.)

EXERCISE PAGE FOR CHAPTER 3

Student Last Name _____

Student First Name _____

Discussion Section _____

On the course web site, use the link on the **syllabus** to find a pdf file of your **assigned reading** for this chapter. The same reading is referenced below. Use the bulleted list of questions to test your knowledge of this reading.

Enard, W.; Przeworski, M.; Fisher, S. E.; Lai, C. S. L.; Wiebe, V.; Kitano, T.; Monaco, A. P.; and Pääbo, S. (2002). Molecular evolution of *FOXP2,* a gene involved in speech and language. *Nature* **418:** 869–872.

- What is the biochemical function of the protein encoded by the *FOXP2* gene, and what kind of phenotype has been observed in humans with mutant alleles of this gene? Why is it important that a mutation in *FOXP2* and similar phenotypes were found in two unrelated individuals?
- What data do the authors show in their Figures 1 and 2, and which additional observation on the evolutionary conservation of the *FOXP2* protein do they mention?
- How do they assess the functional significance of the amino acid substitutions they found in *FOXP2*?
- Do humans show any amino acid polymorphism in the exon 7 protein domain in which the two human-specific substitutions were found?
- What does Figure 2 indicate about the ratio of amino acid replacements over nucleotide changes in exon 7?
- What is the tentative interpretation that the authors propose for their data? Which alternate hypothesis do they also mention?
- Which speculation do they offer (towards the end of their paper) on how the human allele of *FOXP2* gene might facilitate the trait of a spoken language?

Feedback

On a scale from 1 to 10, the being the best, please rank the above reading for

1. Interest and Relevance to the topic (10 being most interesting/relevant) _____
2. Readability (10 being most clearly written, easy to understand) _____

On the back of this page, enter any comments that you may have on this reading.

CHAPTER FOUR

MILESTONES OF HOMININ EVOLUTION

While taxonomic trees can be constructed by comparing extant species, a more detailed picture of their evolution requires the study of **fossils**, which are mineralized body parts of organisms that have lived in a distant past and have been preserved in sand, lava, or other sediments. Fossils and sediments are dated using various physical and biological methods (see Appendix 1). Scientists called **paleoanthropologists** study hominin fossils and artifacts to learn how our ancestors lived: how they moved, what they ate, how they escaped large carnivores, and what their social lives may have been. In this chapter, we will establish a framework of hominin evolution in space and time, which will be fleshed out in subsequent chapters.

A. General Remarks on Hominin Evolution

All paleoanthropologists today are convinced that humans evolved from ancestors shared with chimpanzees and bonobos. Some alternative views known as "creationism" or "intelligent design" are based on an interpretation of the biblical Book of Genesis as a scientific report. However, if instead the Book of Genesis is viewed as a spiritual document, then there is no contradiction to the theory of evolution (see Appendix 2).

The oldest clearly hominin fossils to date were all found in Africa. For this reason, and because our closest nonhuman relatives—chimpanzees, bonobos, and gorillas—are confined to Africa, our common ancestors most likely lived in Africa as well.

Between 10 and 5 mya, a **cooler and more seasonal climate** caused the replacement of much African forest with grasslands or swamps (see Chapter 3). In response to this climate change, the

Panini simply stayed in their familiar tree canopy habitat, their numbers shrinking as the forests receded. In contrast, the *Hominini* tapped into the new resources of their changed habitat by venturing down from the trees at least occasionally. There were new opportunities for them to supplement their diet with meat of herbivorous mammals, which populated the new open habitats. There were also new dangers in the form of large carnivores, including saber-toothed tigers and giant hyenas. Those who managed to grasp the opportunities while avoiding the dangers became our ancestors.

Under these conditions, mutations that promoted humanlike traits (bipedalism, large brain, hand precision grip, etc.) seem to have been positively selected for. These changes allowed hominins to supplement the inherited plant diet by scavenging meat, brains, and bone marrow from the carcasses left by large predators. The bold *transition to a partially ground-dwelling lifestyle* with an increasing component of meat to the diet separated the *Homini* from the *Panini*. This separation probably occurred around 7 mya.

Audiovisual presentations on hominin evolution include a three-part video prepared under the direction of leading paleoanthropologist Donald Johanson (available in the Audiovisual Library of The University of Texas, Austin under call number 4127) and a three-part NOVA series published in November 2009. A website of the Smithsonian Institution (http://www.mnh.si.edu/anthro/humanorigins/) shows many color photos of hominin fossils and reviews ongoing research.

B. Calendar of Hominin Evolution

Humans have not evolved along a straight line. Most hominin fossils represent extinct branches of an evolutionary "bush" (Figure 4.1), and only a few of them are likely to be direct ancestors of modern humans. This means that hominin evolution was precarious, with many forms trying different ways of coping with the loss of their original habitat. Most of these forms went extinct before one species, modern *Homo sapiens,* came to rule the world. Some fossils belonging to different hominin species were found in the same area and the same sediment, suggesting that they may have run into each other.

Following is a dated list of key events and fossils that most paleoanthropologists consider to be close to the line of ancestors who gave rise to modern humans.

10 to 5 mya	Cooler climate in Africa replaces forests with open habitats
7 mya	*Hominini* separated from *Panini* (chimpanzee/bonobo lineage)
3.4 to 1.1 mya	*Australopithecus* species
2.4 to 1.5 mya	*Homo habilis*
1.9 to 0.3 mya	*Homo erectus*
800 to 150 kya	Archaic *Homo sapiens*
300 to 30 kya	*Homo neanderthaliensis*
200 kya to present	Modern *Homo sapiens*
10 kya to present	Agriculture

C. The Earliest Hominins

The search for the oldest hominins, the ones that lived right after the hominin/panin split, has led several teams of paleoanthropologists on a competitive race. Under arduous conditions, they search parts of Africa that were located on the fringes of forests 7 mya and where geological deposits from that time are exposed today.

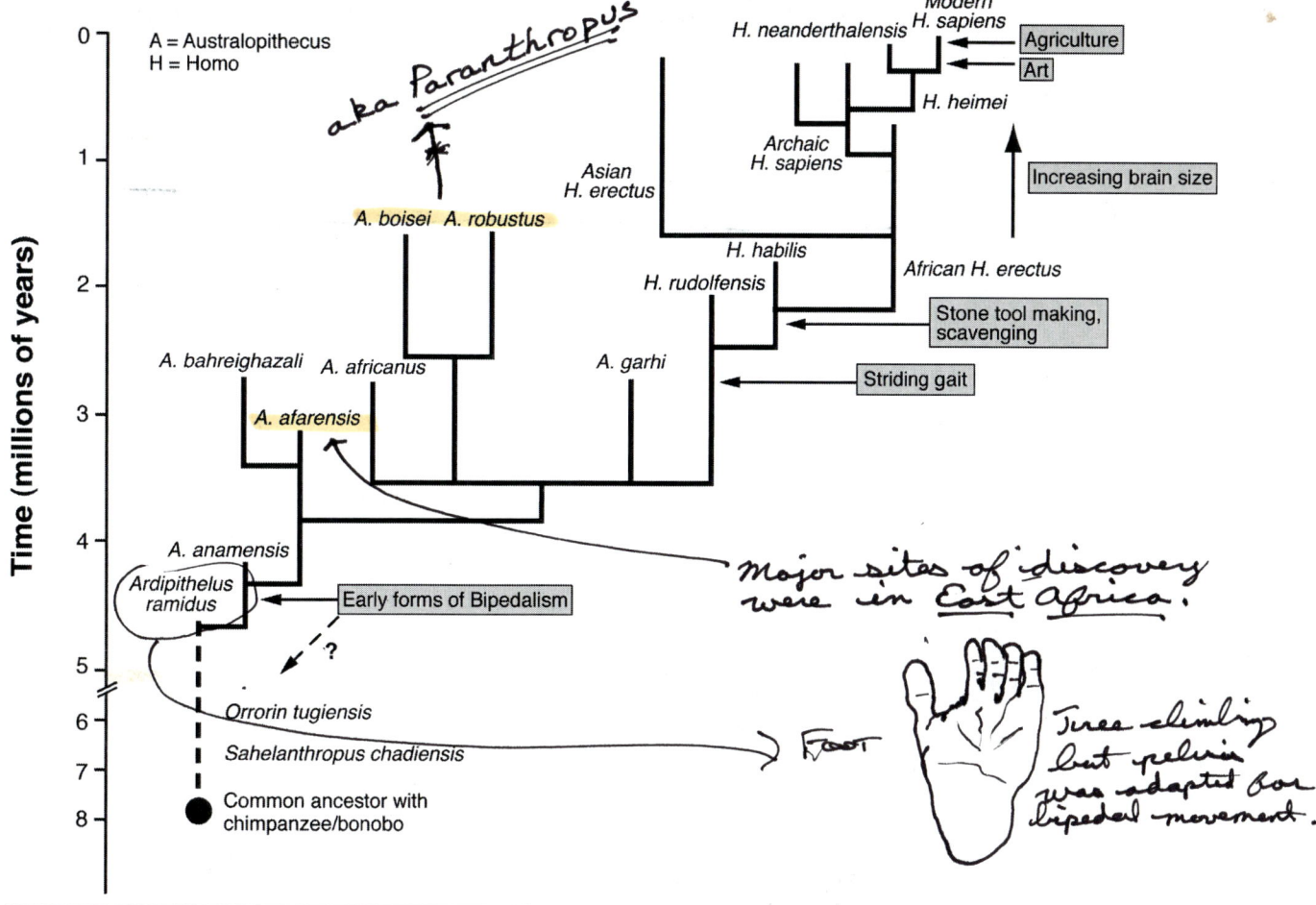

FIGURE 4.1:
Simplified hominin phylogeny

The earliest bona fide hominin fossils, named *Sahelanthropus tchadensis,* was found in what today is the Sahara desert in Chad, Central Africa. Dated 6 to 7 mya, the fossils include a cranium (see Figure 2.7) and fragmentary lower jaws and teeth. *S. tchadensis* has apelike features, including a small cranial capacity. Other traits, such as modest canine teeth, are reminiscent of later hominins. This mixture of traits, along with the fact that the fossils are dated close to the time when the *Hominini* and *Panini* split, suggests that *S. tchadensis* may be close to the shared ancestor.

The second-oldest bona fide hominin, named *Orrorin tugeniensis,* was discovered in the Tugen Hills of Kenya and is dated 6 mya. Parts of femoral (upper leg) bone suggest that *Orrorin* may already have been bipedal. Canine and molar teeth indicate that *Orrorin* ate mostly fruit and vegetables, with occasional meat.

A large collection of hominin fossils assigned to *Ardipithecus ramidus* was found in Ethiopia's Afar desert and dated 3.4 mya. The form of the pelvis, limbs, and the abductible big toe indicate that *A. ramidus* moved bipedally on the ground and quadrupedally in trees. The teeth indicate a nonspecialized, omnivorous diet. The modest size of the canine teeth suggests a lower level of inter-male conflict than in modern chimpanzees.

D *Australopithecus afarensis* and other Australopithecines

A well-investigated group of hominin fossils belongs to the genus *Australopithecus*, a name meaning "southern ape." It alludes to the fact that the first australopithecine fossil was found in South Africa, in a quarry near Johannesburg. A related set of bones, found at Sterkfontein in South Africa and dated 3.5 mya, shows a big toe that could be diverged from the platform of the foot much like an ape's big toe. Both fossils were assigned to the same species, *Australopithecus africanus*.

Fossils assigned to a related species, *Australopithecus anamensis*, were found in Kenya as well as in Ethiopia, and dated about 3.9 and 4.2 mya (Figure S4.a). They include a tibia (shin bone) with shock-absorbing, trabecular bone at the top, indicating that this hominin was already bipedal. Also, the absence of a hollow at the lower end of the humerus (upper arm bone) indicates that *A. anamensis* could not lock the elbow for knuckle walking the way modern apes can (see Figure 2.5)

Other fossils and footprints found in East Africa were dated 3.0 to 3.6 mya and assigned to *Australopithecus afarensis*. The species is named after the Afar triangle in northeast Ethiopia, where cranial fragments, mandibles, and postcranial elements of more than fifty individuals were found. The best-known specimen of this group (Figure 4.2) was named "Lucy" by their finders. At the party following the discovery, the Beatle song "Lucy in the Sky with Diamonds" had been playing over and over.

Actual footprints of australopithecines had been found earlier near Laetoli in Tanzania (Figure S4.b). The footprints were left by two individuals walking side by side over newly deposited volcanic ash dated 3.6 mya. The soft ash hardened like cement, leaving these amazing footprints detailed and well preserved. They differ from modern human footprints by their flat soles and by the divergence of their big toes (see Chapter 5). However, they show the weight distribution of a habitually bipedal organism, rather than that of an ape walking upright for a few yards.

Australopithecines have been dubbed "bipedal apes" because they were similar to modern chimps except that they were bipedal. Their face was apelike with a low forehead, a bony ridge over the eyes, protruding jaws, and no chin (Figure 4.3). Their cranial capacity (400–450 ccm) was only slightly greater than that of chimps (400 ccm). Pelvis and leg bones resemble those of modern hominins, with a short, curved pelvis, legs longer than arms, shock-absorbing shin bones, and femoral necks built for **habitual bipedalism** (see Chapter 5). Of all the traits that set modern humans apart from apes, habitual bipedalism is the first that is clearly documented in the fossil record.

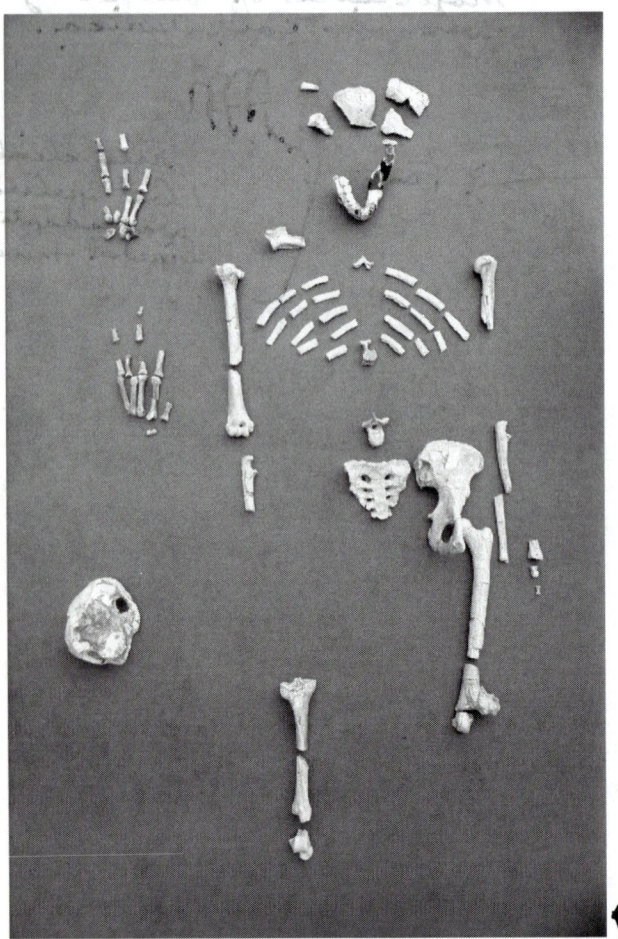

FIGURE 4.2:
Australopithecus afarensis. This specimen, nick-named "Lucy", was found at Hadar (Afar triangle) in Ethiopia and was dated 3 to 3.6 mya.
Image © Alan Nogues/Sygma/Corbis

The australopithecines' bipedalism, however, did not give them the striding gait of modern humans. Their **divergent big toes** would not deliver the same strong push-off as the toes of later hominins. At the same time, australopithecines retained an apelike ability for tree climbing. The curvature of their toes and fingers and the upward orientation of their shoulder joints are also adaptations to tree climbing. This ability was critical for a species that could not outrun its predators on the ground and had not yet acquired the skills and the tools to fend them off.

Later australopithecine species including *Paranthropus robustus* and *Paranthropus boisei* lived as recently as 1.1 mya (see Figure 4.1). They are species with strong jaws, large molar teeth, large zygomatic arches, and sagittal crests on their skulls indicating big chewing muscles (Figure S4.c). *Australopithecus garhi*, dated about 2.5 mya, may be a direct ancestor of *Homo* but could also be a dead end like *Paranthropus*.

The different forms of early hominins are interpreted as alternative attempts to cope with the replacement of forest by open habitats. While the robust species relied on big jaws and teeth for grinding seeds and other hard plant materials found in grasslands, their gracile cousins seem to have built on cognitive abilities and cooperative hunting skills acquired in their traditional rain forest habitat. More importantly, they learned to scavenge herbivore carcasses left by large carnivores.

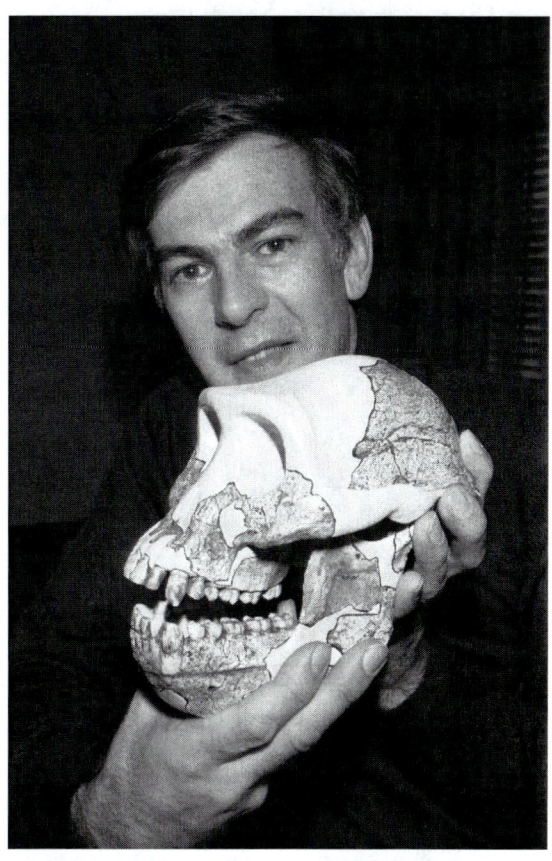

FIGURE 4.3:
Australopithecus afarensis, displayed by anthropologist Donald Johanson, discoverer of "Lucy". Image © Bettmann/CORBIS.

E. *Homo habilis*: Man, the Toolmaker

A different kind of hominin found at Olduvai (Tanzania) and East Lake Turkana (Kenya) was dated between 2.3 and 1.5 mya. The few fossils found so far include hand and foot bones, a lower jaw, and part of a skull (Figure S4.d). The new species was named *Homo habilis* (Latin for handyman) because its fossil bones were found near the oldest known **stone tools** (Figure S4.e). As indicated by the genus name *Homo,* the anthropologist who first evaluated *H. habilis* concluded that this form must be close to the evolutionary line leading to the more recent species of the genus *Homo*.

The supposed toolmakers' hands had at least partially **opposable thumbs**, which means they could rotate their thumbs so that they were facing the palms of their hands instead of just being laterally diverged. Opposable thumbs allow the "precision grip" required for finely manipulating objects, as we modern humans do with a needle or a pair of forceps (see Figure 2.7). The estimated cranial capacity of *Homo habilis* is nearly 650 ccm as compared to about 400–450 ccm for australopithecines.

The importance of stone tools for hominin evolution cannot be overestimated (see Chapter 6). These tools must have greatly enhanced the ability of hominins to obtain meat by **scavenging**, as indicated by the association of stone tools with large assemblies of animal bones showing cut marks from stone tools.

Meat provides protein and fat as major dietary components. Protein from meat is of high quality; it contains in particular "essential" amino acids, which humans cannot synthesize. Fat is a necessary solvent of fat-soluble vitamins and a critical source of essential fatty acids. Fat is also a concentrated source of energy, supplying more calories per gram than protein or carbohydrates. (Being overweight was not a problem for early hominins due to their active lifestyle.) In addition, fat provides phospholipids, molecules that are needed in abundance for the brain.

The brain not only requires proteins and lipids to build and maintain, it also consumes much energy (more than 20 percent of the total energy burned by a modern human at rest). It has therefore been argued that the evolution of a large brain was a luxury that hominins could only afford when they had acquired a substantial meat component to their diet.

F. *Homo erectus:* First Hominin to Leave Africa

The oldest hominin fossils that looked more humanlike than apelike were discovered near Bejing, China, and on the island of Java. Although unearthed far apart, these fossils turned out to be quite similar. The name given to them, *Homo erectus,* was to indicate the **upright posture** of the species. Assessing the age of the *H. erectus* fossils from Java and China turned out to be difficult, but a similar fossil from Dmanisi in Georgia, Asia, was dated at 1.6 to 1.8 mya. In Africa, fossils resembling *H. erectus* have been found in sediments from 1.9 to 0.3 mya. (Some paleoanthropologists assign the African forms to a different species, *H. ergaster,* but we will use the name *H. erectus* for both.)

An almost complete skeleton of a *H. erectus* boy was found at Nariokotome on Lake Turkana, Kenya. Even though the **Nariokotome boy** was dated 1.6 mya, he looks strikingly modern (Figure 4.4). He measured 1.53 m (meters) when he died, but he would have exceeded 1.84 m had he lived to maturity. His tall stature with long legs and slim waist is typical of modern humans adapted to hot, arid environments.

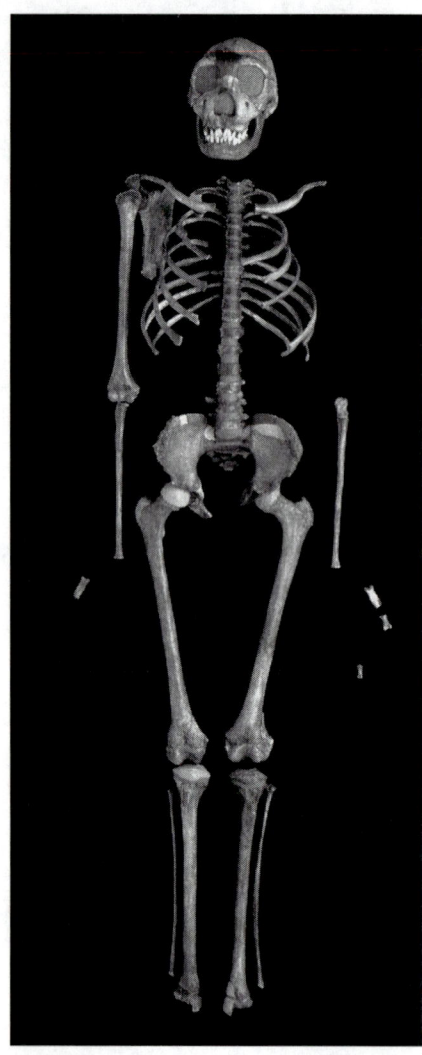

FIGURE 4.4:
Homo erectus, found at Nariokotome on Lake Turkana/Kenya and dated 1.6 mya
Artifact Credit: National Museums of Kenya, Nairobi; Photo Credit: © 1985 David L. Brill, humanoriginsphotos.com

The skull of *H. erectus* shows an elongate braincase, heavy brow ridges, and a mandible without chin (Figure S4.f). The cranial capacity of *H. erectus* varies but averages near 900 ccm, as compared to 400–450 ccm for australopithecines and an estimated 650 ccm for *Homo habilis*. The canine teeth of *Homo erectus* were inconspicuous, and the sexual size dimorphism was modest, indicating that the hominin mating system was evolving away from extreme polygyny (Figure S4.g).

H. erectus was a committed bipedalist, retaining no adaptations for tree climbing. Instead, *H. erectus* had a **striding gait**, as indicated by strong hips, knees, and ankle joints along with a fully aligned big toe and a large heel bone for a strong Achilles tendon. *H. erectus* also had hands with fully opposable thumbs, which may explain why their stone tools exceeded those of *H. habilis* in size and quality (see Chapter 6).

H. erectus probably learned to **tame the fire**, as suggested by burned bones, grains, and clay fragments. In addition to providing warmth for comfort, fire can be used to deter predators, to drive game animals over a cliff to their death, and of course, to cook meals. Cooking tenderizes meat, and it makes plant materials softer and less toxic.

Evidently, having meat and being able to cook allowed *Homo erectus* to survive in climates that did not produce fruit and other soft plant materials year-round. It was probably due to these advances, and to the larger brains sustained by them, that *H. erectus* became able to migrate out of Africa into areas with cooler climates and more sparse vegetations (Figure S4.h). This was not the last hominin to be caught by "Wanderlust" (German for the desire to travel), as we will see later in this chapter, but based on the available fossil record, *H. erectus* was the **first to leave Africa**.

G. Archaic *Homo sapiens*

A heterogeneous group of hominin fossils, dated 800 to 200 kiloyears ago (kya), look more evolved than *Homo erectus* but do not yet show the full set of traits that characterize modern *Homo sapiens*. (Some anthropologists call this group *Homo heidelbergensis* after a lower jaw found near Heidelberg, Germany.) The average brain size of archaic *H. sapiens* was 1,250 cm^3, as compared to 900 cm^3 for *H. erectus* and 1350 cm^3 for modern *H. sapiens* (Figure S4.i).

Archaic *H. sapiens* forms have been *found in Africa, China, India, and in particular in Europe*, where they may have evolved locally from *H. erectus* or represent new arrivals from Africa. Archaic *H. sapiens* fossils found in Africa are dated between 600 and 200 kya. The jaw from Heidelberg is thought to be 500 kya old. The most spectacular collection of archaic *H. sapiens* fossils, dated 800 to 300 kya, was unearthed in the Sierra de Atapuerca of Spain. By 1997 over 1,600 human bones had been excavated, representing at least thirty-two individuals.

Body and limb bones indicate that archaic *H. sapiens* were big, stocky people. Their sexual size dimorphism was modest, suggesting a mildly polygynous mating system. The maxilla (upper jaw bone) of archaic *H. sapiens* had a concave surface, as in modern *H. sapiens*.

Archaic *H. sapiens* living in Europe migrated with the seasons, setting up temporary shelters where the hunting or fishing was good. Foundations of **large huts** found at Terra Amata on the French Riviera range from 26 to 49 feet in length and 13 to 30 feet in width (Figure S4.j). Animal bones and shells left at the sites show that archaic *H. sapiens* ate a **mixed diet** of fish, shellfish, turtles, birds, and on the young of large mammals including deer, wild boar, and ibex.

Archaic *H. sapiens* also made **more diverse stone tools** than those of *H. erectus*, making them useful as knives, scrapers, spear points, and so on (see Chapter 6). Also, archaic *H. sapiens* made **advanced hunting spears** from spruce.

Archaic *Homo sapiens* are thought to be ancestral to both modern *H. sapiens*, which most likely originated in Africa, and *H. neanderthaliensis*, which seem to have evolved in Europe. Molecular data suggest that the lineages of *H, neanderthaliensis* and modern *H. sapiens* split more than 500 kya, with little subsequent interbreeding.

H. *Homo neanderthaliensis*

Archaic *Homo sapiens* gave rise to the latest hominins, modern *Homo sapiens* and *Homo neanderthalensis*. The latter is named after the Neander valley near Düsseldorf, Germany, where one of the first fossils of this species was found. Being discovered in 1856, three years before publication of Darwin's *Origin of Species*, the Neanderthal fossil immediately heated up the debate on human evolution. Since then, more than 400 Neanderthals, dated 150 to 30 kya, have been found in Europe and western Asia.

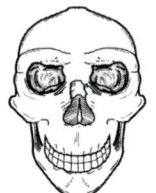
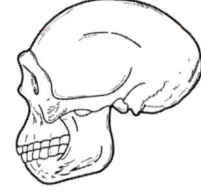

Homo erectus

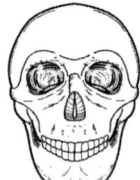
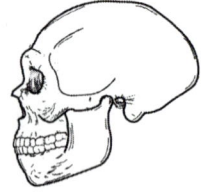

Homo neanderthalensis

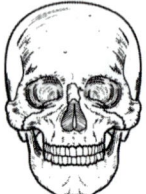
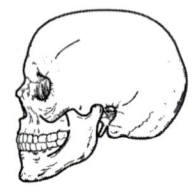

Homo sapiens

FIGURE 4.5
Skulls of *Homo erectus*, *Homo neanderthaliensis*, and modern *Homo sapiens*. Compare braincase, supraorbital ridges, prognathism, and chin.
Image © Kendall Hunt Publishing Company. Adapted from Lewin R. (1993) *The Origin of Modern Humans.* New York: Scientific American Library/HPHLP, p. 149

Neanderthals differed from modern *Homo sapiens* morphologically. Their skull was elongated with a sloping forehead, a projecting nose, and a receding chin (Figure 4.5). Their brain size, at an average of 1400 cm^3, was slightly greater than that of modern humans. Their *body build was stocky and very muscular*, with a barrel-shaped rib cage and bowed, relatively short legs. Many of these traits can be understood as *adaptations to the harsh ice age* in which Neanderthals lived.

The first detailed description of Neanderthal anatomy was based on a nearly complete skeleton of a large male, over 40 years old, and with signs of severe arthritis. An analysis of the fossil by famous paleontologist Marcellin Boulé emphasized the apelike features of the individual, such as his heavy brow ridge and slouching posture, although the latter is readily ascribed to his arthritis. It appears that Boulé wanted to discredit the notion that Neanderthals were human ancestors.

Boulé's writing set the tone for the perception of Neanderthals as brutish and dim-witted creatures, which persists today as the cartoon character of a caveman. However, Neanderthals did have **advanced cultural traits**. They cared for their sick and aged, and they buried their dead ceremoniously, with gifts including flowers. Other grave goods, such as tools, suggest that Neanderthals believed in some form of life after death. A hollowed-out thighbone of a bear with serial holes, found near a Neanderthal fireplace, suggests that its owner may have played the flute.

Neanderthals were found in Europe, Africa, and Asia. In Europe, they disappeared from the fossil record between 45 and 35 kya, in a wave moving from east to west. Their disappearance coincided with the arrival of a new hominin, modern *Homo sapiens*. There has been much speculation on how Neanderthals and modern sapiens related to each other. Did they make love or war? An auxiliary weapon may have given the advantage to modern *H. sapiens* in combat or competition: the spear-thrower, a hand-held stick with a notch into which the end of a spear could be placed to be thrown with increased momentum (see Figure S6.j).

I. Modern *Homo sapiens*

The last 200 kya of hominin evolution are characterized by the appearance of modern *Homo sapiens*, **anatomically indistinguishable from living humans**. This form originated in Africa and eventually spread over the entire Earth.

The oldest fossils assigned to modern *H. sapiens* were found in Ethiopia and dated 200 kya. An almost complete skull from the same area is from 160 kya (Figure 4.6). The overall shape and many anatomical details are characteristic of modern *Homo sapiens*, except that the strong supraorbital ridges are reminiscent of earlier hominins. By 125 kya, anatomically modern humans are found throughout Africa, including Palestine.

The skull of modern *H. sapiens* is distinct from the skulls of both *H. erectus* and *H. neanderthaliensis* (see Figure 4.5).

- neurocranium (braincase) shorter and higher
- supraorbital ridges light or absent

- more projecting chin
- average cranial capacity about 1350 ccm
- viscerocranium (face) is more tucked under the neurocranium

The limbs of modern *Homo sapiens* are more slender, indicating a lighter, less muscular build than that of archaic *H. sapiens* and Neanderthals.

The tools, hunting, and fishing techniques of modern *Homo sapiens* were originally similar to those of archaic *Homo sapiens* but later became much more diverse and efficient (see Chapter 5). Most significant were **sharp and long flint blades**, which could be processed into knives, perforators, hand drills, and so on. These flint tools were used in turn to fashion other implements from reindeer antlers, bison horns, and mammoth tusks. Examples include barbed harpoon tips for fishing as well as needles for sewing together animal hides for clothes and windscreens.

Beginning with the Cro-Magnon culture of *Homo sapiens* in France about 40 kya, we find impressive arts and crafts, including decorated tools, ceremonial batons, beads, and pendants made from teeth, ivory, or shells (Figure S4.k). We also find clay figurines, musical instruments, marvelous **ivory carvings** of humans and animals (Figure 4.7), and spectacular **cave paintings** created between 30 and 10 kya (Figure S4.l). These artworks suggest that Cro-Magnons felt and expressed themselves the same ways we do.

Modern *H. sapiens* wore clothing extensively enough for the human body louse to evolve. They also decorated their bodies with beads and pendants made from ivory teeth, shells, and such, suggesting that they had a strong sense of self.

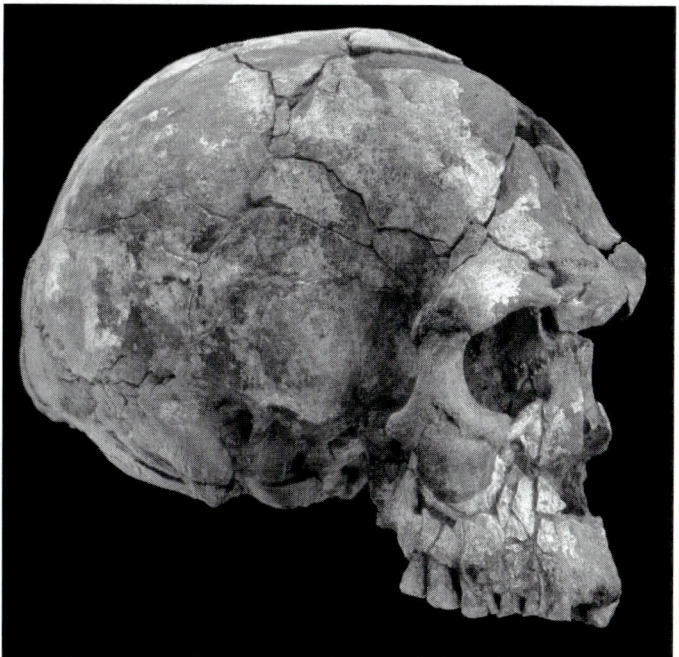

FIGURE 4.6

Modern *Homo sapiens*. Skull of an adult male from Middle Awash/Ethopia dated 160 kya.
Fossil Credit: Housed in National Museum of Ethiopia, Addis Ababa; Photo Credit: © David L. Brill, humanoriginsphotos.com

J. Out of Africa—Again

Fossils of modern *Homo sapiens* were discovered all over the world, with the *oldest* finds dated roughly as follows: 100 kya in the eastern Mediterranean, 50 kya in central Asia, 45 kya in Australia and in Europe, and 25 kya in the Americas. How did they get there? There are two hypotheses. According to the **single-origin hypothesis** (a.k.a. Garden Eden or out-of-Africa hypothesis), modern *H. sapiens* originated once around 200 kya in Africa. They dispersed into subpopulations that started to migrate 50 kya to Europe and Asia, and from there to Australia and as late as 30 to 22 kya to the Americas (Figure 4.8). Wherever they arrived, they

FIGURE 4.7:

Ivory head carved by modern *Homo sapiens* in Laudes, France, dated 25 kya.
Image © Gianni Dagli Orti/ CORBIS

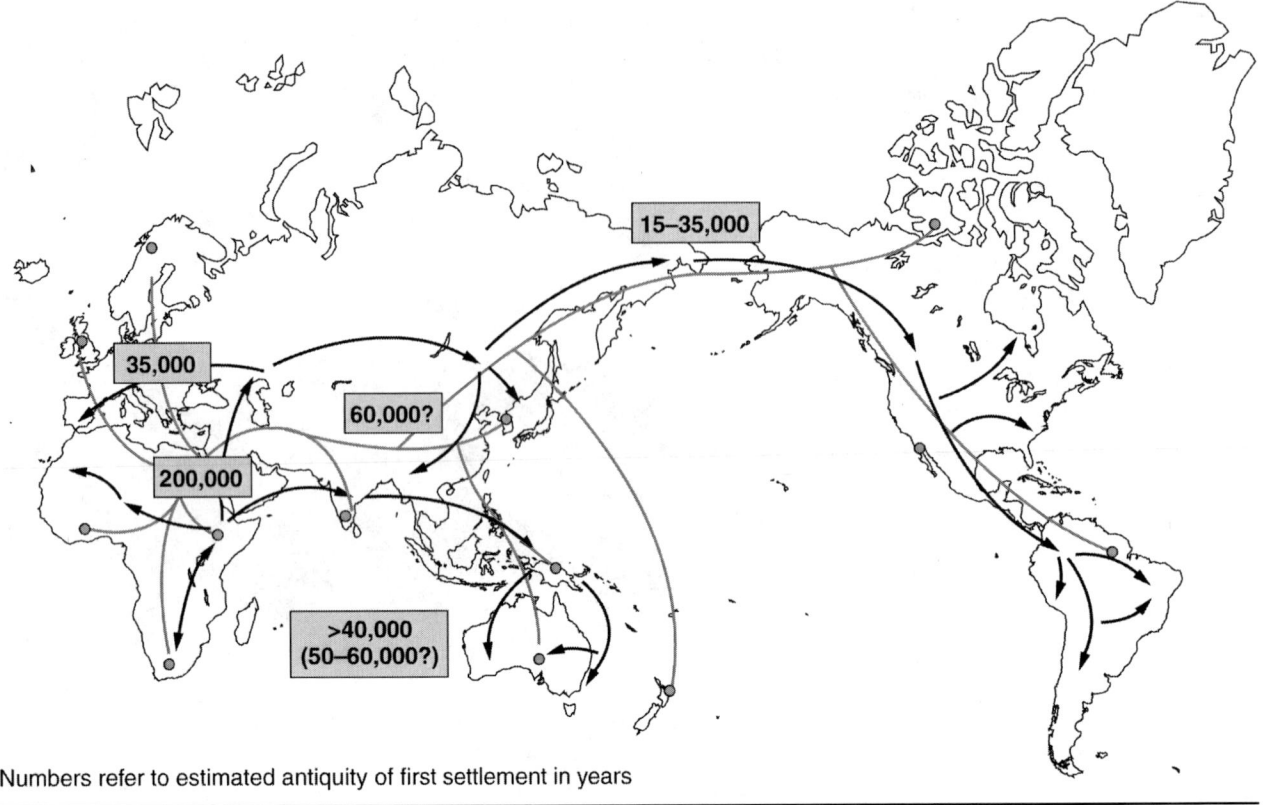

Numbers refer to estimated antiquity of first settlement in years

FIGURE 4.8:

Migration routes of modern *Homo sapiens* (arrows) from Africa over the rest of the world, inferred from the oldest fossil bones or tools found, as well as genetic distance between extant human populations (connected dots based on blood proteins).
Image © Kendall Hunt Publishing Company. Redrawn from Cavalli-Sforza L.L. (1991) Genes, peoples and languages. *Scientific American* **265**: 104–110. p. 107 top

displaced any preexisting hominin populations after periods of coexistence with little or no interbreeding (Figure S4.m). According to the alternative **regional continuity hypothesis**, modern *H. sapiens* evolved multiple times independently from local populations of *H. erectus*, archaic *H. sapiens*, or Neanderthals.

Compelling evidence supporting the out-of-Africa hypothesis has come from molecular studies on extant humans. When DNA sequences of humans from different countries are compared, those of African origin show greater variation among themselves than humans from other continents, suggesting that the DNA of Africans has been accumulating mutations for a longer time. With the aid of computers, one can organize all DNA variants into various trees with branch points representing mutations that could have generated the observed variants. The tree with the smallest number of mutations that can account for all variants must be considered the most likely tree. The best trees from multiple studies on different DNA segments consistently show two major branches, one including only Africans and the other comprising individuals from other locations.

As discussed in Chapter 2, the constant trickle of mutations that accumulate in DNA may be used as a "molecular clock," if the rate of accumulation can be calibrated against time points known from independent evidence. Based on different DNA segments analyzed, and on different assumptions about the sizes of founder populations, the time estimates for the origin of modern *H. sapiens* cluster around

200 kya. This figure matches well with the age of the oldest actual fossils of *H. sapiens* found in Africa, as mentioned earlier. The comparisons of DNA sequences also confirm earlier work on blood proteins.

K. From Hunting and Gathering to Agriculture

Between 12 and 7 kya, humans made a fundamental change from hunting and gathering to agriculture, known as the **agricultural revolution**. Agriculture is defined by the presence of plants and/or animals that have been genetically changed by human breeding (Figure S4.n). The first major center of agriculture, the "fertile crescent," stretching from today's Israel to northern Iraq, was based on wheat, barley, lentils, peas, goats, sheep, and cattle. Another center in Mesoamerica cultivated maize, squash, beans, cotton, and gourds. In the Huanghe and Yangzi basins of China, agriculture flourished on rice, millet, soybean, peas, taro, and pigs.

What caused the agricultural revolution? The answer to this question seems to be complex, with probably many local variations. One driving force was probably the exhaustion of natural resources by overhunting and overgathering around growing settlements. Another probable cause was more of an opportunity: the end of the last Ice Age some 12 kya, which brought a wealth of cultivable plants and herdable mammals to the vast areas that had previously been covered with ice. A third cause could have been a cultural enrichment of life that thrived on arts and crafts, trade, and other social specializations that could not be supported when nearly every adult had to help with hunting or gathering.

Whatever blend of causes may have swayed a community to make the transition to agriculture, there is no doubt that it changed human life fundamentally. Having spent more than 99 percent of their evolution in roving bands of extended families, humans now found themselves in **larger communities** held together by religion and customs rather than by kinship. Along with livestock and fields, the **concept of property** emerged. Societies became governed by complex rules that were made by kings and priests and enforced by soldiers and tax collectors. The food of agriculturalists was less diverse and healthful than the food of hunters and gatherers, as indicated by an increase in tooth cavities and bone diseases in the fossil record.

Agriculture in Eurasia must have been favored by the presence of **large and herdable mammals**. Around 4000 BC, humans in Eurasia already had the "big five" domestic mammals that continue to dominate today: sheep, goats, pigs, cows, and horses. They provided readily available food of high quality as well as mechanical power. In the Americas and Australia, most large mammals became extinct when these continents were settled by modern *Homo sapiens*. Therefore, the agriculturalists on these continents had no herd animals to breed. It has been speculated that large domestic mammals have boosted civilizations in Eurasia, giving them a decisive edge in economical and technological development, which in turn translated into military and political dominance.

The world population, which had been small and fairly constant, began to grow rapidly after the agricultural revolution. Presumably, the growth began with shorter intervals between births because infants could be weaned at an earlier age. With growth came overgrazing and forest clearance, portending some of the serious problems we are struggling with today (see Part V of this book).

Appendix 1: Dating Fossils and Artifacts

Paleontologists use two broad categories of dating methods: absolute and relative. Many absolute dating methods rely on the constant rate of decay of certain radioisotopes.

Radiocarbon dating can be applied directly to organic matter. While most of the atmospheric and organic carbon is the carbon 12 isotope, a small fraction is carbon 14. Upon an organism's death, no new carbon is added to the tissue, and the C14 present decays with a half-life of 5,730 years. Thus, after determining the ratio of C14 to C12 in organic matter, one can compute how long the matter has been dead. This method is most useful for objects younger than 40,000 years.

The potassium-argon method is used for dating the rock surrounding more ancient fossils, 300,000 years and older. About 0.01 percent of all potassium in the Earth's crust is potassium 40, which slowly decays into argon 40. The greater the ratio of argon-40 to potassium is in a rock, the more time has passed since the rock was last at high temperature. Further radioactive decays used for dating rocks are uranium to lead, and rubidium to strontium.

Other absolute methods count electrons trapped by flaws in crystals, flaws that in turn are generated by radioactive elements in rock. For objects between 40,000 and 300,000 years of age, paleontologists may use thermoluminescence (TL) or optically stimulated luminescence (OSL). Both methods measure the light that is emitted when trapped electrons are liberated by heat or light of certain wavelengths. Thus, these methods measure the time elapsed since a stone tool, for example, was heated by fire or exposed to the sun before it was buried.

While TL or OSL can be applied only once to a given sample, electron spin resonance (ESR) can be used repeatedly. Here, trapped electrons are excited by microwaves, and the resulting signal is proportional to the number of trapped electrons.

Yet another method relies on periodic changes in the magnetic field of the Earth, which was reversed 0.7 mya, 2.5 mya, and 4.4 mya. Sediments containing magnetically charged particles represent a fossil compass indicating whether the Earth's magnetic field was like today or reversed at the time of sedimentation.

If elements suitable for absolute dating methods are not available, the paleontologist may rely on relative methods such as *biostratigraphy*. The fossil remains of certain organisms have changed distinctly during their evolution. For instance, the shells of diatomes and the teeth of rodents, elephants, antelopes, and pigs have proved particularly useful as markers of time. Of course, these markers, as well as the geomagnetic changes, were originally calibrated at sites where absolute methods could also be used.

Appendix 2: Evolution and the Book of Genesis

The theory of evolution, especially its application to the human, is often seen as contradictory to religion, in particular to the first book of the Judeo-Christian bible, known as the Book of Genesis. It states that God created the entire world including all plant and animal species in one week, and it implies that species have not changed since then. This controversy is a major issue in U.S. politics, but I believe it is unfortunate and completely unnecessary.

Humans have always been wondering about their origin and their relation to the rest of nature. Religious documents from different cultures give different answers in the form of myths. (The term "myth" is used here not in its derogatory sense of a simply erroneous belief but in its original sense of a legend that relates the beginning and the basic customs of a people.) For instance, myths of Northwest American Indians tell about humans turning into bears or ravens, or vice versa, often implying human admiration for animal traits, such as the bear's strength or the raven's smartness. Some Native American myths are about humans being helped into existence by animals. A much different view is expressed in the Book of Genesis, which says that God created man in His image, and that He gave man dominion over all other creation. Here, human dominance of nature is legitimized as divine order.

It is important to realize that the Book of Genesis may not have been meant to be a scientific report. Rather, it appears to be a spiritual document in which the authors have expressed their thoughts and feelings about the human condition. Supporting this interpretation, the first two chapters of the First Book of Moses describe the creation of humans in different ways. According to 1. Moses 1: 27–28, God created a man and a woman simultaneously, both in His image. According to 1. Moses 2: 7 and 21–22, God first molded a man from earth and then a woman from one of the man's ribs. Thus, it appears that the biblical accounts of creation may be spiritual documents in which multiple authors have expressed their ideas about being part of nature yet having unique attributes, about the relationship between the genders, about having a superior intellect and a knowledge of good and evil, as well as other topics.

Those who take the Book of Genesis as a scientific report are led to believe that species are immutable, and that they have all been created simultaneously. These beliefs are incompatible with overwhelming scientific evidence and are therefore rejected by most scientists. Those who view the Book of Genesis as a spiritual document appreciate it as such and do not perceive it as contradictory to the theory of evolution. They view science and religion as two separate realms of the human mind, each with its own rules and purpose and not as conflicting with each other.

While everyone is free to interpret the Book of Genesis as they deem appropriate, it is worth pointing out that the Catholic Church has given its formal stamp of approval to the theory of evolution. In an official statement sent to the Pontifical Academy of Sciences in October 1996, Pope John Paul II acknowledged that an overwhelming body of evidence supporting the theory of evolution could no longer be ignored. Several Protestant theologians have also stated that the Book of Genesis and other biblical contents should be viewed as spiritual documents, in which the authors reflect on how humans fit into a divine order.

EXERCISE PAGE FOR CHAPTER 4

Student Last Name _____

Student First Name _____

Discussion Section _____

On the course web site, use the link on the **syllabus** to find a pdf file of your **assigned reading** for this chapter. The same reading is referenced below. Use the bulleted list of questions to test your knowledge of this reading.

Cavalli-Sforza, L. L. (1991). Genes, peoples, and languages. *Scient. Amer.* **November 1991:** 104–110.

- Which human populations did the author select for study, and why?
- What data did the author collect, and how did he use them to calculate the genetic distance between human populations?
- Which data from other research groups does Cavalli-Sforza mention as corroborating evidence, and as a way of calibrating his own data with regard to time?
- What are the results shown in the top diagram on page 107? What do the red lines, yellow lines, and numbers mean?
- How does the author explain the resemblance between the linguistic and anthropological data shown on pages 108–109?

Feedback

On a scale from 1 to 10, the being the best, please rank the above reading for

1. Interest and Relevance to the topic (10 being most interesting/relevant) _____
2. Readability (10 being most clearly written, easy to understand) _____

In the remaining space, enter any comments that you may have on this reading.

CHAPTER FIVE

HABITUAL BIPEDALISM AND HAND PRECISION GRIP

While many mammals can stand and walk on their hind legs for short times, only for humans is this the standard way of getting around. This critical ability of ours is technically known as *habitual bipedalism*, but we will just call it **bipedalism** for short. Another human characteristic is the *opposable thumb*, the power to rotate the thumb until it faces the palm and the palmar sides of the other digits. With this **hand precision grip**, the hand is no longer an organ for grasping branches while swinging in a tree canopy but a sensitive tool for manipulating objects. In this chapter, we will review some of the major changes in the human musculoskeletal system that made bipedalism and the hand *precision grip* possible. We will also explore when during hominin evolution these changes arose, and what selective advantages they may have conferred at the time.

A. Comparative Skeletal Anatomy of Humans versus Extant Apes with Regard to Bipedalism

Nonhuman primates are generally quadrupedal, although, from a human perspective, it would be more descriptive to call them quadrumanual. Both the forelimbs and the hind limbs of apes are fairly similar to a human arm and hand. It is the human hips, legs, and feet that have evolved in a special ways. They give us the upright posture and striding gait that set us apart. We will look first at the skeletal features underlying bipedalism.

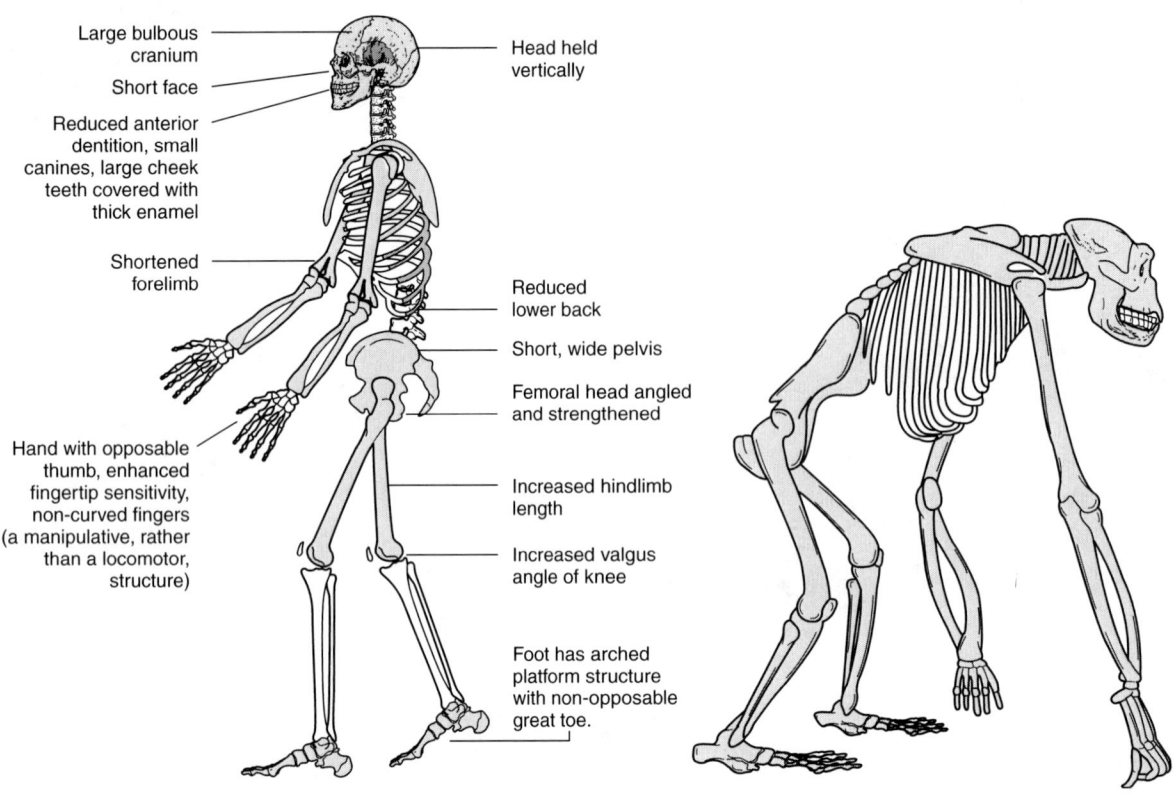

FIGURE 5.1:
Skeletal adaptations of the human to bipedalism
Redrawn after Lewin R. (1993) *Human Evolution: An Illustrated Introduction.* Cambridge: Blackwell Scientific Publications.

If one compares an entire human skeleton and an entire ape skeleton in lateral view, the following distinctions in regard to bipedalism are apparent (Figure 5.1).

- The human braincase is large and balanced atop the spine, in contrast to the smaller braincase of apes, which is sticking out forward. Correspondingly, the articulation between skull and spine, and the associated *foramen magnum* (opening for the spinal cord) are shifted from a posterior to a central position in base of the human skull (Figure S5.a).
- The human spine has a shock-absorbing double-S curvature, while the ape spine follows a simple arched curve.
- The human pelvis is wide and short, bringing the center of gravity close to the point of support at the hip joint. This reduces the tendency of the erect body to tip over.
- Human legs are longer than the arms, allowing longer strides. In nonhuman primates, the arms are longer than the legs, allowing more effective climbing and swinging in trees.
- The human foot forms an elastic platform for walking and running, rather than an anchor for grasping and holding.

The human leg shows some more adaptations to bipedalism that are less obvious. The **femur** (thighbone) can rotate on the **tibia** (shinbone) so as to lock the knee joint in the extended (straight) position, thus reducing the effort of standing upright. Also, the triangular portion of upper human *tibia* is made of trabecular, shock-absorbing bone. Additionally, the **valgus angle** (departure from a straight line) between human femur and tibia allows humans to center the body weight over one foot while the other leg is in motion (Figure 5.2). This facilitates the striding human gait, which contrasts with the waddling gait of nonhuman primates walking bipedally, who have no valgus angle and have to shift their body weight over the supporting foot at each step.

The human pelvis and foot show further adaptations to bipedalism, which will be discussed later in this chapter in the contexts of sexual dimorphism and time of evolution.

B. Functional Anatomy of Pelvis, Femur, and Gluteus Muscles

The most dramatic changes on the evolution of bipedalism occurred in the musculoskeletal system of the hip and the leg. We will focus here on the pelvis, the neck of the femur, and the role of three hip muscles, the **gluteus minimus**, **gluteus medius**, and **gluteus maximus**, which originate on the pelvis and spine and insert on the femur.

Walking consists of alternating **stance and swing phases**, separated by the **heel-strike** and **push-off** with the big toe (Figure 5.3).

During the stance phase, the human pelvis tends to tilt down on the unsupported side from the weight of the trunk (Figure 5.4). However, gravity is counteracted by

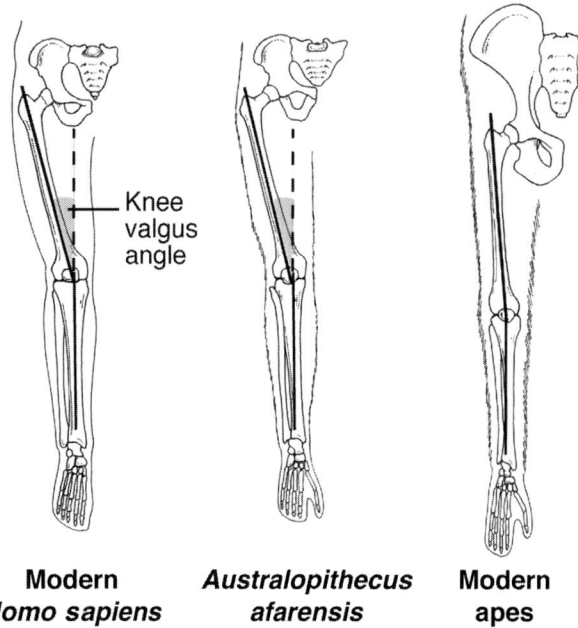

FIGURE 5.2:
Modern humans and australopithecines have a valgus angle between femur and tibia whereas the knee of apes is straight.

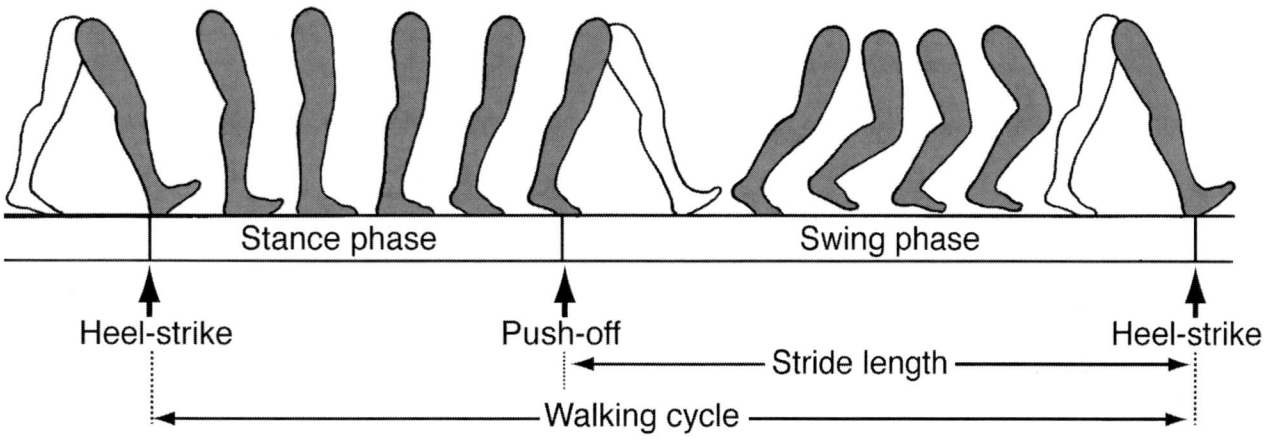

FIGURE 5.3:
Stance Phase and Swing Phase of Human Walking. Right leg is shaded. The stance phase begins with the heel strike, and the swing phase begins with the pushoff.
Redrawn after Lewin R. (1993) *Human Evolution: An Illustrated Introduction*. Cambridge: Blackwell Scientific Publications.

FIGURE 5.4:
During human walking, the *gluteus medius* and *g. minimus* muscles contract on the side that is in stance phase, preventing the pelvis from tilting down on the unsupported side, which is in swing phase.
Redrawn after Lewin R. (1993) *Human Evolution: An Illustrated Introduction.*
Cambridge: Blackwell Scientific Publications.

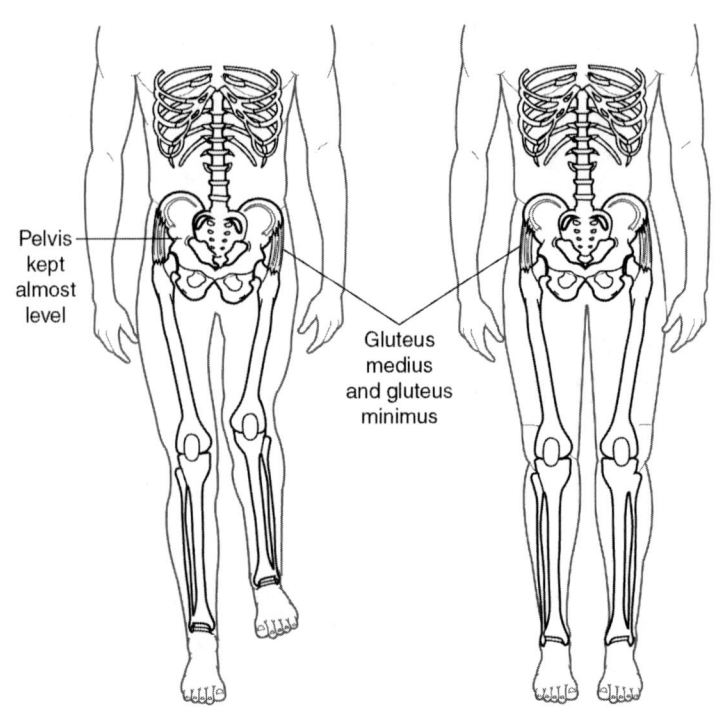

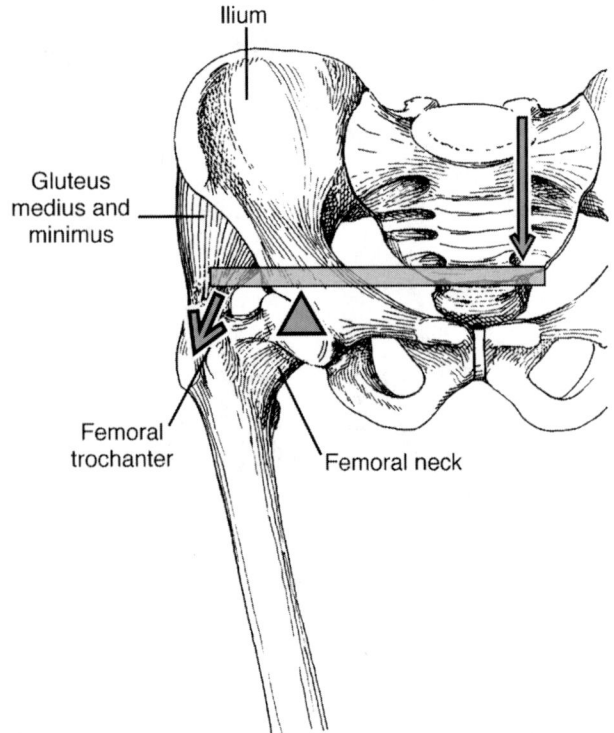

FIGURE 5.5:
In humans, the force (bold arrow) of the *gluteus medius* and *minimus* muscles balances the body weight (light arrow), with the hip joint (triangle) acting as a fulcrum for the pelvis (horizontal bar).
Redrawn after Lovejoy C. O. (1988) Evolution of Human Walking. *Scientific American.* **Nov. 1988:** 118–125

the *g. medius* and *g. minimus*, as they contract on the stance side (Figure 5.5). These two muscles are classified as abductors of the thigh because that is their action if the leg is free to move. By contrast, the corresponding muscles have a different function in apes, where they extend (straighten) the hip joint and thus help to propel the body forward.

The key to understanding the changing role of the *gluteus* muscles in the evolution of bipedalism is the **ilium**, that is, the upper, flaring portion of the pelvis (Figure 5.6). In apes, the *ilium* is flat and almost parallel to the back, whereas in australopithecines and humans the iliac crest (upper margin) is curved around laterally and ventrally, thus moving the origins of the *g. medius* and *g. minimus* muscles from posterior to lateral.

In humans, the *g. medius* and *g. minimus* muscles originate *laterally* from the *ilium* and insert *laterally* on the **greater trochanter** (lateral bony knob on upper end) of the *femur* (see Figures 5.4 and 5.5). Being abductors of the thigh, they stabilize the pelvis during the stance phase of walking. This is in contrast to apes, where the *g. medius* and *minimus* muscles originate *posteriorly* on the ilium and insert *posteriorly* on the trochanter. In this situation, the *g. medius* and *g. minimus* extend the thigh, along with a third muscle, the *g. maximus*, which is the only extensor of the thigh in humans. Because the *g. medius* and *g. minimus* do not function as abductors of the thigh in apes, they do not stabilize the pelvis laterally during bipedal walking. Apes therefore have a waddling or swaggering gait, as they shift the center of their body mass over the new stance leg at each step.

The changed function of the human *g. medius* and *g. minimus* is associated with several other modifications in the musculoskeletal system of the hip and leg.

The *gluteus maximus* muscle is relatively small in apes, where it acts synergistically with the *g. medius* and *g. minimus* as an extensor of the hip joint. In humans, the *g. maximus* is the largest muscle in the body. It originates medially from the *ilium* and the lower parts of the spine, and it inserts posteriorly on the *femur*. The human *g. maximus* acts as an extensor of the hip in climbing, stair stepping, and running. In walking (when the *femur* remains nearly extended at all times), the *g. maximus* does not so much propel the body, but rather it prevents the trunk from pitching forward during the heel strike.

The cross section of the femoral neck reflects the different actions of *g. medius* and *g. minimus* in hominin versus nonhominin primates. The distribution of bone material in this section helps paleoanthropologists to decide whether a fossil primate was already bipedal. The use of this criterion is explained on page 123/124 of your assigned reading. The exercise questions at the end of this chapter focus on this aspect.

C. Sexual Dimorphism of the Human Pelvis: Ultimate and Proximate Causes

In many animal species, the sexes differ not only in their reproductive organs but also in body size, shape, weaponry, or ornamentation. Such differences between adult males and females, are known as *sexual dimorphisms* (see Chapter 2). They often enhance the fitness of their carriers through *sexual* selection while coming at a price in *natural* selection. The costly antlers of deer and the tail feathers of peacocks are well-known examples. Here we will discuss the sexual dimorphism of the pelvis as a uniquely human trait. The same example will also serve to remind us of another basic concept in biology: the existence of *ultimate causes*, which are unique to evolving subjects, in addition to the *proximate causes*, which are familiar from other sciences.

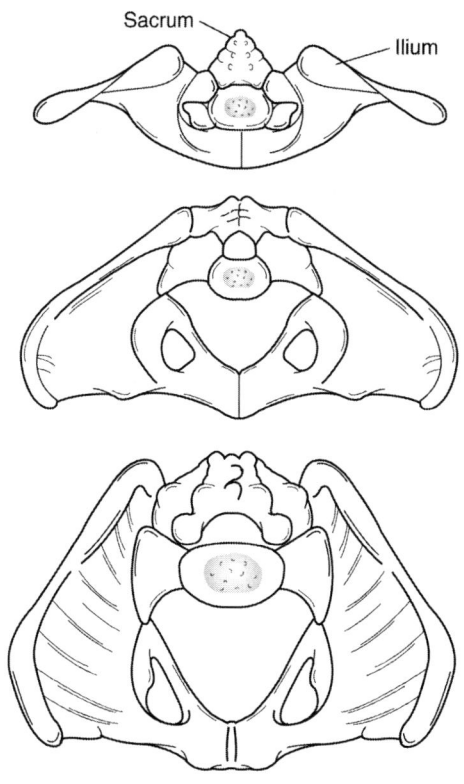

FIGURE 5.6:
Pelvic girdles of chimpanzee (top), *Australopithecus* (middle) and modern human (bottom) shown in view from the top. Note the laterally and ventrally directed bending of the iliac crest during hominin evolution.
Image © Kendall Hunt Publishing Company. Redrawn after Lovejoy C. O. (1988) Evolution of Human Walking. *Scientific American.* **Nov. 1988:** 118–125

The human pelvis is shorter than that of other primates (see Figure 5.1). This facilitates bipedalism by bringing the center of gravity close to the hip joint, as mentioned earlier. The human pelvis is also wide, especially in women, with an opening that is large enough to pass a child's head at birth. The need for a wide pelvis explains a human feature that is unique among primates: the constricted human waistline, or taille.

The fact that human childbirth is still difficult indicates that the width of the female pelvis is a tight compromise between opposing functional demands. On the one hand, childbirth requires a wide pelvis to accommodate the unusually large human skull. On the other hand, walking and running are more effective with a

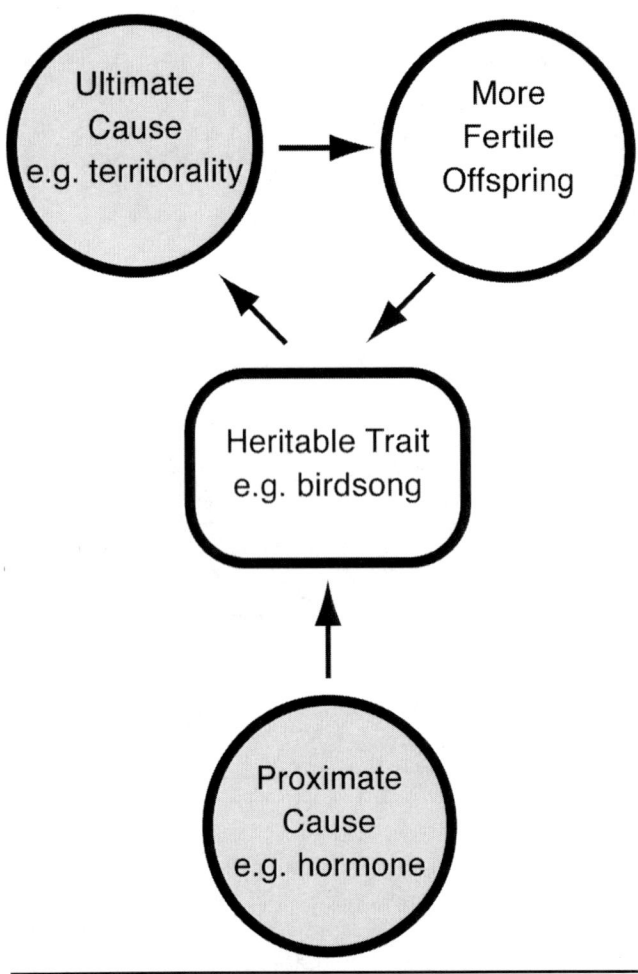

FIGURE 5.7:
Proximate and ultimate causes in biology

narrow pelvis because it provides a short leverage for the trunk weight, which needs to be counteracted by the pull of the *gluteus minimus* and *medius* (see Figures 5.4 and 5.5).

In general terms, the fitness-enhancing effect of a trait, which causes positive selection in evolution, is called its **ultimate cause** (Figure 5.7). Again, ultimate causes are specific to biology because its subjects evolve, while the subjects of other sciences do not. Ultimate causes must be distinguished from **proximate causes**, which are known from physics and other sciences that deal with nonevolving subjects. Proximate causes in biology are usually of a biochemical or physiological nature.

As an example of ultimate and proximate causes, consider the behavior of male songbirds at the beginning of the breeding season. Their singing attracts females while scaring away other males. The singing male will therefore breed successfully, and to the extent that the singing behavior is heritable, his sons will again be singers and successful breeders, and so forth. Thus, the ultimate cause of the male singing is the acoustical marking of a territory. The proximate causes of the same singing behavior include testosterone and its metabolites; this has been shown by experiments involving castration and hormone supplementation (see Chapter 17).

The ultimate causes controlling the width of the human pelvis act differently in the two sexes. For both men and women, a narrow pelvis helps to conserve energy during walking and running. For a woman, a wide pelvis reduces the risk of birth complications, which pose an immediate threat to her and to her newborn, and an indirect threat to any earlier children she may already have. For a man, the same risks are less of a threat to his reproductive success because he has better chances of having more children with other women.

Because of different bundles of ultimate causes in the two sexes, it is not surprising to see that pelvic width is sexually dimorphic (Figure 5.8). The female pelvis is wider, leaving a greater birth canal but also spacing the two hip joints at a greater and ergonomically less favorable distance. The sacrum (lower end of vertebral column) in women is also shorter and wider, and curved more dorsally. The greater distance between hip joints and the curvature of the sacrum affect the way women tend to walk with pelvic rotation (hip sway) and stand with hips relaxed to one side.

Most men are sexually attracted to women showing wide hips and walking with pelvic rotation. Men who prefer such women are likely to have more children because their mates, on average, are less prone to birth complications. Also, the same proximate cause that promotes wide hips (estrogen level) confers greater fertility. To the extent that the male attraction to women with hip sway is genetic, fathers will pass it on to their sons. Thus, in the course of evolution, men attracted to women with wide hips have outreproduced men who are not attracted to this female trait. (This does not mean that men make a *conscious* decision to avoid birth complications when they choose women with pelvic rotation. Most ultimate causes act unconsciously.)

D. Evolution of Hominin Bipedalism and Striding Gait

Of all the characteristics that set humans apart from apes, bipedalism seems to have evolved first. This did not immediately include our striding gait, which came later.

Paleoanthropologists use several criteria to decide whether a fossil hominin species was already bipedal. We have already mentioned the curvature of the spine, the length of legs relative to arms, the position of the foramen magnum in the base of the skull, the short and wide shape of the pelvis, the curvature of the iliac crest, the bone distribution in the neck of the femur, the valgus angle between femur and tibia, and the shock-absorbing bone in the head of the tibia.

Perhaps the single most suitable structure to trace the evolution of bipedalism and striding gait is the primate foot. The foot skeleton of a modern human is strikingly different from the feet of nonhuman primates, both extant and fossil (Figure 5.9). The

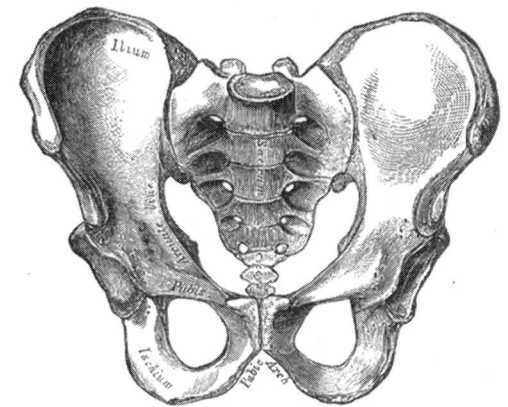

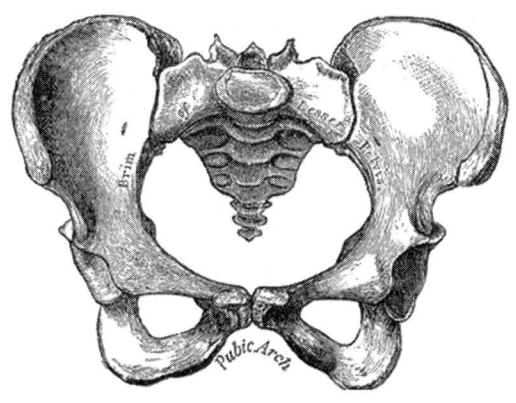

FIGURE 5.8:
Male pelvis (top) and female pelvis (bottom), drawn to the same height. From Gray's Anatomy

latter are all flexible, nimble grasping organs, with an abductible first toe and curved lateral toes, more like human hands than human feet. In contrast, the human foot is specialized for bipedal walking and running. Its tarsal and metatarsal bones are tightly bound by tendons, forming an arched platform. The first toe is big and strong, nonabductible, and nonopposable, suited for a powerful push-off. The other toes are small with limited mobility. The human *calcaneous* (heel bone) is especially large and strong, offering a large insertion for the "Achilles tendon," the tendon of the *gastrocnemius* (calf) muscle.

How old is bipedalism? As mentioned earlier, *Orrorin tugeniensis* may have been bipedal, but the evidence is still scant. *Ardipithecus ramidus* walked bipedally on the ground but used his feet as a second pair of hands when climbing in trees. A wide range of data shows that *Australopithecus afarensis* was a bipedalist.

- The pelvis has iliac crests curved ventrally (see Figure 5.6).
- The insertions for gluteus medius and minimus on femoral trochanter are shifted laterally.
- The neck of femur shows thinning of upper aspect, indicating that gluteus medius and minimus muscles counteracted the bending stress caused by gravity.

- The valgus angle and shock-absorbing bone in tibia are as in modern humans (see Figure 5.2).
- Footprints at Laetioli show bipedal walking.

The early forms of bipedalism, however, were apparently *not* associated with a **striding gait**. Australopithecines could not lock their knee joints in the extended position the way humans can. We therefore think that australopithecines stood and walked with a bent-hip–bent-knee posture. Most importantly, their big toes were not built for the strong push-off required in striding. The Laetioli footprints mentioned in Chapter 4 clearly show that their big toes were abductible (Figure S5.b). The same is indicated by foot bones found in South Africa and dated 3.5 mya, which include the *cuneometatarsal joint*. It connects a tarsal bone, called the *cuneiform*, to the first metatarsus, which in turn supports the big toe. The *cuneometatarsal* joint of *Australopithecus* shows a distinct curvature indicating that the first metatarsus, and with it the big toe, could be diverged from the rest of the foot (Figure S5.c).

Did *Homo habilis* have the striding gait? A set of foot bones assigned to this species shows a metatarsus I that is almost fully aligned with metatarsals II to V (Figure S5.d). The terminal bone of a big toe found in a younger sediment of the same area is large enough to deliver a strong push-off. Based on this limited evidence, it is assumed that *Homo habilis* had a striding gait, or was at least well underway towards this advanced stage of bipedalism.

Homo erectus, with fully modern feet, legs, and pelvis must have had a fully striding gait and apparently was an excellent runner.

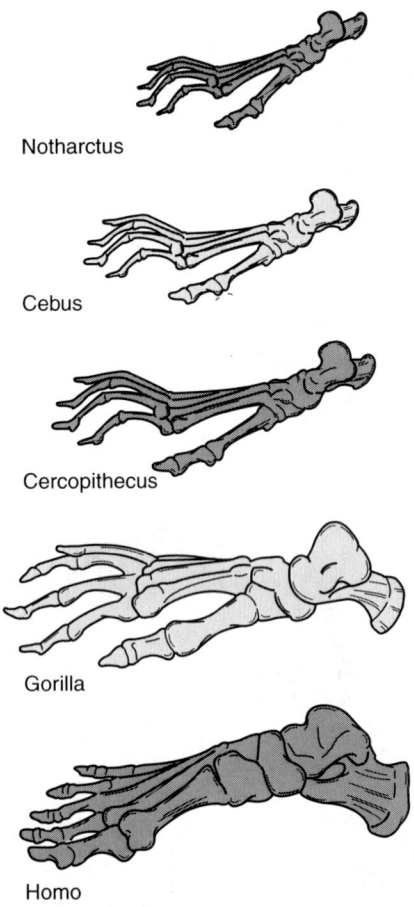

FIGURE 5.9:
Foot skeletons of the human and four non-human primates. *Cercopithecus* and *Cebus* are Old World and New World monkeys, respectively. *Notharctus* is a fossil primate, dated 40 mya.
Image © Kendall Hunt Publishing Company. Redrawn after Harrison G.A. et al. (3rd ed. 1998) *Human Biology*, Oxford University Press.

E. Possible Benefits and Risks of Bipedalism to Early Hominins

Why would random mutations that promoted bipedalism have been positively selected for? The current hypotheses were developed under the assumption that bipedalism evolved in savannas or open grasslands. This assumption may have to be revised depending on the habitat(s) in which the oldest hominins are found.

According to the **"better view" hypothesis**, bipedalism evolved because it affords a better view over long distances. Bipedalists moving in tall grass can survey their surroundings constantly and not just periodically, as nonhuman primates do when they pause to stand up on their hind legs in order to raise their heads. Thus, bipedalism could mean a critical improvement in predator avoidance.

The **agility hypothesis** emphasizes a bipedalist's ability to quickly swivel around the vertical axis. This ability allows him/her to quickly recognize, evade, or fight a predator coming from behind. Such agility is a major advantage over a large quadrupedal attacker, as any bullfight demonstrates.

The **"man-the-provider" hypothesis** focuses on the ability of bipedalists to carry large objects on their arms and shoulders. Thus, males could have carried carcasses obtained by scavenging to "base camp," where they would trade the prized meat for sex with females. Because meat is more nutritious than plant food, the females could raise more young, many of which would bear the providers' genes.

Another potential advantage of bipedalism may have been more efficient travel between shrinking food patches. Indeed, the walking of modern humans is **more energy efficient** than the quadrupedal moving of extant chimpanzees. Of course, the important question is whether the bipedal walking of early *Hominini* was more energy efficient than the locomotion of early *Panini*, and this question is much more difficult to answer.

Finally, bipedalism may be viewed as a **thermoregulatory adaptation**. A large mammal foraging during the day in an open habitat must cope with the heat from solar radiation. During midday, a hominin with an upright stance exposes a smaller surface to the sun than an ape of the same body weight with a quadrupedal stance. A bipedalist therefore needs less water for sweating or can forage during more hours of the day.

Evidently, the adaptive value of bidedalism depends on several variables including the distance between food patches, threat from predators, time spent in trees versus out in the open, temperature, length of time needed for foraging each day, and the availability of drinking water. For most of these variables, a cooler, drier, and more seasonal climate would seem to tip the scales in favor of being bipedal. As discussed earlier, such a climate change has indeed taken hold early during hominin evolution. It is therefore plausible that the bipedalism of early hominins, even though it was less than perfect, was positively selected for and became a stepping stone for further evolutionary events.

F. Evolution of the Hand

While the human foot is a specialized organ, extensively remodeled from the foot/hand of early primates, the human hand conserves a primitive state, with a few modifications that render the thumb fully opposable and make the terminal phalanges broader and stronger.

The human hand is a very flexible organ, which can assume a range of positions from fully extended ("high five") to forming a cup to closing into a fist. Correspondingly, the digits (fingers) can be stretched, curved, or curled. This is in contrast to apes, which use their hands mostly with curved fingers. Also in contrast to nonhuman primates, the terminal phalanges of human fingers are broader, in accord with their more versatile use in touching and holding.

The skeleton of the human hand shows eight *carpal* (wrist) bones, five elongated *metacarpal* bones, and five digits (Figure S5.e). The first digit (**pollux**, or thumb) consists of two phalanges and the other fingers of three. The joint between one of the carpals, called the **trapezium**, forms a saddle joint with the first metacarpal. This joint allows rotation within a cone of about 45 degrees.

The muscles that move the thumb include the strong **abductor pollicis** and **adductor pollicis**, which diverge the first metacarpal—and hence, the thumb—or align it with the other metacarpals, respectively. The **opponens pollicis** muscle sweeps the thumb across the palm. It originates from the *trapezium* and inserts on the radial side of the palmar surface of the first *metacarpal*. Along with the general mobility of the human hand, the anatomy of the thumb permits the power grip and the precision grip (see Figure 2.7).

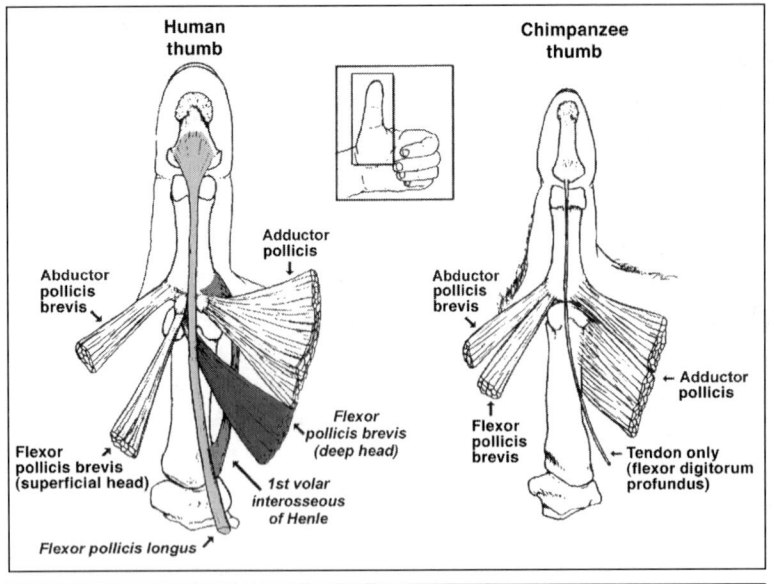

FIGURE 5.10:
Muscles that move the thumb. The shaded muscles with italicized names flex the human thumb and are not present in apes.
Redrawn after Sussman R. L. (1994) Fossil evidence for early hominid tool use. *Science* **265:** 1570–173

Humans also have three muscles that flex the thumb and are not present in apes (Figure 5.10). They are known as **flexor pollicis longus, flexor pollicis brevis**, and **1st volar interosseus of Henle**. Corresponding to the greater force exerted by these additional thumb flexors, the first metacarpal bone is stronger with a broader head in humans as compared to chimps and bonobos. The first metacarpal's head width/length ratio can therefore be used as a proxy for hominin hand evolution. According to this criterion, the characteristics of human hand bones were fully developed in *Homo erectus*, and at least partially in *H. habilis*, while the hand of *Australopithecus afarensis* was still apelike. The last point is confirmed independently by the long, curved phalanges of *A. afarensis*, which were still adapted to tree climbing.

The appearance of human hand features in *Homo habilis* coincides approximately with the age of the first hominin stone tools, dated 2.5 to 2.7 mya (see Chapter 6).

Although the evolutionary steps from the early primate hand to the modern human hand were smaller than those observed for the foot, hand evolution followed foot evolution rather than preceding it. Indeed, bipedalism may have had a "coattail" effect on positive selection for a strong opposable thumb. Such a thumb could be a hindrance for a nonbipedal species swinging in the tree canopy, since it would get in the way of quickly hooking a hand over a branch.

More generally, it appears that bipedalism has "freed" the hominin forelimb to evolve into a new organ no longer constrained by the functional requirements of walking or tree climbing. The evolution of wings in the reptilian ancestors of birds seems to be an interesting parallel. In the case of the hominins, the evolution of the hand is closely associated with the appearance of the first stone tools, which will be discussed in the next chapter.

EXERCISE PAGE FOR CHAPTER 5

Student Last Name _____

Student First Name _____

Discussion Section _____

On the course web site, use the link on the **syllabus** to find a pdf file of your **assigned reading** for this chapter. The same reading is referenced below. Use the bulleted list of questions to test your knowledge of this reading.

Lovejoy, O. (1988). Evolution of human walking. *Scient. Amer.* **Nov. 1988:** 118–125.

- What are the mechanical forces imposed by the body weight on the upper aspect and on the lower aspects of the femoral neck? To which of these forces is bone tissue more resistant?
- At the same time, what force is placed on all aspects of the femoral neck by contraction of the gluteus medius and minimus muscles?
- If you superimpose the forces from body weight and gluteus action, what is the overall stress pattern placed on the femoral neck of the human?
- How does the overall stress pattern imposed on the femoral neck of a chimp differ from that of the human, and why?
- How should the different stress patterns placed on the femoral necks of humans and chimps be reflected in the distribution of bone tissue as seen in cross sections?
- How does the femoral neck of an australopithecine look in cross section, and what does this tell us about bipedalism in these early hominins?

Feedback

On a scale from 1 to 10, the being the best, please rank the above reading for

1. Interest and Relevance to the topic (10 being most interesting/relevant) _____
2. Readability (10 being most clearly written, easy to understand) _____

In the remaining space, enter any comments that you may have on this reading.

CHAPTER SIX

TOOLMAKING OF MONKEYS, APES, AND HOMININS

Since humans lack natural weapons, such as claws and large canine teeth, and because they cannot outrun large carnivores, their survival depends on **tool use,** such as throwing rocks and branches (Figure S6.a). An advanced step is **tool making,** that is, the purposeful modification of natural objects to render them more effective as tools. Tool making is not a uniquely human trait because some nonhuman primates also make tools. However, these are much simpler, and they are made from perishable materials, such as plant shoots or vines. Only hominins have left stone tools, and they reveal a lot about how their makers lived. In this chapter, we will compare the tool kits of chimpanzees and some of the most primitive people living today. We will also explore what the fossil record tells us about hominin stone tools, and indirectly, about the human brain.

A. Tool Use and Toolmaking of Captive Macaques and Chimpanzees

Humans have always wanted to know whether human activities, from drinking beer to painting pictures, could also be observed in nonhuman primates. As models for human tool use and toolmaking, researchers have studied these capabilities in apes and monkeys kept in experimental stations.

A classical study, published by Wolfgang Köhler in 1927, was done on chimpanzees at the Anthropoid Station of the Prussian Academy of Science on Tenerife, Canary Islands. The chimps were given **human implements,** such as ladders or boxes, which they could use to reach bananas or other treats suspended overhead. Köhler's star student was a male named Sultan. He was given a narrow bamboo stick that would fit snugly into a wider stick, so that together they would be long enough to reach a banana. Sultan worked on the sticks for over an hour (Figure S6.b). When he had finally stuck them together, he was delighted and immediately used the composite tool to retrieve the food. Sultan also learned to stack boxes to reach a treat.

Köhler and others who reviewed his results noted typical mistakes made by the chimps. They would lean a ladder flat against a wall, apparently for the **optical effect** of maximizing its height, while ignoring the instability of the ladder's position. Similarly, boxes to be used as stepping stools were positioned on one corner, apparently again for the optical effect of maximum height but very unstable. When looking for an elongated object to be used as an angling stick, the chimps tried to use a rope, not anticipating that it would be too limp for the purpose. Another time, the experimenters pulled up bananas using a rope and a pulley, placed the metal ring on the free end of the rope, and put the ring over a nail in the wall. The chimps would simply have to take the ring off the nail to release the bananas. Instead, they tried to bite off the rope.

In a more recent study by Tokida and coworkers on macaques (Old World Monkeys), individuals were challenged to retrieve an apple from the middle of a horizontal, transparent plastic tube. First, they were offered a prepositioned, hooked stick in the tube. They quickly pulled on it in order to retrieve the apple. Subsequently, they learned to bring the stick to the tube from increasing distances, and to select an appropriate stick from a selection of sticks with varying usefulness. One female, Tokei, was particularly adept. Eventually, she learned to pull up a shrub from the ground, pluck the leaves, and bite off most of the roots so that the stick could be used as a retrieval instrument.

Tokei then demonstrated a remarkable ability for **problem solving**. She discovered that she could throw a stone from one end of the tube to drive the apple out the other end. When Tokei's consorts pilfered her apple as it emerged from the tube, she learned to throw the stone with less force so that she had time to fend off the pilferers before they got the apple. Finally, Tokei brought her infants to the tube, pushed them in, and got them to fetch the apple for her.

How can apes and monkeys be so clever, even creative, in using sticks, stones, and their children but so inept in using human implements such as boxes and ladders? They do not spend much time just tinkering around with objects when there is no immediate incentive to do so. This is very much in contrast to human children, who spend much of their lives **exploring** the look, feel, and mechanical properties of things by banging their toys against furniture and the like. As a result, humans are later able to tell just from the braided look of a rope that it will not be stiff enough for angling. Of course, apes do not grow up with human implements. Another set of studies has therefore compared the tool kits that chimpanzees and humans made for themselves in their natural habitats.

B. Comparison of Human and Chimpanzee Subsistence Techniques

To get an idea what kind of tools the common ancestors of humans and nonhuman primates may have used, William McGrew and coworkers compared the tool kit of one of the most primitive human societies known today (Tasmanian aborigines) with the tool kit of a group of chimpanzees living in a similar natural habitat in Tanzania.

The tools of the Tasmanian aborigines were very simple. Their most complex artifact was a blind for catching birds. It consisted of four components: sticks, grass, bait, and a stone, which the bird catcher sitting inside the blind would use to knock down the bird. All other aboriginal tools consisted of single components. Straws to suck up sap; sticks with fire-hardened tips to be thrown as spears; torches made from bark; or baskets to carry shell fish. These aborigines used stones as they found them to chop down and sharpen sticks. However, they did not modify stones.

The chimpanzees were proficient at using sticks and vines to "fish" for termites and ants, which are a good source of protein (Figure 6.1). Fishing sticks were modified by stripping off bark and leaves and then biting the sticks to the right length. Sometimes a chimp would prepare several fishing sticks and set them aside for later use (Figure S6.c). Prepared sticks were poked into termite mounds and held for several seconds, while termites "attacked" the stick. The stick was then withdrawn and the attached termites licked off or stripped off by hand. In some chimp communities, the ends of fishing sticks were brushed up by biting, so that more termites would attack it. Chimps also used twigs to clean their teeth, both self-directed and socially. Additionally, chimps wadded up leaves to sponge for ants and fruit pulp.

FIGURE 6.1:
Chimpanzee fishing for termites by poking a twig into their nest. Termites attacking the tool are pulled out and licked (or stripped) off. Image © DLILLC/Corbis.

Other investigators observed chimps using unmodified stones to crack nuts. Recently, chimps in Senegal were seen preparing spears by biting the tips of sticks. The spears were jabbed into tree holes where bush babies (small primates) sleep during the day. Speared bush babies were pulled out and eaten. Most of the chimps using this hunting technique were females. Their method differs from the way males hunt in groups: They ambush their prey and kill it by beating and biting.

In comparison, the tools of the Tasmanian aborigines are more advanced than the chimp tools, but only modestly so. It seems reasonable to assume that a common ancestor had toolmaking capabilities somewhere in this range and that hominin toolmaking evolved further from there.

How many of the tools described in this section would have shown up in the fossil record and would have been recognizable as intentionally made?

C. Paleolithic Tools

Fossil tools are, for the most part, *stone* artifacts—hence, the time period when such tools were used has been named the **Paleolithic** (from Greek *paleo-*, "old," and *lithos*, "stone"). Such tools are made from lava, chert, and flint, which give sharp edges when they brake. Only from relatively recent times do we find tools made from softer materials like bones, antlers, and tusks. Wooden tools are only rarely preserved.

Stone tools are classified according to their apparent use as hammer stones, choppers, scrapers, and so on. Such names are based in part on microscopic inspection of the tools' wear and tear marks. The names also reflect the experiences of contemporary paleontologists who survive for months in the field using only self-made stone tools.

Fossil tools are also classified in reference to the period of time when they originated and according to the way they were made. Such tool **industries** are often named after the **type sites** where they were *first found*, which are not necessarily the sites where the tools were *first made*. Table 6.1 lists some prominent industries and the hominin species with which they are associated.

Table 6.1: Stone Tools

Period	Industries	Time (approx.)	Hominin species
Lower Paleolithic	Olduwan	2.5 to 1.2 mya	*Homo habilis*
	Acheulean	1.6 to 0.1 mya	*Homo erectus*
Middle Paleolithic	Levallois	350 to 100 kya	Archaic *Homo sapiens*
	Mousterian	200 to 30 kya	*Homo neanderthaliensis*
Upper Paleolithic	Many, world wide	50 to 10 kya	Modern *Homo sapiens*

D. Olduwan Stone Tools (Early Lower Paleolithic)

Olduwan stone tools are named after the site of their first discovery, the Olduvai Gorge in Tanzania. Dated as early as 2 mya, they are representative of the oldest hominin tools we know (Figure S6.d,e). They were made mostly from lava and more rarely from chert, a fine-grained aggregate of silica.

One type of Olduwan tool was an egg-shaped stone that was used without modification as a **hammer stone**. The other Olduwan tools are stones modified using the **hammer-percussion technique:** the toolmaker ("stone knapper") uses a hammer stone held with one hand to strike a **core stone** held in the other hand (Figure 6.2). First, the knapper hits the core stone so as to create a platform for future strikes. Next, the knapper delivers a forceful hammer stone blow near the edge of the striking platform. The percussion angle should not be perpendicular to the platform but somewhat acute (75 to 80 degrees).

A well-aimed stroke releases a thin, sharp-edged fragment called a **flake**. Such flakes are effective tools for cutting animal hides, literally opening up a most valuable food source: meat from animals. Hammer stones, of course, can also be used for cracking open skulls and bones to obtain highly nutritious brain and bone marrow.

By striking off flakes repeatedly, core stones can be modified into larger tools named according to their shape or apparent use. Cores with sharp edges running part way around the circumference are known as **choppers**. They are suited for recovering dried meat from a scavenged carcass, and for chopping off tree branches, which can then be used to dig up plant roots and tubers. Similar but flatter cores are called **scrapers** because they are best for removing fat from hide or for fashioning a point on

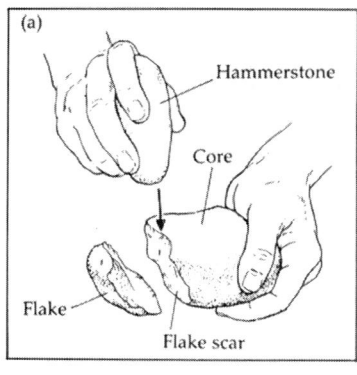
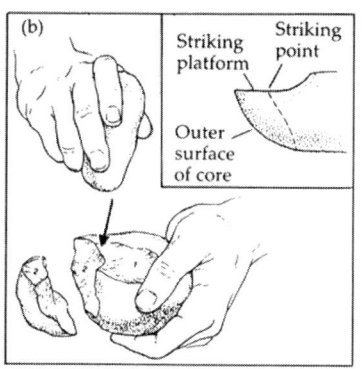
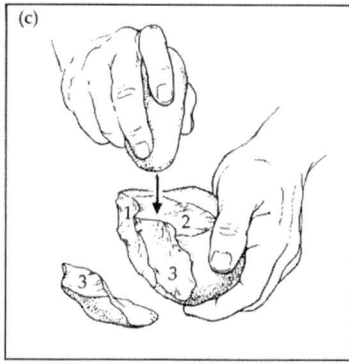
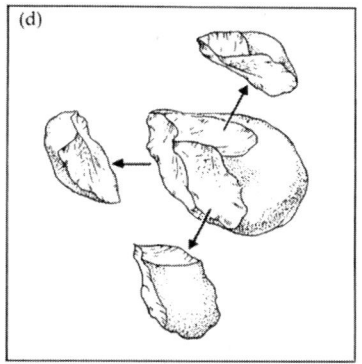

FIGURE 6.2:

Hammer percussion technique used to make the oldest known hominin stone tools From *Human Evolution* by Roger Lewin. Copyright © 2005 by Blackwell Publishing Ltd. Reproduced with permission of Blackwell Publishing Ltd.

a stick. Pointed sticks could be used to fend off predators, such saber-toothed tigers, as well as scavengers, such as hyenas, which would contest hominins for the carcasses of animals left partially eaten by predators.

Olduwan tools and *Homo habilis* bones were found in the same location, referred to as Bed I of the Olduvai Gorge, dated about 2 mya. Australopithecines were also around at the same place and time, but their hands, with nonopposable thumbs and curved, tapering fingertips seem less conducive to toolmaking than the hands of *H. habilis*. The latter hominins also had the larger brains. For these reasons, it is generally assumed that the Olduwan stone tools were made by *H. habilis*.

Similar stone tools, dated 2.5 mya, were found later in Gona, Ethiopia. Most of the edges of these tools are sharp, indicating that their makers were expert stone knappers. However, the Gona stone tools have not been associated with any hominin fossils so far.

Stone tools were of utmost importance for hominins trying to make a new living outside of their traditional habitat, the tree canopy. Living on the ground, they could tap into a rich new source of nutrition: the meat of the herbivores populating the new open landscapes. Stone tools gave them access to this meat, first by scavenging and later by hunting. However, in doing so, hominins made themselves vulnerable to large predators, such as saber-toothed tigers, and competing scavengers, including hyenas. Unable to outrun these enemies, and lacking natural weapons, early hominins were probably saved from extinction by their ability to throw and jab pointed sticks.

Incidentally, throwing objects and knapping stone tools both require the same types of movement: extending the lower arm and flicking the wrist with force and precision. Children and baseball players spend long hours practicing just these movements.

E. Acheulean Stone Tools (Late Lower Paleolithic)

Another prominent and long-lasting stone tool industry produced the **Acheulean Stone Tools**. They are named after Saint Acheul in Northern France, where they were first found. However, these tools were originally associated with *Homo erectus* in Africa. Compared with their Olduwan predecessors, the Acheulean stone tools are larger (compare Figures S6.d,f) and reveal a much higher degree of craftsmanship. Made from flint, they are trimmed bifacially, so that no part of the original core stone's surface is left unmodified. The tool's sharp edge extends over much of its contour, which may be ovoid or pear-shaped (Figure 6.3).

The signature tool of the Acheulean industry is the **hand axe**, which can be held with one hand grasping the rounded end of the tool while its pointed end is sticking out. It could be used for a variety of tasks including:

- chopping and cutting wood
- slicing animal hide

FIGURE 6.3:

Acheulean cleaver (a: left) **and hand axe** (b: right), bifacially trimmed from flint stone.
Fig. 6.3a: Artifact Credit: Courtesy Denise de Sonneville-Bordes, Centre Francois Bordes, Institut du Quaternaire, Batiment de Geologie, Universite de Bordeaux I; Photo Credit: © David L. Brill, humanoriginsphotos.com
Fig. 6.3b: Artifact Credit: Institut de Quaternaire, Batiment de Geologie, Universite de Bordeaux I; Photo Credit © David L. Brill, humanoriginsphotos.com

- dismembering and defleshing bones
- cutting bones
- as a weapon

Because of its multiple uses, the Acheulean hand axe has been dubbed the "Swiss army knife of the Paleolithic." However, the Acheulian tools were found in about a dozen different sizes and shapes, adapted for different uses. They have been made for about 1.5 million years, making them the dominant technology for the longest time in human history.

F. Middle Paleolithic Tools

In the Middle Paleolithic stone tool industries, the emphasis shifted from core tools, like the Acheulean hand axe, to flake tools that were larger and more advanced than their Olduwan forerunners.

The **Levallois technique**, associated with archaic *Homo sapiens*, works in four steps. First, the edges of a cobble are trimmed into a rough shape. Second, the upper surface of the core is trimmed to produce a convex surface, like a turtle shell, which anticipates the shape of the flake to be produced eventually. Third, a flake is removed from one end of the core to produce a striking platform for the blow that will detach the desired flake. Finally, the ultimate blow of the hammer stone is delivered to the prepared platform site, driving off a large flake that is shaped as anticipated during the preparation of the convex surface (Figure 6.4). Thus, the shape of the final flake could be adapted to its eventual use.

The **Mousterian industry** is named after its type site in Le Moustier, a rock shelter in the Dordogne region of France. Mousterian tools have been found with fossil remains of *Homo neanderthaliensis* and early populations of modern *Homo sapiens* in Africa and the Middle East. These tools are made from flint or chert using refined versions of the Levallois technique; the flakes were processed for special uses. For instance, some have a square tab protruding from the base, which apparently was used to haft the flake to the end of a spear.

In addition to stone tools, several wooden spears were recovered from a brown coal mine at Schöningen, Germany, and dated about 400 kya. The spears were expertly crafted from spruce, with the tips coming from the hard base of the trees. All spears were well balanced with the center of gravity about a third of the way from the tip. Up to 2.3 m long and 47 mm in diameter, these spears must have been thrown by

FIGURE 6.4:
Levallois technique. The core has been flaked deliberately around its edge before the large flake at the center could be struck off without damage to its tip.
Artifact Credit: Courtesy Denise de Sonneville-Bordes, Centre Francois Bordes, Institut du Quaternaire, Batiment de Geologie, Universite de Bordeaux I; Photo Credit: © 1985 David L. Brill, humanoriginsphotos.com

big and strong people—as we think archaic *Homo sapiens* were. The spears were found among the remains of wild horses, presumably the spear throwers' game.

G. Upper Paleolithic Tools

The tools of the Upper Paleolithic are the tools of modern *Homo sapiens* after 50 kya, which is also the time when modern *H. sapiens* began migrating out of Africa (see Chapter 4). Their tools are characterized by spectacular variety, inventiveness, and refinement.

The hallmark of upper paleolithic tools are long **blades** struck from flint cores. The key to making these blades was the use of a punch, which allowed the stone knapper to deliver the force of a hammer blow precisely to the right spot of the core (Figure S6.g). These blades were generally elongated in shape, and they had sharp, straight edges (Figure 6.5). They were processed into knives, scrapers, perforators, drills, arrowheads, spear tips, and other tools.

Perhaps the most versatile derivatives made from blades are tools called **burins**, which are used for gouging and engraving. They were made by trimming one end of a blade to a sharp angled tip similar to a modern box cutter or utility knife. Burins were used to fashion tools from wood, bones, antlers, and similar materials (Figures S6.h,i). Such tools included arrowheads, spearheads, and barbed harpoon tips, which could interchangeably be hafted to wooden projectiles. Needles were also important tools, used for sewing together animal hides into clothes, shoes, backpacks, and such.

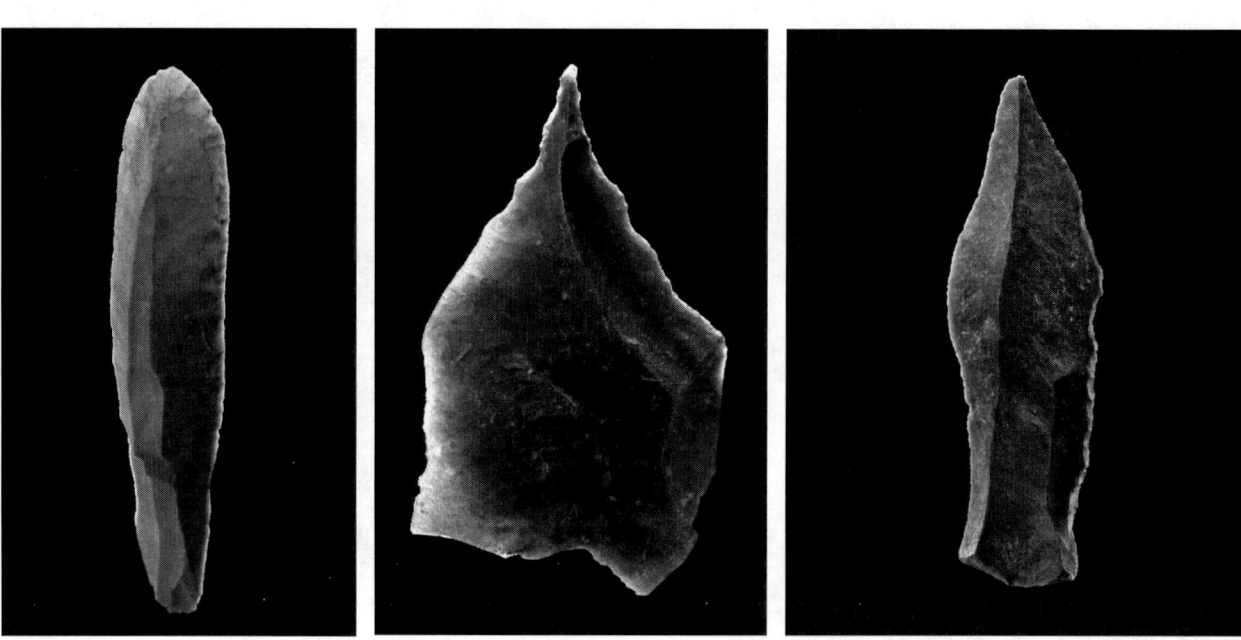

FIGURE 6.5:

Flint blades. Cro Magnon stone knappers were experts in striking long blades from flint cores. Such blades were often modified for use as scraper (a: left), perforator (b: middle), or back knife (c: right). Fig. 6.5a and c: Artifact Credit: Courtesy Denise de Sonneville-Bordes, Centre Francois Bordes, Institut du Quaternaire, Batiment de Geologie, Universite de Bordeaux I; Photo Credit: © 1985 David L. Brill, humanoriginsphotos.com

Fig. 6.5b: Artifact Credit: Courtesy R. Deffarge, Institut de Quaternaire, Batiment de Geologie, Universite de Bordeaux I; Photo Credit: © 1985 David L. Brill, humanoriginsphotos.com

To fashion a needle, converging tracks were gouged into a bone or antler until a blank for the needle could be lifted out. The blank then had a hole drilled into it with a pointed flint blade and was ground with sandstone to a fine sharp needle.

Spear-throwers, known as *atlatls* by native American peoples, were hand-held sticks of wood or antler, cupped at one end. Functioning as extended arms, they helped to impart more energy on thrown arrows and spears. Bow and arrow became the weapon *par excellence* for bringing down game (and unfriendly neighbors) from a safe distance but with great precision (Figure S6.j).

Upper Paleolithic tools, such as the barbed harpoon tip shown in Figure 6.6, are known mostly from Europe. However, similar harpoon tips were found by the Semliki River in the Republic of Congo—in accord with the hypothesis that Modern *Homo sapiens* originated in Africa (Figure S6.k).

H. How Much Skill Does It Take to Make a Stone Flake?

How much skill does it take to make a sharp stone flake? This question was addressed by trying to train a bonobo named Kanzi to do just that. Kanzi was a star student: He had been raised in captivity, understood spoken English, and had learned to communicate using a set of tile symbols (see Chapter 7). He also had star trainers including Sue Savage-Rumbaugh, the primatologist who had trained Kanzi to communicate with tile symbols, and Nicholas Toth, an experimental anthropologist specializing in stone tools.

FIGURE 6.6:
Upper Paleolithic harpoon tip carved from bone. Artifact Credit: Courtesy R. Deffarge, Institut de Quaternaire, Batiment de Geologie, Universite de Bordeaux I; Photo Credit: © 1985 David L. Brill, humanoriginsphotos.com

First, Kanzi was given an incentive to make stone flakes by showing him how they would gain him access to favored food items. These were originally enclosed in a box with a plexiglass lid that was secured with a string, which could be cut with a stone flake. Later, the food items were placed in drum-shaped containers with thick, transparent membranes that had to be sliced open, an operation that mimicked an early hominin's task of having to cut an animal hide in order to get to the meat underneath.

Then the trainers modeled the hammer percussion technique for Kanzi, who was an eager participant (Figure S6.l). Next Kanzi was left by himself with various lava and chert stones provided to him. He used the hammer percussion technique, but the flakes he produced were small and with stepped edges. He picked the larger flakes for use, indicating that he recognized their better quality. However, he did not learn to prepare a striking platform for subsequent hits at an acute angle.

Unexpectedly, Kanzi developed a technique of his own. He learned to produce flakes by throwing stones against hard objects, such as the laboratory floor or other stones, and he perfected his technique to where he would hit a stone lying 1 m away. However, the quality of the flakes produced by the throwing technique was poor.

Kanzi demonstrated his **problem-solving abilities**, but he did not acquire the stone knapping skills of *Homo habilis*. Whether his preference for an inferior technique reflected cognitive or anatomical limitations, or whether he simply sought to avoid the risk of hitting his core-holding hand with a hammer stone, is not clear.

The attempt to train Kanzi as a stone knapper has pointed out a possible gap in our knowledge about the evolution of stone tools. Whereas flakes and cores left behind by Olduvan stone knappers are clearly recognizable as artifacts, the products of Kanzi's work could be mistaken for stones damaged by natural causes. This means that hominins older than *Homo habilis*, such as the australopithecines, could have made coarse, unconspicuous stone tools using Kanzi's throwing technique, and they could have used these stone tools to fashion implements from wood and other perishable materials, but we would not know anything about it.

I. Were Early Hominins Right-Handed?

Some archeological sites show evidence of extensive stone knapping. By evaluating the production on such sites quantitatively, and by doing some experimental stone knapping himself, paleoanthropologist Nicholas Toth arrived at a surprising conclusion: Most of the ancient stone knappers were already right-handed (Figure 6.7 and assigned reading below).

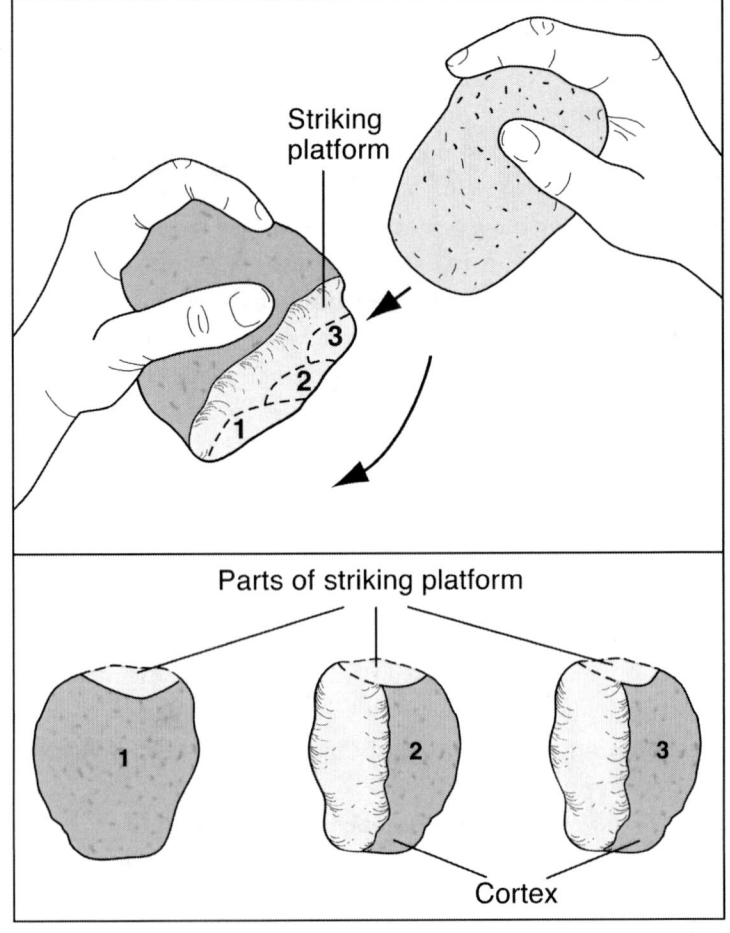

FIGURE 6.7:
A right-handed stone knapper will hold the hammer stone in the right hand. He will tend to rotate the core stone in the left hand *clockwise* between hammer stone blows. In the second and subsequent flakes, the cortex is to the right when the striking platform is pointing up. Redrawn from Lewin R. (1993) *Human Evolution: An Illustrated Introduction.* Cambridge: Blackwell Scientific Publications. p. 123

The conclusion of Toth extends many observations made earlier. Extant humans are, for the most part, right-handed. Historical drawings, paintings, and sculptures of people using tools or weapons indicate that most people were right-handed from 5 kya to present. In caves with Paleolithic art, hand prints on the walls are mostly from left hands, indicating that the right hand was used for painting.

In modern humans, handedness is part of a more general phenomenon known as **brain lateralization**, with the control of certain activities occurring predominantly in one brain hemisphere. Toth's evidence suggests that brain lateralization occurred early in hominin evolution. We will continue the topic of brain lateralization in the following chapter.

EXERCISE PAGE FOR CHAPTER 6

Student Last Name _____

Student First Name _____

Discussion Section _____

On the course web site, use the link on the **syllabus** to find a pdf file of your **assigned reading** for this chapter. The same reading is referenced below. Use the bulleted list of questions to test your knowledge of this reading.

Toth, N. (1985). Archaeological evidence for preferential right-handedness in the Lower and Middle Pleistocene, and its possible implications. *J. Human Evol.* **14:** 607–614.

- What does Toth call a "right-oriented flake" and a "left-oriented flake"?
- What does a right-oriented flake suggest about its original position on the core stone, relative to the previously struck flake?
- When a stone knapper turns the core stone clockwise between successive blows with the hammer stone, will the resulting flakes be right-oriented or left-oriented? To find out, see Figure 6.7 and flake an apple with a knife. What flakes are produced when the core stone is rotated counterclockwise?
- What kind of relationship does Toth postulate between a stone knapper's handedness and the way he or she turns a core stone during stone knapping? How does Toth support his contention?
- Thus, should there be a bias in the types of stone flakes produced by right-handed versus left-handed knappers?
- What stone flake collections did Toth examine for the frequency of right-oriented and left-oriented flakes?
- What does Toth conclude from his data?
- What is the significance of the study mentioned in the "postscript"?

Feedback

On a scale from 1 to 10, the being the best, please rank the above reading for

1. Interest and Relevance to the topic (10 being most interesting/relevant) _____
2. Readability (10 being most clearly written, easy to understand) _____

In the remaining space, enter any comments that you may have on this reading.

CHAPTER SEVEN

LANGUAGE AND BRAIN EVOLUTION

A complex spoken language is one of the uniquely human characteristics. Chimpanzees and bonobos can learn sign languages. If they interact with humans on a daily basis, they also come to *understand* spoken English, but they do not learn to *speak* it. So are there any special attributes to the human neck anatomy that make us superior speakers? It turns out there are. A critical one is the space above our voice box, which enhances the clarity and articulation of our speech. But because this space is a common passageway for *air and food*, it comes with a price tag: the hazard of choking. Evidently, improved speech and choking hazard have been a package that was positively selected for in hominins but not in other primates. Why in us and not in them? Evolution acts on whole organisms, and the space above our voice box evolved in a context that other primates did not have: bipedalism, toolmaking, and a big brain.

A large and complex brain is another uniquely human trait. It enhances many of our capabilities far beyond the level of apes, from superior toolmaking to creating an inner world of memories and plans for the future. But again, our brain comes at a price: It costs a lot of energy to build and maintain, it has made birthing difficult, and it has made child rearing a long and arduous task. In this chapter, we will explore how speech and other complex human activities may have provided the ultimate cause for the evolution of our expensive brain. We will also use language and brain as examples to illustrate the general concept that human traits may have evolved as an interactive system.

A. Definition of Language

The minimal requirements of any **language** are symbols, referents, and a connecting code. Each language has a set of **symbols**, such as spoken words or the gestures of sign language. The symbols are the vocabulary that one needs to memorize in order to learn a language. Each symbol is associated with a **referent**, which may be an object (apple), action (eating), or state of mind (hunger). The **code** defines the relationship between referents and symbols, such as the "meaning" of words. Code is normally learned by growing up with elders and peers. In special situations, such as a military operation, participants may agree to use a secret code, in which "apples" may refer to the enemy's ammunition, and "eat" may mean to destroy. Clearly, then, for a language to work, speaker and listener must apply the same code.

In addition to symbols, referents, and code, spoken human languages have *syntax* and *grammar*, which need to be taken into account in encoding and decoding. "Dog bites man" has a different meaning from "man bites dog." Complex relationships between events and activities, such as conditional relationships, are mediated by auxiliary verbs and conjunctions as in "I *may* eat the apple *when* I am hungry."

Each of us has spent years on learning our first spoken language, and many of us have learned additional languages, spoken ones or others. Generally, humans are adept at coordinating complex series of fast and precise movements, not only in speech, but also in athletics, in dance, or in playing a musical instrument. We have great respect for those who excel in languages, sports, or performing arts. Thus, we can take language as an example of a more *general human ability to learn to do just about anything with speed and precision, for survival or just for fun.*

B. Do Animals Have Language?

Do social animals, especially nonhuman primates, use spoken language in their natural environment? Vervet monkeys (*Cercopithecus aethiops*) live in groups of about twenty-five individuals in savannahs of East Africa (Figure 7.1). When they see predators, they sound **alarm calls** that alert the rest of the caller's community. The type of alarm call given depends on what kind of predator the caller is seeing. A loud barking call is sounded for leopards, a short snore for eagles, and a ratcheting sound for snakes. The response of other monkeys to each type of call is appropriate to evading the predator that triggered the call (Figure S7.a). When the leopard call is heard, the monkeys run up the next tree; the eagle call makes them look up in the air and seek cover; hearing the snake call has them standing up on their hind legs and looking around in the grass.

Are these alarm calls language? Play-back experiments with audiotapes have shown that the calls alone trigger the appropriate responses; the caller's body language is dispensable. The calls have no similarity to sounds that the predators make, so each

FIGURE 7.1:
Vervet monkeys give specific alarm callls at the sighting of their main predators: leopards, eagles, and snakes. The responding monkeys' escape behaviors differ accordingly.
Image © Paul Souders/Corbis

call is a symbol rather than an imitation of the predator. Sounding an alarm call is not part of the evasive action taken by the caller himself, so that the response of the group members involves a decoding process rather than being a simple imitation of the caller's behavior. Finally, at least some of the callers *know* what kind of response their alarm calls will trigger, because they can *lie*. When two troupes of vervet monkeys clash, a member of the losing party sometimes sounds a fake alarm call, thus breaking up the fight and preventing more serious injury to his or her buddies.

Alarm calls have also been observed in other primate species, but distinct calls warning of different predators seem to be rare. Chimpanzees have alarm calls that can be heard from two miles away, but they do not seem to be specific as to the kind of threat. However, chimps bark differently when they encounter an abundance of food, and others will heed the call to join the feast. Youngsters shriek when they are lost or in some other predicament, and mothers will rush to the rescue. On close range, chimps communicate by displays, body language, and facial expressions. Males intent on affirming their rank fluff up their hair, throw rocks, beat up vegetation, and burst into wild charges. For reconciliation after fights, they hug and kiss. If they need help, they will approach others holding out an open hand. They smile when they are relaxed and friendly, and they press their lips together when they are ready to attack.

In summary, the alarm calls of vervet monkeys fulfill the criteria of a simple language because the callers use symbols encoding specific referents, and the callers expect that the listener will decode them accordingly. Overall, spoken language is not nearly as prominent and sophisticated among nonhuman primates as it is in humans. Instead, social primates rely more on body language, which plays a big role in their lives.

C. Primates Can Learn Sign Languages

In order to compare how human children and young apes learn languages, psychologists have trained chimpanzees and bonobos in American Sign Language and in the use of tile symbols with geometric patterns.

In a landmark experiment, Allen and Beatrice Gardner trained an adopted female chimpanzee, Washoe, to use **American Sign Language (ASL)**. Keeping Washoe in their household, they raised her like a human child. In Washoe's presence, the Gardners communicated only by ASL, and they taught Washoe to use ASL by modeling the signs for her and by holding her hands and putting them through the appropriate movements. During 51 months of training, Washoe learned to use 132 signs correctly and consistently (Figure S7.b). On occasion, Washoe would spontaneously combine ASL signs to refer to novel objects. For example, she signaled "water/bird" when she saw a swan. The Gardners concluded that Washoe had at least some understanding of the symbolic meaning of ASL signs.

Washoe then came to live with another trainer, Roger Fouts, who later joined her with four other chimps who had been taught ASL independently. To his surprise, the chimps started to communicate in ASL even when no human was present. Evidently, they were eager not only to learn ASL as a way of communicating with humans, but also to use it among themselves.

In another well-documented experiment, Sue Savage-Rumbaugh trained Kanzi, a young male bonobo, to communicate using geometric symbols called **lexigrams**. First, he was taught a vocabulary of lexigrams, which referred to objects or actions like "melon" or "tickle." He learned to recognize 109 out of 194 lexigrams all of the time and another 40 most of the time. Next, Kanzi had to match the lexigrams with words spoken to him over headphones (Figure S7.c). In double-blind tests, he pointed out the correct lexigrams, and missed only 3 words out of 109. Chimpanzees trained similarly learned to operate a keyboard with 100 lexigrams. For example, if they wanted M&M candy, they had to punch in five keys in this sequence: Please/Machine/Give/M&M/./

After years of interacting with humans, chimps and bonobos understand spoken English language well enough to carry out composite instructions. For instance, when Kanzi was asked to "give an onion to Panshiba," another bonobo, he went outside, pulled an onion out of the ground, came back inside, and gave it to Panshiba.

These results show that apes can master large sets of symbols and code to the point of combining them for new referents. With time and in a friendly environment, they learn to follow composite verbal instructions spoken in English language. They are eager to use their new communication skills not only with humans but also with peers. Still, their language abilities are modest as compared to those of human children.

D. Unique Position of the Human Larynx

The inability of nonhuman primates to speak human language does not seem to be rooted in cognitive problems. What else could be the limiting factor?

The critical human trait is the *position* of the voice box, or **larynx**. It contains the vocal folds, which vibrate to produce our voice. The larynx sits on top of the windpipe through which air passes on its way to and from the lungs (Figure 7.2). Human adults have a critical space above the larynx, known to anatomists as the **supralaryngeal space (SLS)**. Located behind the base of tongue, the SLS extends from the pointed end of the soft palate down to the larynx. The SLS has two critical functions. First, it serves as a conduit for passing *air* from the nasal cavity to the larynx and further down the windpipe to the lungs. Second, the same SLS passes *food and drink* from the mouth to the esophagus, the tube that runs behind the windpipe and ends in the stomach.

The dual function of the SLS as a passageway for air as well as food and drink has two problems. The first one is common to all mammals: the passageways for air and food/drink are *crossing over* in the SLS. Note that the nasal cavity lies over and

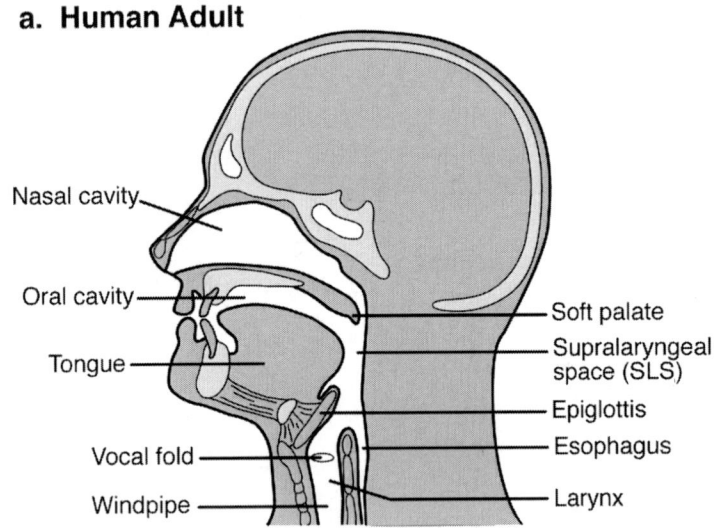

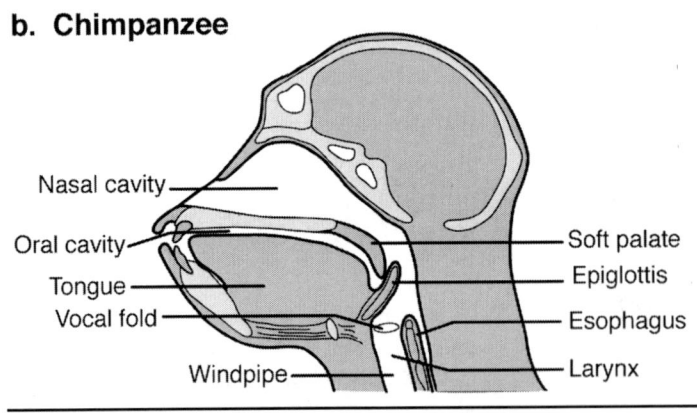

FIGURE 7.2:

Supralaryngeal space. Human and chimpanzee heads shown in median section. The low position of the human larynx creates a large supralaryngeal space, which does not allow the epiglottis to lock behind the soft palate.

behind the oral cavity, whereas the rest of the air passageway (larynx and windpipe) lies in front of the esophagus. Thus, food or drink might run down the "wrong way" into the windpipe were it not for a lid known as the **epiglottis**. It folds down when we swallow, closing off the larynx so that food passes over the epiglottis into the esophagus. At all other times, the epiglottis stands up, allowing air to pass through the larynx.

Ingenious a construction as the epiglottis is, it does not avoid the second problem of the SLS, which is found only in humans: the hazard of choking when a bolus of food gets stuck in the SLS. This is a life-threatening situation that must be ended quickly by expelling the bolus upward, and the Heimlich maneuver is designed to accomplish this. Why do we rarely hear about similar choking accidents in nonhuman primates? It turns out their SLS is too short to cause this problem.

A chimpanzee's larynx is positioned *high* so that the epiglottis can reach behind the soft palate, thus connecting the larynx directly to the nasal cavity (see Figure 7.2). Indeed, fluid can run around the erect epiglottis laterally into the esophagus, allowing chimps to drink and breathe simultaneously (Figure S7.d). Human *babies* can do the same, and it turns out their larynx and epiglottis are positioned high, as they are in chimps and other mammals of any age. In other words, the *low* position of the larynx, and the resulting length of the SLS, are unique to humans *beyond their early childhood*.

Why do we have such a hazardous situation in our necks, and why are we discussing this oddity in the context of language? The long human SLS develops during childhood, beginning around 1.5 years of age. This is the age when children begin to speak, and indeed, the SLS modulates the sound produced by the vocal folds, giving tone and clarity to the human voice. The SLS also gives our tongue more flexibility to move, which greatly enhances the articulation of our speech. Evidently, the speech-promoting functions of the SLS have outweighed its choking hazard in hominin evolution.

E. Advantages of Having a Language

While the timing of language evolution is still being debated, there is no doubt that it provides great advantages, especially in conjunction with a better brain. Language and memory create an inner world of symbols that can be communicated in the absence of their referents. While the exact role of language in personal development and societal interactions is a wide and controversial field, a few simple scenarios and observations will illustrate possible ultimate causes of having a complex spoken language.

If a guard from village A sees two scouts from village B, the guard can *transmit his fleeting observation*, which may never repeat itself, to the elders of village A. They will combine the guard's information with memories of earlier raids staged by men from village B, and they will put village A on high alert. This scene has probably played out millions of times during hominin evolution, and communities with an effective language had a better chance of survival.

In addition to communicating signs of imminent danger, language can *conceptualize experiences* and reduce them to their essence. Take a simple sentence like "Game is best approached from downwind." Here, "game" refers to any wild animal worth hunting, rather than a particular species. The phrase "from downwind" means that wind should be blowing from the game to the hunter, not the other way, because that would carry the hunter's smell and noise to the game. Coining the term "downwind" reduces the rule for the hunter to its essential aspect and makes it independent of the wind's geographical direction.

Language is involved in developing the *concept of self*, as is indicated by the way children refer to themselves first by their name and only later by "I" or "me." This concept is critical for individuals to see themselves as both subjects and objects of personal interactions. This ability in turn greatly facilitates any planning of group activities in hunting, defense, and cultural ceremonies.

Finally, language would seem to be indispensable to *defining and reinforcing complex norms of behavior*. All cultures have rules about sex, marriage, sharing food, and cooperation in foraging and defense. Since agriculture, societies have needed more rules about property, taxes, and inheritance. Good rules reward productivity and cooperation while discouraging laziness and selfishness. Societies with effective rules outcompete those with ineffective rules, and much effort is therefore spent on educating children and adolescents about the norms of their community. Among adults, gossip and advertising of opinions continue the process of defining and reinforcing acceptable behavior. These vital processes could not proceed without language.

F. Dramatic Encephalization in Hominin Evolution

One of the most dramatic events in hominin evolution is the increase in size and complexity of the brain, a process known as **encephalization**. From australopithecines to modern *Homo sapiens*, the size of the brain has tripled from about 450 to 1350 cm^3 (see cover art and Figure S7.e). In contrast, the great apes have stayed at a fairly constant brain size of 400–450 cm^3. In addition to overall size, the human brain differs from the chimpanzee brain especially in the **cerebral cortex**, the folded layer of gray matter (cell bodies of neurons) that covers the brain's surface and is involved in memory, planning, and other complex mental functions. Due to the greater number of folds, the cortex of the brain is more convoluted in a human than it is in a chimpanzee. If flattened, a human's cerebral cortex would cover four letter-sized pages: a chimp's would cover only one; a monkey's would cover a postcard; and a rat's would cover a postage stamp (Figure S7.f).

Genes that promote the growth of the cerebral cortex are currently being investigated. Humans with mutant alleles of these genes are afflicted with *microcephaly*, a severe reduction in the size of the cerebral cortex associated with mental retardation. One of these genes, known as *ASPM*, has undergone rapid evolution in hominins but not in other primates. The same gene also shows a lot of variation within and between ethnic groups, indicating that human evolution with regard to brain cortical size is still going on.

The cerebral cortex consists of two lateral hemispheres, which are connected by the **corpus callosum**, a broad band of white matter (neuronal axons). The right hemisphere receives sensory input from and provides motor control over the left half of the body, and vice versa. Each hemisphere is subdivided into four lobes (Figure 7.3). The **frontal lobe** is involved in activating skeletal muscles and in the planning and coordination of body movements including speech. The lateral boundary of the frontal lobe is a groove called the **lateral sulcus**; it separates the frontal lobe from the **temporal lobe**, which is involved in smell, hearing, and related associations. The posterior boundary of the frontal lobe is another groove known as the **central sulcus**; behind it is the **parietal lobe**, which integrates sensory input of temperature, taste, and touch, as well as the **occipital lobe**, which processes visual information.

G. Plasticity in Response to Injury and New Tasks

Along with size, one of the most amazing properties of the human brain is its **plasticity**. This term refers to the brain's ability to adapt to changed circumstances, to find new divisions of labor, sometimes after an injury or a stroke, but more commonly when a person acquires a new skill.

A dramatic illustration of brain plasticity after injury is the story of Phineas Gage, a railroad worker who suffered a horrible accident. When tamping down powder to blast some rock, he apparently forgot to cover the powder with sand. When his tamping iron struck a spark, it set off the charge, and the iron shot back through his head. The iron was 1.25 inch in diameter and 3.5 feet long. It entered Gage's head under his left cheekbone and went out through the top of his head, destroying most of the front part of the left side of his brain. Amazingly, Gage survived and was released from the hospital after ten weeks. He could walk, talk, and take care of himself. However, his company would not give him his former job back because his personality had

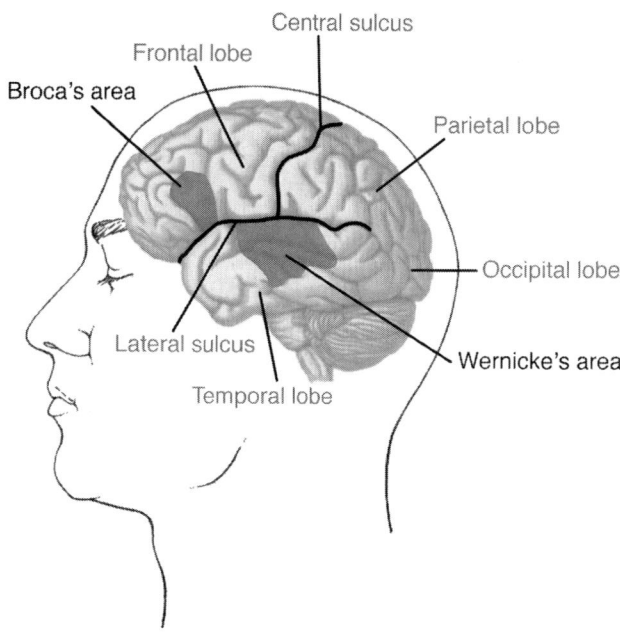

FIGURE 7.3:
Human cerebral cortex shown in left lateral aspect. Frontal, parietal, occipital, and temporal lobes are demarcated, in part, by two grooves, the central sulcus and lateral sulcus.

changed. Before the accident he had been their most capable foreman, disciplined and with a well-balanced mind. After the accident he had become fitful, irreverent, and grossly profane. He never worked as a foreman again, and instead made a living as a stable hand. Twelve years after the accident, he began to have epileptic seizures and died soon thereafter.

Gage's recovery shows that the brain can shift activities from one area to another. This has also been demonstrated under more normal circumstances using modern methods of brain imaging. In the **somatosensory cortex**, the strip of parietal lobe right behind the central sulcus (see Figure 7.3), specific areas receiving touch and temperature sensations from the leg, trunk, arm, and other body parts have been mapped. The map looks distorted, with sensitive areas like the hand and face occupying much larger areas of cortex than the thigh or trunk. More to the point, this map changes according to usage. For example, violin players use the fingers of their left hand to manipulate the strings of their instrument, and some players practice several hours a day. In doing so, the fingertips of their left hand get stimulated more than the fingertips of their right hand, or the fingertips of an average person. Correspondingly, the somatosensory cortex area for the left-hand fingertips is increased in violin players. This change is greater if they start playing in childhood, but still significant if they begin as adults.

H. Lateralization, Handedness, and Sequencing

Another remarkable feature of the human brain is the **lateralization** of certain brain functions, meaning that these functions are carried out predominantly or exclusively in one lateral half of the brain. Two of these areas are involved in speech, and both are limited to the left half of the brain in most people. One is known as **Wernicke's area**, located in the left temporal lobe above and behind the ear (see Figure 7.3). It is an area of integration for sensory input from reading and hearing speech. If damage occurs to Wernicke's area, a person is unable to understand any written or spoken information. Such a person can speak fluently, but the words make no sense. Another brain area critical to speech is **Broca's area**, located in the left frontal lobe behind the temple. Persons with damage to this area can read and understand the speech of others, but their own speech is laborious and slurred.

The lateralization of Wernicke's and Broca's areas was demonstrated by simple tests given to human patients who had their **corpus callosum**, which connects the right and left cerebral hemispheres, severed to alleviate epileptic seizures. In a typical test, the patient was asked to describe an object placed in her *right* hand without looking at it. Usually this presented no problem, but the same patient *could not* describe the same object when she held it in her *left* hand. Sensory information from the left hand is projected to the right sensory cortex and vice versa. (This crossing over occurs in the brainstem and was not affected by the surgery.) However, after severance of the corpus callosum, the speech centers lateralized to the left cortex could no longer communicate with the right somatosensory cortex, which receives sensory projections from the left hand.

The lateralization of speech control centers depends on the handedness of people. While more than 95 percent of all right-handed people have their speech control in the left brain, most non-right-handed people have some language abilities on both sides of the brain. On the one hand, this lack of lateralization seems unfavorable because left-handed people are more likely to suffer from stuttering and migraine headaches. On the other hand, there seems to have been some positive selection for left-handed people, because, due to their lower frequency, they were more difficult opponents for right-handers than vice versa in sports and in combat.

Speech is a fast and complex activity. It seems adaptive to lateralize the control of such activities to one half of the brain, because then all participating neurons are close together, and the shorter connections result in faster communication. Is the control of other fast and complex movements also lateralized? Not in an obvious way. The neurons that activate the skeletal muscles are located in the **motor cortex**, the strip of frontal lobe in front of the central sulcus (see Figure 7.3). The neurons for the muscles on the right side of the body are located in the left motor cortex and vice versa, but both sides of the motor cortex have equivalent functions.

There may be a degree of asymmetry in clusters of neurons called **sequencers**, which coordinate the activity of different neurons in the motor cortex. Such sequencers are critical to the execution of complex movements that need to be both precise and powerful, such as speech or throwing a spear. Like Broca's area, other sequencers may also be localized in the left frontal lobe. The right hemisphere seems more involved in face recognition, picking up emotional qualities of speech and music, and processing many kinds of information in parallel.

I. Possible Ultimate Causes of Encephalization

The human brain is a marvelous organ, but it is also expensive in terms of both metabolism and child rearing. Although the brain represents only 2 percent of an adult's body weight, it consumes 18 percent of the energy of a human at rest. During pregnancy, the brain of the growing fetus places great demands on the mother's metabolism and on the placenta's capacity for transporting nutrients and oxygen. Due to these constraints, and because of the limited size of the birth canal, human babies are born with a brain that has reached only 28 percent of its adult size and undergoes most of its growth after birth. (For comparison, chimpanzees are born at 43 percent of their adult brain weight.) Correspondingly, human babies are extremely helpless and have very little control over their movements, as compared to ape babies, who can at least hold on to their mothers.

Encephalization has also had major implications for human family structure. The vulnerability of human babies extends into their childhood, when humans are still more dependent than apes on support and protection. The enormous investment required to raise a human child has made it adaptive for both parents to share the effort, which explains why virtually all human societies have the institution of marriage. It includes a covenant between husband and wife to forego extramarital sex, so that the husband can be confident that the children he helps to raise are biologically his. Several unique characteristics of human sexuality seem to have evolved because they stabilize marriage, as discussed in Chapter 2.

Why did humans evolve to have such large brains, which are expensive to build and maintain, and make birthing and child rearing more difficult than in any other mammal? Several ultimate causes of human encephalization have been proposed; only two of them are outlined below.

The first argument hinges on the **utility of sequencers**, the clusters of neurons that program other neurons for complex patterns of fast firing, as mentioned earlier. In addition to speech, humans can learn to do many other things with speed and precision, such as knapping stone tools. However, this takes a lot of coordination and practice, and anybody who has watched babies and infants knows how jerky their hammering movements are initially. The same holds for other human activities, like throwing a spear or playing a musical instrument. Once these activities are in use, there is a constant selection pressure for carrying them out with greater speed and/or more precision, which require brains with better sequencers. Model experiments with computers indicate that the required size of sequencers increases rapidly with growing demands on power and accuracy of the movements they control. Thus, large brains may have been positively selected for because they accommodated better sequencers.

Another hypothesis on the evolution of the large human brain is centered on **social intelligence**. Hominin communities probably ranked in complexity between the simplest human societies today and chimp or bonobo societies. The latter are already complex. In particular, male chimpanzees spend much effort on achieving two conflicting goals: On the one hand, they *compete* with one another for top rank because the alpha male gets all the food and females he wants. On the other hand, males need to *cooperate* for hunting and defense. Balancing competition and cooperation is not an easy task, and chimp males are busy forging alliances, appeasing opponents, winning over their opponents' friends, and so on (see Chapter 21). When the brains of thirty social primate species were analyzed, the volume of the cerebral cortex, relative to the overall volume of the brain, was positively correlated with the group size of the species. Thus, hominin encephalization may have been driven by the rewards of smart social behavior.

FIGURE 7.4:
Baby chimp (a) **and adult chimp** (b, not to scale). Redrawn from Gould S.J. (1977) *Ontogeny and Phylogeny.* Cambridge: Belknap.

Baby Chimp   **Adult Chimp**

J. Paedomorphosis as a Proximate Cause of Encephalization

Having considered possible *ultimate* causes of encephalization, that is, why the owners of big brains may have had more offspring, we will now turn to a plausible *proximate* cause, meaning a genetic and developmental mechanism that could have boosted the development of our large brain. We obtain a clue by comparing head development in humans versus chimpanzees (Figure 7.4). While the head of a *newborn* chimp shows an uncanny similarity to the head of a human, the head of an *adult* chimp is dominated by the larger size of the face. In other words, humans and chimps share a similar growth pattern of the head up to birth, and then humans continue with this pattern, while chimps change to another pattern in which the face grows more than the braincase. The visual impression is borne out by quantitative data (Figure 7.5). If brain mass is plotted versus body mass during mammalian development, the two grow proportionally to a point of inflection when the mass of the brain levels off. Primates reach the point of inflection later than other mammals, but the human brain grows the longest.

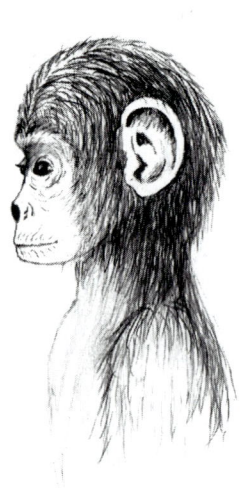

FIGURE 7.5:
Brain weight and body weight. Straight inclines represent periods of proportional growth. This period lasts the longest for the human. Redrawn from Gould S.J. (1977) *Ontogeny and Phylogeny.* Cambridge: Belknap.

Such an extension of an early growth patterns into a later period of development is so common in evolution that scientists have coined a special word for it: **paedomorphosis**, meaning the retention of juvenile features. Another example are the axolotls, amphibians that maintain as adults the body plan of typical amphibian larvae with external gills and other adaptations to aquatic life.

The growth of specific tissues or body regions is generally regulated by signals that act on target genes, which in turn control the cyclic division of cells. Indeed, the human *ASPM* gene promotes the proliferation of *neuroblasts*, the cells that give rise to neurons. As mentioned earlier in this chapter, the function of this gene makes the difference between an *Australopithecus*-sized and a modern human brain. There is only a handful of known other genes that have a strong enough effect on brain size to be detectable by analyzing human family trees. Thus, it appears that activity changes in a small number of genes could bring about the dramatic encephalization that occurred in hominin evolution.

K. Human Characteristics Are Mutually Reinforcing

Some 50,000 years ago, there was a rapid increase in the variety and sophistication of stone tools made by modern *Homo sapiens* (see Chapter 6). Around the same time, archaeological findings begin to indicate trade networks, sculpture, cave painting, body ornaments, musical instruments, and grave provisioning (Figure S7.g). Some anthropologists call this phenomenon the **great leap forward** and ascribe it to a genetic change that caused a brain reorganization. This may have enabled our ancestors to develop *advanced languages*, which would have improved their ability to make elaborate plans and create the myths that unite communities. The new linguistic talent could have enhanced economic and cultural activities to new levels, which then allowed these talented people to colonize the rest of the world and displace all preexisting hominin populations.

Instead of ascribing the great leap forward to the acquisition of *one new capability*, one could also explain it as a cooperative effect in an **interactive system**, in which a small improvement in any one trait enhances the odds for further improvements in other traits to occur (Figure 7.6). For instance, hominin bipedalism may have promoted the evolution of opposable thumbs and better toolmaking in *Homo habilis*, which in turn could have allowed them to forego abductable toes and acquire a striding gait. Better tools and striding gait made for more effective scavenging, resulting in diets with more meat. These in turn supported encephalization. Larger brains improved toolmaking, speech, concept of self, art, and social skills, until it all came together in the great leap forward.

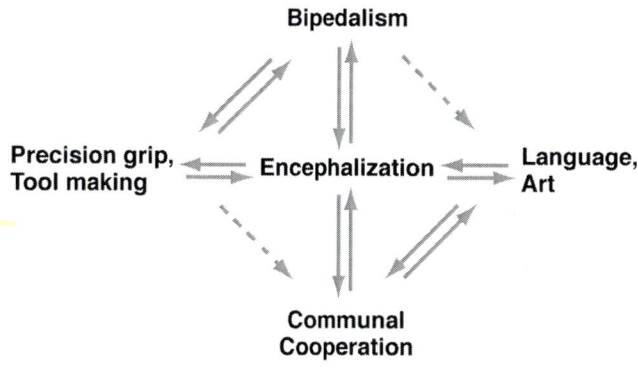

FIGURE 7.6:
Human characteristics are mutually reinforcing.

EXERCISE PAGE FOR CHAPTER 7

Student Last Name _____

Student First Name _____

Discussion Section _____

On the course web site, use the link on the **syllabus** to find a pdf file of your **assigned reading** for this chapter. The same reading is referenced below. Use the bulleted list of questions to test your knowledge of this reading.

Evans, P. D., et al. (2004). Adaptive evolution of *ASPM,* a major determinant of cerebral cortical size in humans. *Human Molecular Genetics* **13:** 489–494.

- What is the phenotype of human patients who do not have a wild-type allele of the *ASPM* gene?
- What seems to be the function of *ASPM,* based on knowledge of similar genes in *Drosophila* and mice?
- Define the K_a/K_s ratio plotted in Figures 1 through 3 of your reading. What do these data show?
- What is the major conclusion that the authors draw from their data? What alternate explanation of their data do the authors mention but consider "highly unlikely"?

Feedback

On a scale from 1 to 10, the being the best, please rank the above reading for

1. Interest and Relevance to the topic (10 being most interesting/relevant) _____
2. Readability (10 being most clearly written, easy to understand) _____

In the remaining space, enter any comments that you may have on this reading.

CHAPTER EIGHT

GERMS, COLONIALISM, AND DARWINIAN MEDICINE

Humans, like other organisms, are hosts to a variety of parasites. As parasites move from one host to the next, they cause infectious diseases, which reduce the hosts' vitality or may even kill them. Sometimes a host acquires a mutation in a critical gene that helps to fend off a parasite, thus allowing the host to raise more offspring. Humans also defend themselves by cultural means, such as hygiene, medications, and vaccines. In turn, parasites evolve through mutations that enable them to subvert their hosts' defenses. These "arms races" of measures and countermeasures shape the course of epidemic diseases. Such diseases have played major roles in human history, and they are a continuing threat in an ever-growing world population.

Despite improved hygiene and vaccinations, about 25 percent of all deaths worldwide are still caused by infectious diseases, primarily AIDS, malaria, and tuberculosis. One of the underlying reasons is that many parasites have become drug resistant. Many antibiotics that have reliably cured bacterial diseases in the past may no longer work. This experience is causing a new emphasis on the evolutionary aspect of infectious diseases, a trend sometimes called *Darwinian medicine*. While traditional medicine is focused on surgery and medication, Darwinian medicine tries to harness symptoms of human disease as natural defenses while trying to minimize the problem of microbial resistance.

A. Classification of Parasites According to Size and Development

Parasites are classified by their size and according to the complexity of their life cycles. The size classification distinguishes between microparasites and macroparasites. **Microparasites** are invisible to the naked eye. Colloquially known as "germs," they include viruses, bacteria, fungi, and protists.

A virus has a coat of proteins surrounding a core of genetic material, which consists of either DNA or RNA. All viruses are parasites because, due to their simple construction, they can only multiply inside of host cells. Viral coat proteins are particularly suited to breaching the walls and membranes of their host cells. Once inside, the viral genes commandeer the host's molecular machinery to produce more viruses, until the host cells bursts and the cycle repeats itself. Examples of human viral diseases are AIDS, influenza, Ebola fever, hepatitis, herpes, and yellow fever.

A bacterium is a prokaryotic cell showing all characteristics of life. While most bacteria are free-living in soil and water, others exist in mutualistic relationships with larger host organisms. Typically, the host provides nutrients and a habitat for the bacteria, while the bacteria deliver other services to the host. For instance, bacteria in the human gut digest food that would otherwise go unused, and they produce some vitamins that we cannot synthesize. Only a minority of bacteria causes diseases, usually by toxic substances they release. Human bacterial diseases include the bubonic plague, cholera, diphtheria, food poisoning, leprosy, syphilis, tetanus, tuberculosis, and typhoid fever.

Other human microparasites are fungi and protists, both consisting of eukaryotic cells. Fungi are normally found on decaying plants and animals but some occur in living tissues, including those causing vaginal yeast infections and "athlete's foot." Diseases from parasitic protists are amoebiasis, leishmaniasis, malaria, and sleeping sickness.

Macroparasites are visible to the naked eye. They include blood-sucking flies, mosquitoes, fleas, lice, and ticks. There are also different types of worms living in humans. The roundworm causing trichinosis lives in the gut and flesh of its hosts. Humans contract the parasite by eating undercooked meat, often from pig, bear, and other meat-eating game. Tapeworms, which may grow several meters long, live in the guts of their human or animal hosts. While the worm's head stays anchored in the intestinal lining with hooks and suckers, the hind end releases segmentlike units that are filled with eggs and leave the host with feces.

Independently of size, parasites are also classified according to the complexity of their life cycles. Most of the parasites discussed so far show **direct development**. This means they transfer directly between host individuals of the same or similar species. Other parasites show **indirect development**. They have complex life cycles involving one primary host (the human or another vertebrate) and one or more intermediate hosts called **vectors** (carriers).

An example of a microparasite with indirect development is *Plasmodium,* a protist causing malaria in about 500 million people. Malaria occurs only in areas that are inhabited by the mosquito *Anopheles,* which acts as the parasite's vector (Figure 8.1). When *Anopheles* bites humans to suck blood for nourishment, it prevents blood coagulation by injecting saliva into the wound. If the saliva contains *Plasmodium,* then the human is infected. From the blood serum, the parasites enter the liver, then

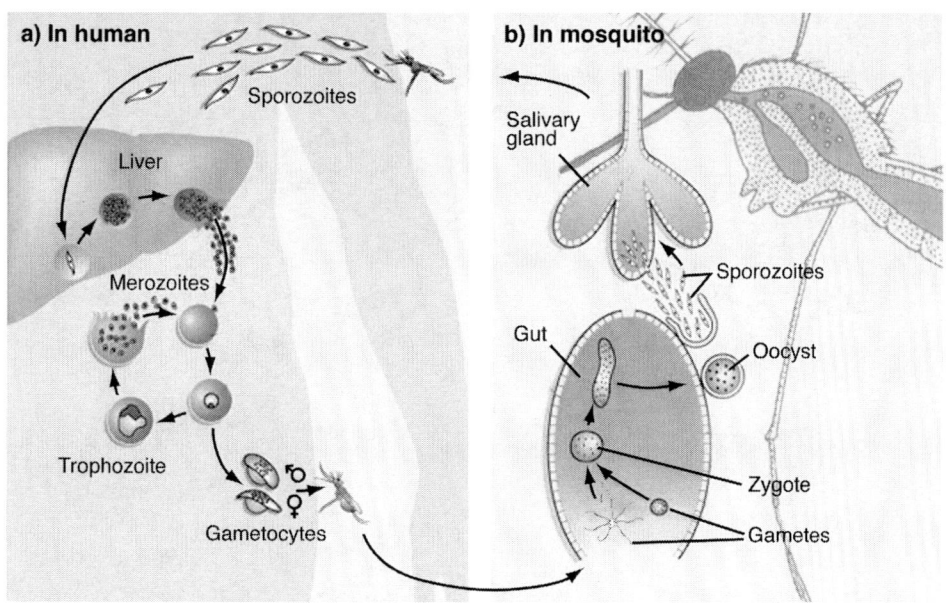

FIGURE 8.1:
Life cycle of *Plasmodium*, a unicellular parasite that is transferred by a mosquito, *Anopheles,* and causes malaria in humans.

red blood cells, and possibly other host cells. Depending on host cells, the parasite cells change their morphology and are named differently. During these stages of development, *Plasmodium* multiplies asexually by cell division, killing its host cells and releasing toxic waste products. As a result, infected humans suffer from fever, headache, and general malaise. About a million each year die.

Sexual reproduction of *Plasmodium* begins with the formation of gametocytes in human blood cells. If bitten again by *Anopheles,* the parasites get into the lumen of the mosquito's gut, where they form gametes that undergo fertilization. The zygotes penetrate the wall of the gut, from where their daughter cells reach the salivary glands. From here they transfer to another human host when the vector bites again, and the life cycle repeats itself. *Plasmodium vivax* is common in tropical as well as subtropical zones. *P. falciparum* is limited to the tropics but more deleterious because it eventually enters the host's brain cells.

Similarly complex life cycles are shown by the microparasites causing sleeping sickness, bubonic plague, and typhus, which are carried to new hosts by flies, fleas, and lice, respectively.

Examples of macroparasites with indirect development include the blood flukes, worms that infect some 200 million people in tropical and subtropical areas (Figure 8.2). Adult flukes live in blood vessels of the gut and urinary bladder of humans and other vertebrates, causing pain, anemia, and dysentery. Fertilized eggs exit with the host's feces or urine. In water, the eggs develop into ciliated larvae, which infect snails as intermediate hosts. Here they multiply by budding, producing a tailed kind of larva that leaves the snail. These larvae penetrate the skin and enter the blood vessels of animals or humans exposed to contaminated water. The bloodstream carries them to various internal organs, where they mature to adults, which mate in blood vessels of the gut or bladder, from where the life cycle repeats itself.

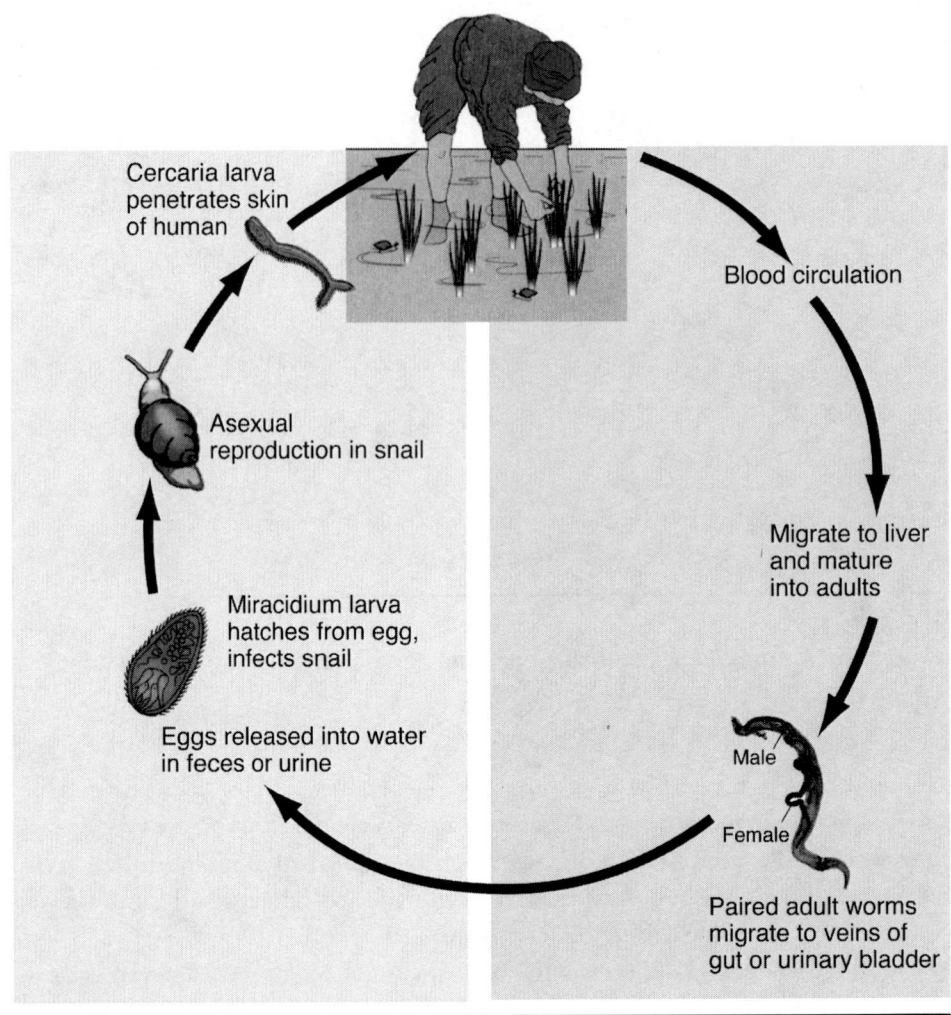

FIGURE 8.2:
Life Cycle of a Blood Fluke *(Schistosoma mamsoni).* Infected people (about 200 million) suffer anemia, dysentery, and pain.

B. Coevolution of Humans and Their Parasites

The often bizarre life cycles of parasites are testimony to the opportunistic ways in which parasites and their hosts have undergone a stepwise process of **coevolution**, in which mutations of one change the selection of itself and the evolutionary partner. Parasites mutate fast because they invest most of their energy in reproduction: Viruses and bacteria multiply within minutes or hours, while macroparasites produce huge numbers of offspring. This excessive multiplication improves not only the odds of finding a new host but also the rate of mutation, that is, the probability that the next generation will include a genetic variant that improves the parasite's viability.

Parasites use a wide range of "strategies" to find new hosts. Some simply wait until their host is eaten by the next host: We humans contract salmonella by consuming spoiled eggs, or trichinosis by eating undercooked meat. Other parasites are very sensitive to signals emitted by their hosts. Ticks, for example, perch on vegetation until they smell a host, at which time they let go of their perch to drop onto their new host. Other parasites employ vectors to reach their next host. Some parasites even affect their hosts' behaviors so that the hosts will spread the parasite. A dramatic case is the rabies virus, which gets into the saliva of mammals and at the same time affects their brain so that they will bite what will be the next host.

Hosts have physiological reflexes to protect themselves. Coughing, sneezing, vomiting, and diarrhea rid humans and animals of parasites. Fever reduces the duration of viral and bacterial infections. Synergistically with fever, the level of iron in human blood serum drops dramatically during a bacterial infection, a response that deprives bacteria of a key mineral. The human revulsion toward the odor of rotting meat prevents ingestion of many parasites.

While most host defenses against parasites are biological, some are learned. Grooming and bathing go a long way towards avoiding external parasites. An acquired distaste for the odors of urine and feces causes us to leave these excretions away from our living areas. Some societal rules reinforce innate tendencies. While most people seem to have an inborn aversion against close contact with obviously ill fellows, ostracizing such people as "unclean" adds force to the aversion.

C. The Mammalian Immune System

The most sophisticated natural defense of mammals against microparasites is the **immune system**. It attacks molecular fragments, collectively called **antigens**, which it recognizes as foreign, or not belonging to the normal inventory of the healthy host. Such antigens include exposed segments of viral coat proteins, bacterial cell walls, and surface proteins of protists as well as cancer cells.

Antigens floating freely in blood serum or lymph fluid may be recognized by white blood cells called *B cells*. Each B cell carries on its surface a specific protein called an **antibody** (Figure 8.3). There are millions of different types of antibodies, and each B cell carries only one type. If a B cell's antibody matches a particular antigen, so that the two bind tightly, then this B cell is stimulated to divide, creating a large clone of B cells, all producing the same kind of antibody. Most of these cells mature into *plasma cells*, which go on to release the same antibody that the clone's founder cell had carried on its surface. This means the host's blood is now flooded with antibodies that will latch on to the matching antigen wherever they find one. Thus, microparasites or cancer cells displaying the antigen are tagged with antibodies, which attract other blood cells that engulf and destroy the antigen carriers. Importantly, some cells from each B cell clone are set aside as *B memory cells*, which remain "on file" for many years, often for life, so that a faster and more vigorous response can be mounted if the same antigen enters the body again. This type of defense is known as **antibody-mediated immunity**.

A complementary defense, known as **cell-mediated immunity**, targets microparasites that spend most of their time inside of host cells where they are not accessible to B cells. Such infested cells are engulfed and digested by blood cells called **antigen-presenting cells (APCs)** because they present antigens from the digested microparasite on their cell surfaces. The presentation is done by **major histocompatibility (MHC) proteins**, which are incorporated into the outer APC membrane along with the antigen. An antigen-MHC combination on the APC's surface is recognized by a matching **helper T cell**, which—similar to a B cell—is stimulated to divide clonally. Most of the daughter cells become *cytotoxic T cells*, which

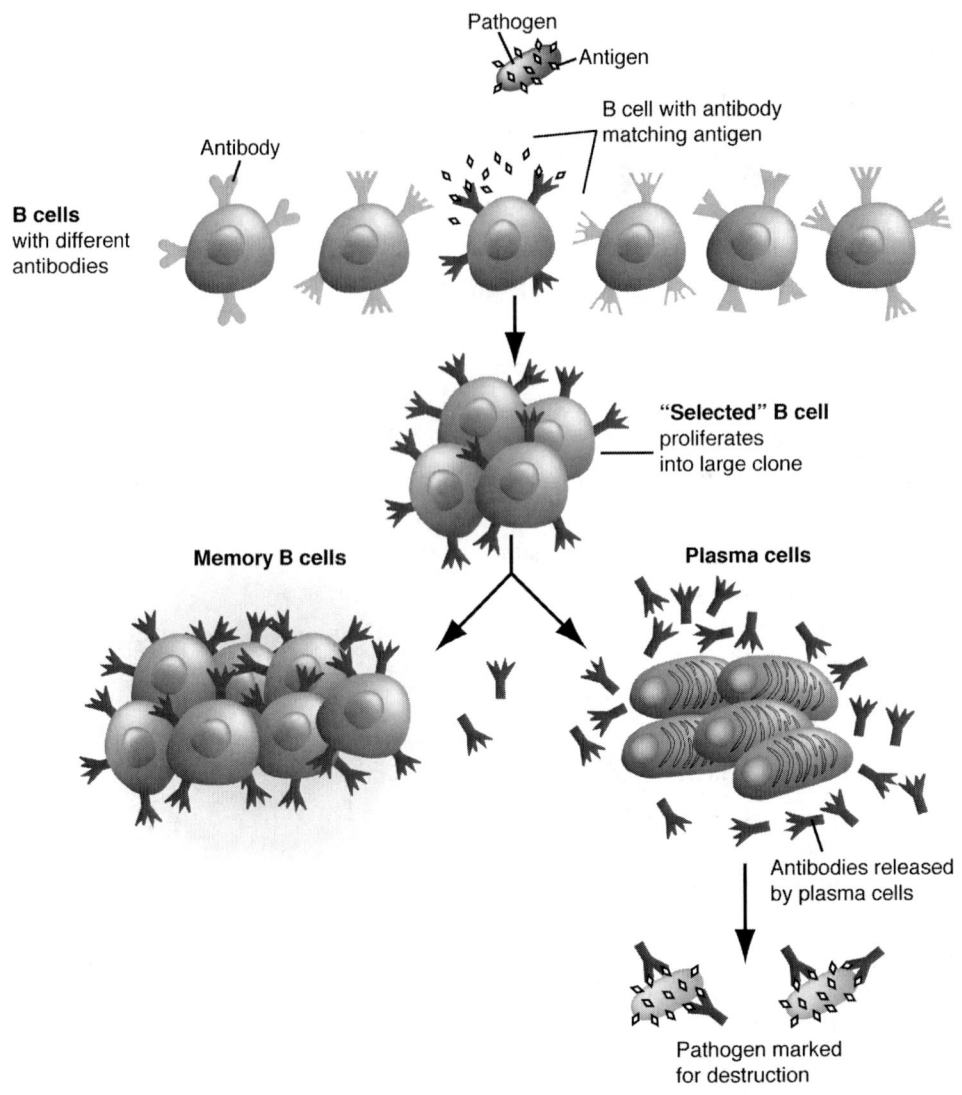

FIGURE 8.3:
Antibody-mediated immunity
Redrawn after Campbell N.A. and Reece J.B. *Biology*, 6th ed. 2002

bind and destroy foreign or abnormal cells that display the matching antigen. Again in parallel to B memory cells, some daughter cells of each helper T cell clone are set aside as *T memory cells* for future use.

Vaccines work by "priming" the immune system with a dead or nonpathogenic version of an expected microparasite. Such "decoy" parasites elicit an immune response including the formation of memory cells, thus saving valuable time and immune system capacity when the "real" (pathogenic) parasite is encountered. Of course, the vaccination is only effective if the decoy and the real parasite share the same antigen(s).

As discussed earlier, parasites and their hosts coevolve in an ongoing war of measures and countermeasures. Many microparasites escape their hosts' immune systems—and any attempts to develop a vaccine—by changing their antigenic coats. For instance, many attempts to develop vaccines against malaria have been unsuccessful because *Plasmodium* rapidly changes the antigens it presents.

D. Costly Defenses Against Parasites

The impact of a parasite on the evolution of its host is illustrated dramatically by defenses that are costly to the host. Well-known examples are genetic variants of *hemoglobin*, the protein complex in red blood cells that ferries oxygen and carbon dioxide. The common form of this protein in human adults is called **hemoglobin A (HbA)**. A mutant variant is **hemoglobin S (HbS)**, so called because it may lead to **sickle-cell anemia**, a condition in which red blood cells become sickle shaped (Figure 8.4). The deformed cells tend to clog blood vessels and break up, causing life-threatening crises. HbS also does not transport oxygen and carbon dioxide as efficiently as HbA. Blood transfusions and medications can save a patient's life, but there is no cure.

HbS differs from HbA by a single amino acid substitution in position 6 of one of its polypeptide chains, β-globin. The difference is encoded by a mutation from the normal β-globin gene to its mutant allele (designated $ß^A$ and $ß^S$, respectively). Given the dire consequences of sickle-cell anemia, why has the $ß^S$ allele not been eliminated by natural selection? As it turns out, the mutant globin gene leaves homozygotes prone to sickle-cell anemia but confers resistance against malaria to heterozygotes (Figure 8.5). If two parents are both heterozygous for $ß^A$ and $ß^S$, then half of their gametes carry only the $β^A$ allele, while the others have only $β^S$. Their chances of fertilization and development are the same, so that statistically one out of four children will be homozygous for $β^A$, one will be homozygous for $ß^S$, and two will be heterozygous like their parents.

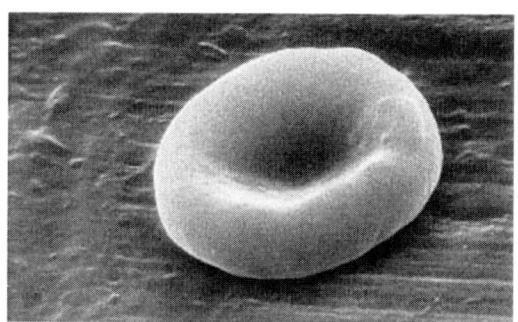
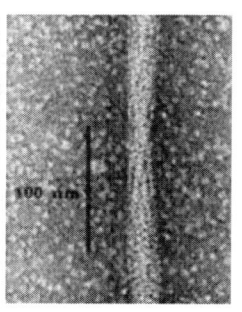
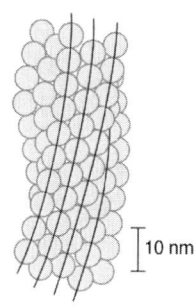
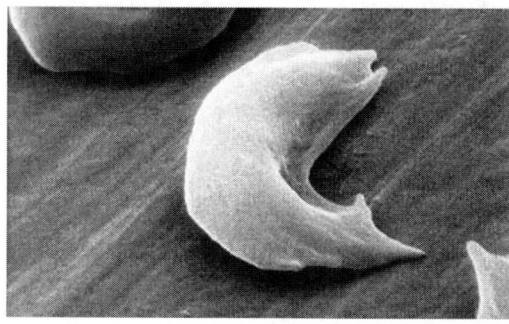

FIGURE 8.4:
Sickle cell anemia caused by abnormal self-assembly of hemoglobin molecules in humans homozygous for the globin $ß^S$ allele **[a]** normal red blood cell **[b]** sickled red blood cell **[c, d]** photograph and interpretative diagram of hemoglobin S self-assembled into a rod-shaped complex
From *Analysis of Biological Development* by Klaus Kalthoff. Copyright © 1996 by McGraw-Hill. Reproduced with permission of The McGraw-Hill Companies.

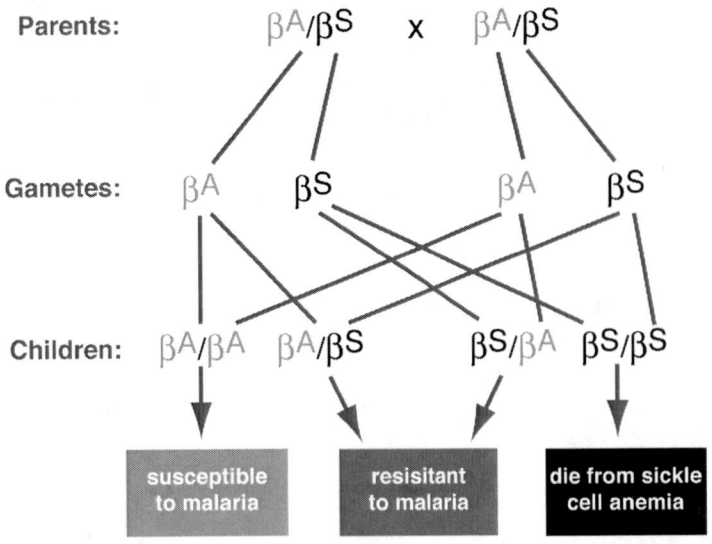

FIGURE 8.5:
Sickle cell anemia and resistance to malaria

Depending on the allele combination they inherit, the children will fare differently in regard to malaria and sickle-cell anemia (Figure 8.5). The β^A/β^A homozygotes will make only normal hemoglobin, rendering them susceptible to malaria. The β^S/β^S homozygotes will synthesize only hemoglobin S, leaving them at risk to die from sickle-cell anemia. The heterozygotes will not be prone to sickle-cell anemia because red blood cells sickle only if *all* their hemoglobin molecules contain the mutant β-globin. The same heterozygotes will be resistant to malaria because *Plasmodium* does not develop in red blood cells in which *many* of the hemoglobin molecules have the mutant β-globin. Thus, in the presence of *Plasmodium*, the heterozygotes will do best.

In theory, in a human population in which the β^S globin allele is present, up to 50 percent of all persons could become resistant to malaria for the price of leaving another 25 percent prone to sickle-cell anemia. In real populations, the frequency of the β^S allele is lower, depending on the selection pressure exerted by malaria. If *Plasmodium* is not present, as in temperate and cold climates where *Anopheles* does not live, then the β^S allele is rare or nonexisting. If there is a high risk of *Plasmodium* infection, the frequency of the β-globin^S allele or similar globin variants is significant (Figure S8.a). In some areas of central Africa, the frequency of the β^S allele is greater than 10 percent. In African Americans, the frequency of the β^S allele is still about 5 percent, even though *Anopheles* has been nearly eradicated in the United States.

People who live in malaria-infested areas, unless they know that they carry the β^S allele, should sleep under mosquito nets impregnated with insect repellent. People who travel in such areas should do the same and use prophylactic drugs.

A parallel case of costly countermeasures against a parasite seems to be cystic fibrosis (CF). It is caused by mutations in the gene for a chloride transporter protein in the cell plasma membrane. Mice heterozygous for such a mutant allele are resistant to cholera. This protection, if present in humans, would explain why CF is frequent (1 in 2,500 live births) among people from Europe, which has a history of cholera epidemics.

E. Some Parasites Can Cause Epidemics

Some parasites cause **epidemics**, that is, diseases that spread quickly in a population. Influenza, AIDS, measles, mumps, and tuberculosis are epidemic diseases. Quantitatively, a parasite can be characterized by its **basic reproductive rate, R_0**. This is the average number of new infections caused by the first infected individual in a population. The epidemic threshold is at $R_0 = 1$, above which an infectious disease becomes an epidemic and below which it will eventually die out. Most epidemics are caused by microparasites, which usually elicit an immune response. Thus, a typical epidemic will spread quickly through a population, leaving its members either dead or immune. Because the parasites can spread only to new hosts that are alive but not immune, epidemics tend to break out in crowded places (big cities, public schools) and in hosts that travel a lot (migratory birds, modern humans).

The reproductive rate of a parasite is not correlated with its **mortality rate**, that is, the percentage of infected hosts killed by the parasite. For instance, rabies is not an epidemic but has a high mortality rate. Conversely, the common cold spreads epidemically but with low mortality. The most dreaded diseases are epidemics with high mortality rates, such as the smallpox epidemic that struck Rome in AD 165 and killed about 25 percent of the population. The bubonic plague, which is caused by bacteria transmitted from rats to humans by fleas, killed about 25 percent of Europe's population around AD 1350, with death tolls up to 70 percent in some cities.

The only human epidemic that has been truly eradicated is smallpox. Other epidemics, such pneumonia and tuberculosis, have been fought off with hygiene and medication but are coming back. One cause of their resurgence is the evolution of mutant strains that are resistant to medication, an alarming trend that will be discussed later in this chapter. Another phenomenon that drives up the risk of epidemics is the relentless growth of the human world population (see Chapter 22). More humans on Earth, more than half of them living in big cities, will make epidemics a continuing threat.

F. The AIDS Epidemic

A frightening new epidemic known as **acquired immunodeficiency syndrome (AIDS)** is caused by the human immunodeficiency virus (HIV). The principal ways of becoming infected are from mother to fetus, by shared injection needles, and by unprotected sex. HIV attacks *helper T cells*, which are critically involved in the immune response to viral infections as discussed earlier. AIDS patients therefore die from various infectious diseases or cancers that they are unable to fight off. There is currently no cure, and the costs of life-prolonging treatment are high. Because of the nature of the disease, people with AIDS easily contract and spread other infectious diseases.

HIV is a dramatic example of a parasite subverting host defenses. Not only does the virus attack the heart of the immune system, it also evolves so fast that all attempts to develop an effective vaccine or drug have been thwarted. Just how fast the virus evolves became widely known through a Louisiana criminal court case in which a physician was sentenced to 50 years in prison for attempting to murder his ex-girlfriend by injecting her with blood drawn at his office from an HIV-infected patient. The conviction was based on comparisons of HIV genes prepared from blood samples of the patient, the physician's ex-girlfriend, and other HIV-infected people from Louisiana. Due to the high mutation rate of HIV, the viral genes from the patient and the ex-girlfriend were clearly more similar to each other than to the corresponding genes from other local HIV carriers. This similarity could not have been detected if HIV evolved more slowly so that samples from the different carriers would be more alike.

AIDS has probably claimed more than 20 million lives so far. About 40 million adults and children are thought to be infected with HIV, and this number increases steadily. The prevalence of HIV infection varies greatly around the world, from 0.02 percent in Japan to more than 30 percent in some South African countries (Figure 8.6).

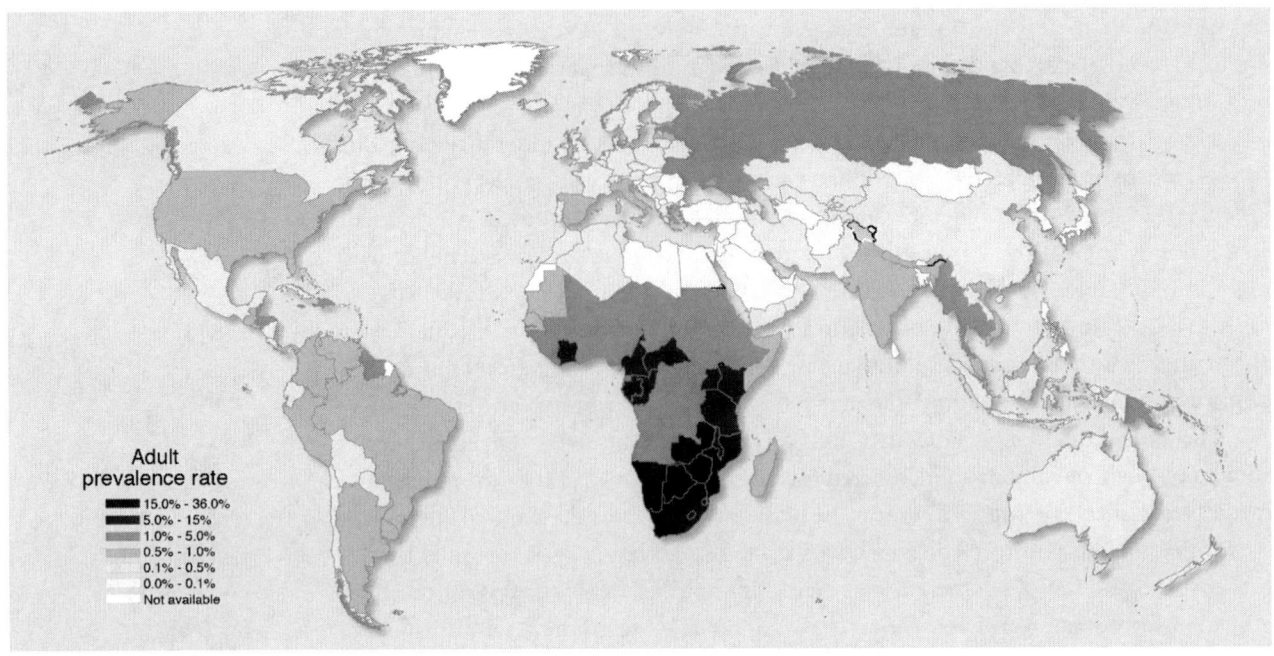

FIGURE 8.6:
Prevalence of HIV infection in defferent countires. In 2005, about 40 million people worldwide lived with HIV. Estimated prevalence rates among adults (15-49 years) was <0.1% in some countires but >30% in others *(2006 UNAIDS Report)*. Reprinted by permission of UNAIDS.
From http://data.unaids.org/pub/GlobalReport/2006/2006GR-PrevalenceMap_en.pdf

Fueled by the social stigma associated with AIDS, there have been scientific and political controversies about whether AIDS is indeed caused by HIV, and consequently, what the most promising courses of prevention and treatment may be. While new hypotheses are the lifeblood of science, it would be irresponsible to abandon the preventative measures that are based on the overwhelming evidence that AIDS is caused by HIV: Don't share injection needles, avoid unsafe sex, and wear latex gloves if you have to touch other people's blood.

Individual persons and ethnic groups differ in their susceptibility to the HIV virus, and in the rate of progression of an HIV infection to AIDS. For example, people deficient for the CCR5 receptor, which along with the CD4 plasma membrane protein initiates HIV entry into T helper cells, are resistant to HIV (Figure 8.7). Likewise, people with extra copies of the gene for CCL3L1 protein, which blocks CCR5, are more resistant. Thus, as in the case of malaria, it can be expected that the HIV virus will affect the genetic constitution of humans living in areas where AIDS is common.

G. Parasites May Evolve to Change Their Hosts

The rapid evolution of parasites often provides them with the **ability to change hosts**. Several infectious human diseases seem to have evolved in animals before they "jumped" to human hosts. For example, the human measles virus is very similar to the rinderpest virus of cattle and the canine distemper virus of dogs.

Wild chimpanzees may carry a virus very similar to human HIV without becoming very sick, suggesting that the virus may have evolved in chimps before a mutation allowed it to survive in humans. A likely way humans could have contracted the virus is by eating chimpanzee meat.

Hunting apes and monkeys for "bush meat" has been customary in African communities, and unfortunately, the habit persists despite the associated health hazards and the fact that most primate species are endangered. Another viral disease, **Ebola fever**, is also contracted by eating the meat of gorillas and chimpanzees. Symptoms include fever, headache, diarrhea, vomiting, and internal as well as external bleeding. The mortality rate is high, and people taking care of patients often become infected themselves if they do not wear protective gear meticulously.

There is currently concern about a strain of the influenza virus known as **H5N1**, which is spread by wild and domestic birds. The virus was originally found in birds of Southeast Asia but has spread to China and Europe. H5N1 has transferred from birds to humans, with a mortality rate of 59 percent in 447 laboratory-confirmed human infections reported through 21 December 2009. The transfers have apparently occurred through feces and blood of diseased birds who shared areas with playing children or who were handled by adult workers in poultry farms.

The greatest concern is that the bird virus may exchange genetic information with viruses causing the relatively harmless seasonal human flu. Such an exchange could lead to the formation of a new virulent strain of flu virus that passes directly from human to human, causing a pandemic (worldwide epidemic) faster than a vaccine could be produced for large numbers of people.

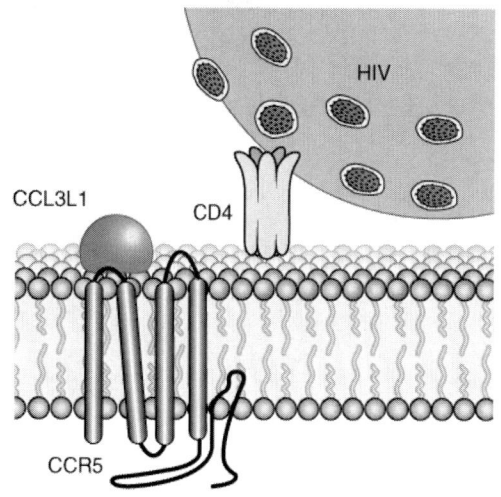

FIGURE 8.7:
Resistance to HIV. CCR5 receptor and CD4 plasma membrane protein act as co-receptors for HIV to enter human cells. Resistance to HIV is conferred by genetic alleles causing loss of CCR5 or over-expression of CCL3L1, a protein that blocks CCR5.
Image © Kendall Hunt Publishing Company. Redrawn after *Science* **307:** 23 credited to K. Sutliff/Science

H. Pathogens in the Colonization of America

The coevolution of humans and their parasites has had profound effects on human history. For instance, the colonization of America was helped decisively by European germs, which devastated American natives while the conquerors were immune. It is estimated that only a small percentage of the American natives survived the invasion by Europeans. The others died not so much from military action but from disease, especially from **smallpox**, a deadly viral epidemic.

When Hernan Cortez led 600 Spanish soldiers to conquer the Aztec empire in 1519, he lost two-thirds of his force during his first onslaught and was beaten back. When he returned two years later, smallpox had killed half of the Aztecs and demoralized the other half. By 1618, Mexico's population had plummeted from 20 to 1.6 million. Likewise, when Francisco Pizarro landed on the coast of Peru with only 200 men to conquer the Inca empire, smallpox had already arrived before him and had killed much of the Inca population including the emperor, unleashing a civil war that weakened his successor (Figure S8.b). In North America, the Indians of the Great Plains and Northwest were decimated, upon the arrival of European explorers, by infectious diseases including smallpox, influenza, and measles, as well as by alcohol abuse. European germs also wreaked havoc on the aboriginal populations of Australia, New Zealand, and Hawaii.

Most infectious diseases were carried by Europeans to indigenous peoples, while the colonists did not seem to pick up new diseases from the locals. Why has the spread of germs been such a one-way street? If one compares the genes underlying the immune system between people from different geographic areas, Europeans and Asians show the greatest diversity. This holds in particular for the genes encoding *MHC proteins*, which present antigens so that they elicit a cell-mediated immune response. The more different alleles of the MHC gene there are in a population, the more likely it is that some individuals will produce an MHC-antigen combination that stimulates a strong immune response. Such individuals will pass on their MHC alleles, so that their descendants will be better prepared to present recurring antigens in combination with effective MHC proteins.

Then why was it the Europeans and Asians who evolved to have more diverse MHC alleles? Probably because they lived for a longer period of time with a wider range of **domesticated mammals**. About 5000 BC, Europeans and Asians already had the "big five": horse, cattle, pig, goat, and sheep. American natives had only guinea pig and llama since most large mammals had become extinct when the first humans reached the Americas, and because only few mammalian species are suited for domestication. The Eurasian domestic animals were also easier to herd, which made them better hosts for infectious diseases. As the pathogens jumped from herd animals to humans, the latter were prompted to diversify their immunity genes.

I. The Problem of Overusing Antibiotics

Several epidemic diseases that had almost disappeared from developed countries are now coming back with a vengeance. One of the underlying reasons is that many bacteria have become resistant to the most effective drugs used against them: **antibiotics**. Typically, they interfere with the ability of bacteria to synthesize proteins or to make their cell walls.

The discovery of the first antibiotic is credited to Alexander Fleming. Having spent years searching for methods to treat soldiers with bacterial wound infections, he was well prepared to recognize the importance of a serendipitous observation he made in 1928, while cleaning out stacks of dishes used to grow bacteria in laboratories: A mold that had contaminated his dishes was killing the surrounding bacteria! Since the mold was identified as *Penicillium notatum,* the bacteria-killing substance released by the mold was named penicillin. Thus, the first antibiotic was a natural product made by a microorganism to kill other microorganisms. Modern antibiotics are produced synthetically, and most of them are still directed against bacteria.

An unsettling experience that physicians and their patients have made over the past decades is that antibiotics lose their effectiveness because the targeted bacteria become resistant. In patients treated for nose or ear infections, the first prescribed antibiotic is often ineffective and only a second or third antibiotic eliminates the disease. Many strains of *Mycobacterium tuberculosis,* the bacterium causing tuberculosis, have become **multi drug resistant (MDR)** and now kill more than 2 million people per year. Other strains have become **hyper-transmissible**, that is, transmissible by very casual contact, like standing together outdoors with an infected person. In poor countries, some patients are having infected lung tissue removed surgically because they cannot afford the high cost of the few antibiotics that might still be able to fight MDR tuberculosis.

The rapid evolution of drug resistance and transmissibility in bacteria is rooted in their fast reproduction. Each cell division involves a round of DNA replication, and thus, a chance for random mutations to occur. Almost all of these will be deleterious so that the mutant will be outgrown by the *wild-type,* that is, the carriers of the prevailing combination of genes, which has been optimized by millennia of natural selection. However, a patient taking an antibiotic changes the rules of the game. Any mutant bacteria that are able to sidestep the antibiotic's action, even if they grow only slowly, can now flourish because the antibiotic has killed their wild-type competitors. This means the patient has become host to a new bacterial strain that is resistant to the antibiotic taken. Over time, additional mutations can bring the resistant strain up to speed. Worse, the resistant strain can use the general ability of bacteria to transfer DNA segments for conferring its resistance to other bacterial species.

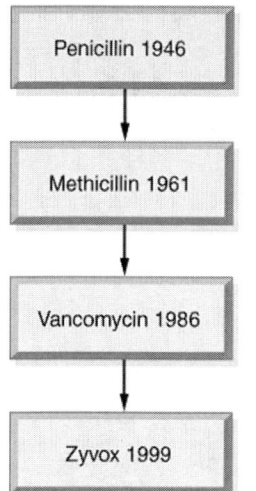

FIGURE 8.8:
Rapid succession of antibiotics needed to treat life-threatening infections with *Staphylococcus aureus.* Years indicate when *Staphylococcus* strains resistant to each drug were first observed. Image © Kendall Hunt Publishing Company. Redrawn after Palumbi S.R. (2001) Humans as the world's greatest evolutionary force. *Science* **293**: 1786–1790

Because each treatment with an antibiotic is a potential selection for resistant bacterial strains, the latter are accumulating in hospitals. Each year the bacterium *Staphylococcus aureus* causes a million hospital-acquired infections in the United States, killing tens of thousands of patients. Some strains are now resistant to nearly all antibiotics (Figure 8.8).

J. Darwinian Medicine

The frightening experience of bacterial resistance is causing a new emphasis on the evolution of infectious diseases, a trend called **Darwinian medicine** in contrast to traditional medicine, which is focused on correcting symptoms. For example, the conventional treatment of a bacterial infection may include aspirin to lower fever and a dietary supplement to correct low iron in blood serum. An evolutionary perspective views the same symptoms as resulting from a coevolution of humans and their parasites. This view helps to appreciate the potential usefulness of the symptoms as defense mechanisms. A Darwinian physician may therefore allow the patient's fever and low blood iron to take their course because both are probably more harmful to the bacteria than to the patient. The Darwinian doctor will also keep in mind that an antibiotic may bring the patient fast relief but also entails the risk—to the patient and others—of helping the evolution of a bacterial strain that has become resistant to yet another antibiotic.

A Darwinian perspective on public health also takes into account the role of domestic animals in the evolution of human infectious diseases. Modern agriculture often involves keeping large numbers of animals in limited space. To head off epidemic diseases, many farmers are routinely adding antibiotics to the feed of *healthy* animals. As a result, the agricultural use of some antibiotics exceeds their medical use by a factor of 100 or 1,000. This has the effect of breeding resistant animal pathogens, which may subsequently jump to human hosts as discussed earlier.

Evolutionary considerations are critical for proper management of serious epidemics. Some drugs can reduce the duration of viral infections, provided they are administered within 48 hours of symptom onset. Most of these drugs were developed for the treatment and prophylaxis of seasonal influenza and are marketed to "people who have no time for flu." However, it is virtually certain that seasonal flu strains resistant to these drugs will evolve. Then, if a carrier of such a strain is also infected with H5N1, the two viral strains may exchange parts of their genetic information, giving rise to a deadly bird flu strain that is already resistant to the antiviral drug developed against the less harmful seasonal flu.

The same Darwinian thinking applies to the evolution of resistance in agricultural pests. Farmers in the United States lost about 7 percent of their crops to pests in the 1940s. The answer was the liberal application of chemical herbicides and insecticides in the 1950s and 60s. But the problem got worse. The losses to pests have doubled and more pesticides are being used. This is a huge problem, but the pests are only following the rules of evolution: the best-adapted survive. Every time chemicals are sprayed on a crop or lawn to kill weeds or ants, a few members of the pest population may survive and create a new generation that is poison-resistant. Modern farmers and gardeners therefore try to outsmart the pests by using a variety of methods. They introduce natural enemies of pests, or they sidetrack the pests with meals that are even more tasty than the vulnerable crops. Pesticides are only used as a last resort if the other methods fail.

EXERCISE PAGE FOR CHAPTER 8

Student Last Name _____

Student First Name _____

Discussion Section _____

On the course web site, use the link on the **syllabus** to find a pdf file of your **assigned reading** for this chapter. The same reading is referenced below. Use the bulleted list of questions to test your knowledge of this reading.

Levy S.B. (1998) The challenge of antibiotic resistance. *Scient.Amer.* **Mar. 1998:** 46-53

- What are antibiotics? How do they affect bacterial growth?
- What physiological mechanisms make bacteria resistant to antibiotics?
- How do bacteria acquire the genes for antibiotic resistance?
- Does antibiotic resistance come with a "price" to bacteria?
- When human patients take an antibiotic medication, what happens to their pathogenic bacteria, to other bacteria, to the patient as a bacterial host, and to other humans living with the patient?
- How do antibiotics work together with a patient's immune system? What happens if a patient does not take a full course of antibiotic medication?
- Which human pathogenic bacteria are already untreatable or poised for resistance to all available antibiotics?
- What can consumers, physicians, and policymakers do to manage the problem of antibiotic resistance?

Feedback

On a scale from 1 to 10, the being the best, please rank the above reading for

1. Interest and Relevance to the topic (10 being most interesting/relevant) _____
2. Readability (10 being most clearly written, easy to understand) _____

In the remaining space, enter any comments that you may have on this reading.

PART III

HUMAN GENETICS AND GENOMICS

The scientific study of the transmission and expression of heritable traits is called genetics. It began with crossing experiments on garden peas by the Austrian monk and naturalist Gregor Mendel. From his observations, he concluded that phenotypic traits, such as shape and color of seeds and pods, are controlled by heritable "factors," which are passed on by the reproductive organs of the flower. Today we refer to Mendel's factors as genes, and we know they are segments of DNA. We have also learned how genetic information flows from DNA to RNA to protein, and how proteins control the structure and function of cells and organisms. In particular, we have come to appreciate the central role of gene regulatory proteins called transcription factors, which tailor gene expression according to tissue and developmental stage.

The powers of genetics were boosted by advanced techniques for the analysis of DNA. In DNA cloning, one makes many identical copies from a DNA segment. In DNA sequencing, one determines the exact sequence in which the four different nucleotides (A, T, C, and G) are lined up in a DNA molecule. These techniques, along with computer software to store and manipulate sequencing data, have made it possible to decipher the entire human genetic information, or *genome*. This monumental effort, known as the human genome project (HGP), will be presented in Chapter 9.

Information from the HGP is now revolutionizing every area of the life sciences and their applications in law, medicine, and pharmacy. Individual variations in genomic sequence make it possible to obtain a unique "DNA profile" of any person, which can be obtained from miniature amounts of tissue, such as blood, hair, or semen left in the course of a crime. Because DNA profiles are passed on from parents to children, such profiles can also be used to prove or exclude paternity. In the United States alone, DNA profiles are used in about 10,000 criminal court cases per year to convict a perpetrator or to free an innocent person. In addition, more than 200,000 civil cases, mostly paternity suits, are being decided each year on the basis of DNA profiles.

Another major benefit of the HGP has been the identification of disease genes, that is, genetic alleles that predispose their carriers to certain diseases. Once such a disease gene is identified, it is possible to devise laboratory tests that can tell a person whether he/she carries the gene. Similar tests for other genetic alleles can help a physician to predict a person's response to certain medications.

The use of genomic tests is also enhancing the prospects of in vitro fertilization (IVF). Traditionally, the development of the fertilized eggs is monitored under the microscope, and embryos that are free of visible abnormalities are implanted in the prospective mother's uterus. This form of "quality control" is now being extended by a procedure called pre-implantation genetic diagnosis (PGD). For this purpose, one cell is removed from each embryo, a procedure that allows the rest of the embryo to develop normally. The DNA from the removed cell is tested for genetic abnormalities, and only embryos passing this test are implanted.

Non-implanted embryos from IVF clinics can be used to generate embryonic stem (ES) cells. As we will discuss in Chapter 10, human ES cells are prized for their active proliferation in culture and for their ability to form any type of adult cell. Continued research on ES cells may therefore lead to cell replacement therapies for patients suffering from lack or malfunction of a particular cell lineage, as in type 1 diabetes, Parkinson's disease, or heart disease.

ES cells are also of major interest in conjunction with another kind of new medicine called gene therapy, to be discussed in Chapter 11. At the heart of many human diseases lies a defect in a single gene, and many of these genes are known. Replacement of the defective gene with its normal allele would seem to be the ideal cure. There have been successful cases of human gene therapy, but more research is needed before gene therapy can live up to its promise. Such research may converge with recent studies on induced pluripotent stem cells (iPS cells), which are similar to ES cells. Such iPS cells can be made from small biopsies of patients by adding a few genes encoding transcription factors characteristic of ES cells. If all goes well, future patients will be able to get replacement cells made from their own iPS cells, which will be immunologically compatible and genetically corrected as needed.

If we keep replacing our defective cells with new ones, will we live forever? Probably not. In Chapter 12, we will explore a progressive loss of DNA from the ends of our chromosomes, a process known as cellular senescence. A lineage of cells exempt from this process is the germ line, which is potentially immortal because it gives rise to eggs or sperm. All other cells, collectively called somatic cells, are bound to die.

All currently approved trials for human gene therapy are for somatic cells, so that even fully cured patients cannot expect any benefits for children they may have after gene therapy. The currently available techniques for genetically modifying eggs or sperm are not safe, and even if they were, they are fraught with ethical problems. Of particular concern are attempts to create "designer babies," because most genes are pleiotropic, that is, affecting more than one trait. The importance of this basic fact became clear from an attempt to make leaner pigs by giving them extra genes for growth hormone. Their meat had less fat, as intended, but the pigs were also suffering from unforeseen ailments including gastric ulcers, arthritis, cardiomegaly, dermatitis, and renal disease. Nobody should be allowed to risk side effects of this magnitude for their children.

By luck or good intuition, Mendel chose to study traits of peas, such as flower color or seed shape, that were controlled by single genes. Meanwhile we have learned not only that a single gene may control more than one trait, but also that a given trait may be affected by multiple genes (see Figure 3.9). In addition, we have come to appreciate that genetic and environmental factors may interact in shaping a trait. Indeed, in Chapters 13 to 15 we will learn about methods for *estimating the extent* to which traits such as intelligence or homosexuality are inherited and how much they can be molded by the environment.

Human genomics and advanced reproductive techniques have provided us with unprecedented powers to cure diseases and to control the genetic endowment of our children. The scientists who founded the HGP have foreseen this responsibility, and they have set aside 3 percent of each annual budget for considering the ethical, legal, and social implications of their research. Their idea was that a framework of laws and policies for our vast new options would emerge from a broad societal dialogue. It is hoped that Part III of this book will help this process along.

CHAPTER NINE

THE HUMAN GENOME PROJECT

The **human genome project (HGP)** was a big science project to decipher the entire human genetic information. The scientists who initiated the project had various goals. Some wanted to learn about the causes of human genetic disorders. Others were especially interested in *polygenic traits*. Many were fascinated by the challenge of the project, which was as ambitious as putting a man on the moon. Actually, compared to space exploration, HGP has been much more relevant to life on Earth and much less expensive.

The DNA molecules from one set of human chromosomes, when stretched out and joined end-to-end, would be about 1 m long and comprise more than three billion nucleotide pairs. Printing only one of the two complementary DNA strands would be a sequence like ...TACGCTGAC... that would fill 500 large books of 1,000 pages each. This human DNA sequence is now available to anyone on the Internet. In addition, complete DNA sequences of more than a hundred bacterial, plant, and animal species are known. This vast information is revolutionizing every discipline in biology, along with many areas of medicine, law, pharmacy, and agriculture.

In this chapter, we will briefly recount the history of the HGP before we explore its major goals and methods. Next we will review some of the results of the HGP, especially in terms of medical applications. The latter include in particular the identification of disease genes and the beginning of individualized medicine, which tailors medical treatment to the individual patient's genetic predisposition. Finally, we will discuss some of the ethical, legal, and social implications of the HGP, especially the needs for privacy and the responsible use of personal genomic information.

A. Brief History of the HGP

The HGP became possible with the development of two basic techniques in molecular biology. One of them, known as **DNA cloning**, produces any number of identical copies from a DNA segment of interest (Figure S9.a). This can be done *in vivo*, using suitable host cells, or *in vitro*, by letting the key enzyme of DNA replication do its work in a test tube.

Another technique basic to the HGP is called **DNA sequencing,** which means determining the exact nucleotide sequence of a DNA segment (Figure S9.b). This procedure used to be laborious, so that in an early debate about the feasibility of the HGP the sequencing part was recommended—tongue in cheek—as punishment for scientists who had committed scholastic dishonesty. Today, sequencing is done by robots at incredible speeds.

In 1988, James D. Watson (codiscoverer of the DNA double-helix) became director of a new U.S. National Center for Human Genome Research (NCHGR). He is credited for bringing the HGP on track during its formative years. In 1991, U.S. Congress approved the HGP, projected to run through 2005 and to cost a total of $3 billion. (For comparison, NASA received $5.9 billion for 1995 alone.) The major goals of the HGP were to:

- construct high-resolution maps of all human chromosomes.
- develop automated sequencing techniques and computer programs to assemble huge amounts of sequencing data.
- sequence the entire human genome.
- identify all human genes.
- obtain corresponding data for model organisms.
- explore the HGP's ethical, legal, and social implications.

Much of HGP has been carried out by a public, nonprofit consortium managed by the NCHGR, the U.S. Department of Energy, and the British Wellcome Trust. The consortium worked in its own laboratories and through grants awarded to other institutions. One of the consortium's central policies was that participating scientists could not acquire patents to the human DNA sequences they were discovering. One who disagreed with this policy, Craig Venter, left the consortium and founded a for-profit company, Celera Genomics Corporation. Celera and the consortium then competed until both published preliminary sequences on the same day, 12 February 2001, in the online versions of the two general scientific journals, *Nature* and *Science*. The finished version of the human genome sequence was published in April 2003. This "canonical" sequence is a composite based on the DNA from several individuals. Complete genomic sequences of individuals from different ancestries are now following.

B. Physical Chromosome Maps

Sequencing is done on relatively short DNA segments. To sequence an entire human chromosome, one has to string together the sequences of a million overlapping DNA segments in correct order. One of the first objectives of the HGP was therefore to establish detailed maps of all human chromosomes showing the locations of known genes and other characteristic DNA sequences. Modern chromosome maps combine the results of two basic mapping procedures known as *physical mapping* and *genetic mapping*.

Physical maps indicate the location of genes on a chromosome in physical units, such as the percent length from one chromosomal end or the number of nucleotide pairs from a neighboring gene. The first physical maps were based on **deletion mapping**, a method that correlates abnormal phenotypes with the absence of chromosome fragments large enough to be seen under the microscope. For example, the human gene for the enzyme acid phosphatase was mapped to one end of chromosome 2 by tracking a missing form of the enzyme in a child to a chromosomal translocation in the mother (Figure 9.1).

Another way of physical mapping, known as **in situ hybridization**, is based on the property of single-stranded nucleic acids to bind tightly, or *hybridize,* so long as their nucleotides are *complementary* (A paired with T and C with G). If chromosomes are spread on a glass slide and their DNA is made single-stranded, segments of their DNA will hybridize to complementary **probes**, that is, single-stranded DNA or RNA molecules tagged with a dye or other visible marker. If such a probe is prepared from a known messenger RNA and added to a chromosome spread, then the probe will "home" to the DNA segment from which it was transcribed. Thus, the probe will map the position of its cognate chromosomal gene. Figure 9.2 shows photographs of human chromosome 21 after in situ hybridization with fluorescent probes for four different genes.

While deletion mapping and in situ hybridization can show the approximate location of a gene on a chromosome, the resolution of these methods is limited. The physical maps constructed as part of the HGP are much finer. They are based on **contigs**, that is, sets of cloned overlapping DNA segments. In a contig spanning an entire chromosome, each DNA segment is millions of base pairs long (Figure 9.3). Each segment in turn is spanned by a sub-contig of smaller DNA segments, and so forth. Eventually, a gene is mapped physically by the position, in the nested set of contigs, of the smallest DNA clone(s) containing the gene. For each human chromosome, such contigs are kept in freezers, from where they can be replicated and shipped to researchers all over the world.

C. Genetic (or Linkage) Maps

Physical mapping requires some molecular information about the gene to be mapped, such as the nature of the gene product or a

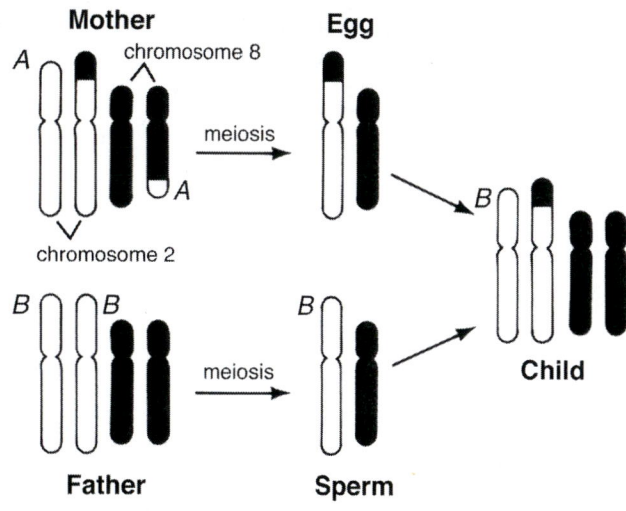

FIGURE 9.1:
Deletion mapping. The human gene for acid phosphatase exists in two alleles, A and B. A child for homozygous parents (A/A and B/B) had a chromosome translocation, which mapped the gene's locus to the tip of the short arm of chromosome 2 (white).
Redrawn after Sutton H.E. *Introduction to Human Genetics,* Saunders, (4th ed. 1988)

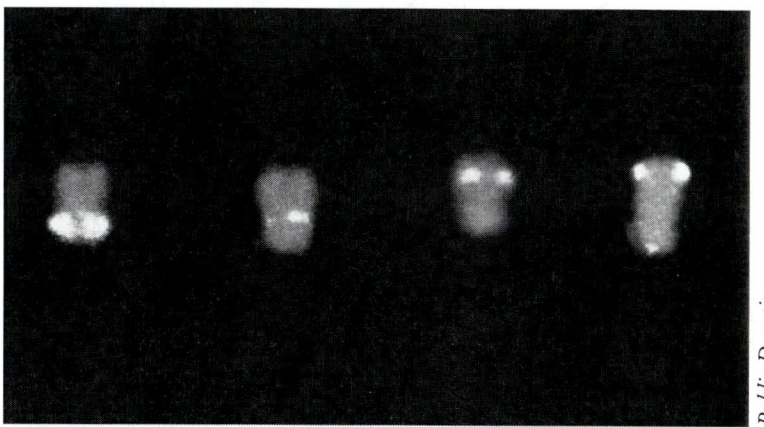

FIGURE 9.2:
Physical mapping of chromosomes by in situ hybridization with fluorescently labeled probes.
Image © Copyright Los Alamos National Security, LLC. All rights reserved.

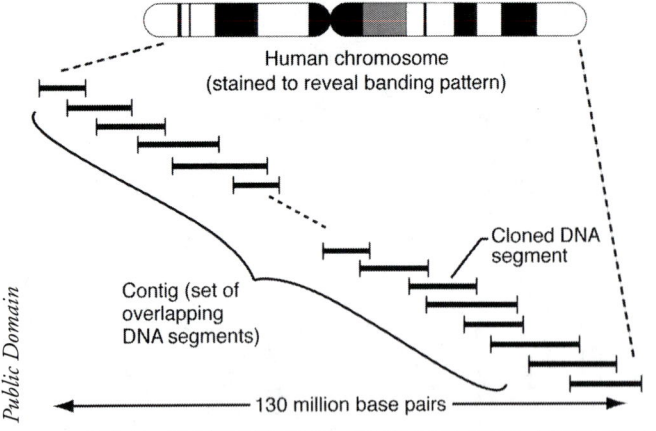

FIGURE 9.3:
Physical mapping of chromosomes by contigs (sets of overlapping segments of cloned DNA)
Image © Copyright Los Alamos National Security, LLC. All rights reserved.

partial nucleotide sequence of the gene. However, in many cases, no such information is available. Here is where genetic mapping comes in.

Genetic mapping is also called **linkage mapping** since it reveals groups of genes that are located on the same chromosome. Because linked genetic alleles travel together during cell division, traits caused by such alleles often show up together in relatives. Figure 9.4 shows the pedigree of a family with *nail-patella syndrome (NPS)*, a dominant genetic disorder characterized by abnormal finger- and toenails as well as small patella (kneecap) size. Individual I-2 in this pedigree shows NPS as well as blood type BO (type B with an underlying I^B/I^O allele combination for gene I, which encodes the A/B/O blood antigen). Her husband has blood type OO, being homozygous for the I^O allele. Half of their children also show the combination of NPS and BO blood type, suggesting that the NPS-causing allele travels together with the I^B allele. This means that the *NPS* gene is probably located on chromosome 9 because gene I was mapped to the same chromosome previously. If *NPS* were located on another chromosome, then about half of the children with NPS should have inherited the maternal I^O allele and should have blood type OO.

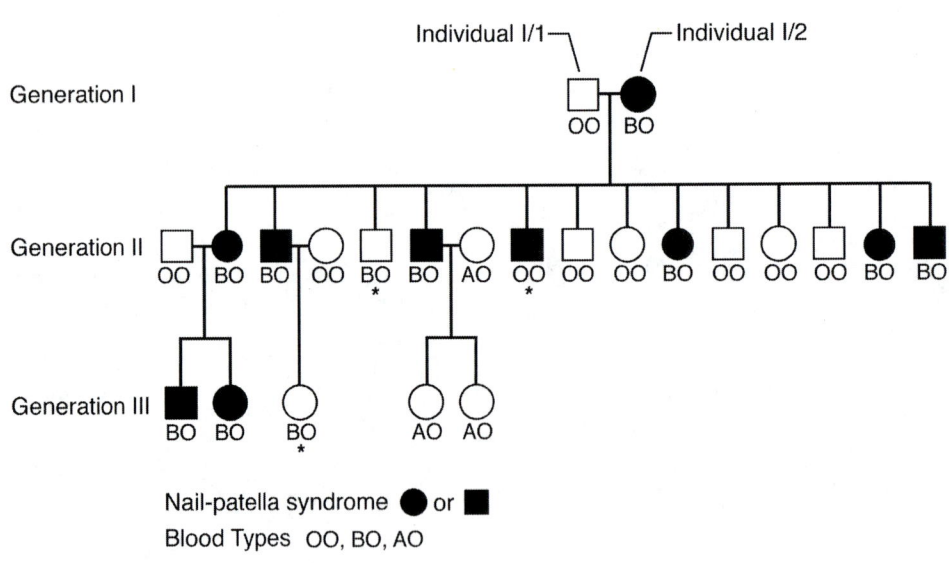

FIGURE 9.4:
Genetic mapping. This human pedigree reveals a linkage of the locus for nail-patella syndrome (NPS) and the locus for the A/B/O blood types. Squares represent males, circles females. The blood type designations reflect the underlying allele combinations. For example, "BO" individuals have one I^B and one I^O allele. Family members with NPS also show the BO blood type, except for individuals marked with an asterisk. The frequency of these recombinants is indicative of the distance between the loci for NPS and A/B/O.

The individuals marked with an asterisk in Figure 9.4 are exceptional in that the prevailing linkage between NPS and the BO blood type is broken. This is explained by the well-known phenomenon of crossing over, or **crossover** for short. It occurs regularly during **meiosis**, the specialized cell divisions involved in making haploid gametes (eggs or sperm). This is because, early in meiosis, pairs of **homologs** (corresponding maternal and paternal chromosomes) are aligned in parallel. This configuration facilitates the exchange of homologous chromatid segments by cutting their DNA strands and crosswise rejoining them (Figure 9.5).

Crossover events can account for the occasional separation of genetic alleles that usually travel together on the same chomosome. Figure 9.5a represents a homologous pair of number 9 chromosomes as it would occur in individual I/2 of Figure 9.4. Each homolog consists of two sister chromatids, and all four chromatids are aligned in parallel. The chromosome pair shows the two genes that are of interest here: the blood antigen gene, which is present the two alleles I^B and I^O in an individual with blood type BO, and the gene controlling NPS. Because NPS is a dominant disorder and only half of the offspring of individual I/2 has NPS, she must have only one copy of the NPS-causing allele (*N*), while the homologous allele is normal (*n*).

If a crossover occurs between non-sister chromatids, as shown in Figure 9.5b, then the previous linkages between the I^B allele and the *N* allele, and between the I^O allele and the *n* allele, may be broken. The resulting four egg genotypes are shown in Figure 9.5c. An egg that inherits the recombinant chromatid with the I^B and *n* alleles, if fertilized by a sperm from individual I-1, will become a child showing blood type BO but no NPS, as is the case for individuals II/5 and III/3 in Figure 9.4. An egg that inherits the other recombinant chromatid, carrying I^O and *N*, when fertilized by a sperm from individual I-1, will give rise to a child with blood type OO and NPS, as seen in individual II/8.

Note that in Figure 9.5 we assumed that the *N* allele is originally linked to the I^B allele. This was the case in the family pedigree shown but is not necessarily so. There are other families in which the *N* allele arose in a chromosome 9 with the I^A allele, so that most of the chromosome carrier's descendants with NPS had blood type AO or AB, depending on the genotype of the other parent.

Crossovers occur almost randomly during meiosis. This observation allows us to make the statistical argument that is critical to genetic mapping: *The further apart two genes are on a chromosome, the more likely it is that a crossover will separate them.* In other words, the distance between any two linked genes can be estimated from the frequency of recombinant gametes produced.

The linkage group of genes assigned to a chromosome, and their relative distances, constitute the **genetic map** of a chromosome (Figure S9.c). In honor of Thomas Hunt

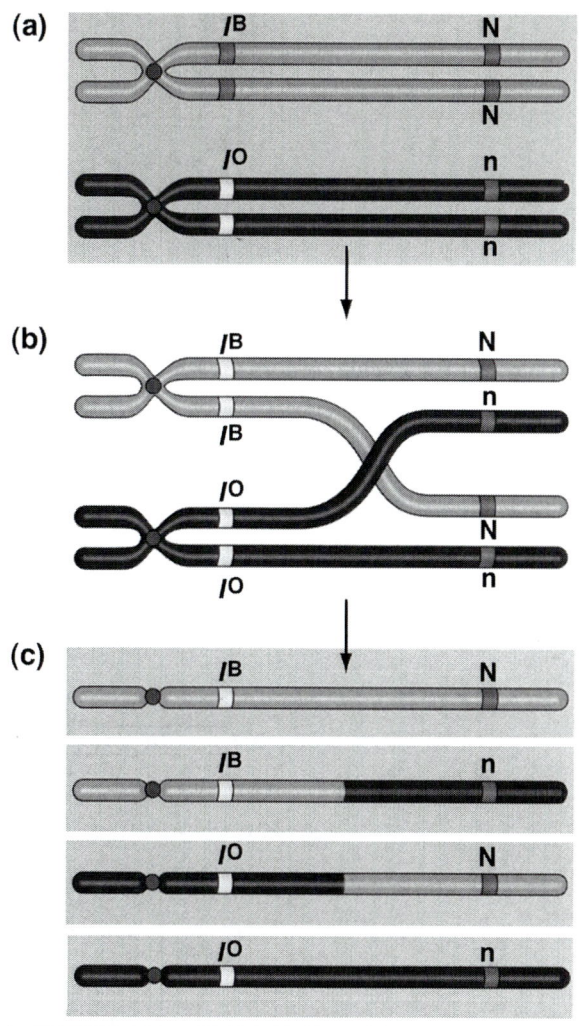

FIGURE 9.5:

Crossing over between non-sister chromatids during meiosis I can create gemetes with novel allele combinations (*Bn* or *ON*) in the same chromosome.

CHAPTER NINE: The Human Genome Project • 115

Morgan, the founder of modern genetics, distances on genetic maps are expressed in **centimorgans (cM)**. A distance of 1 cM means that 1 percent of all gametes produced are recombinant for alleles of these two genes. Based on 23 pairs of human chromosomes and an average of 33 crossovers observed per meiosis, the mean genetic length of a human chromosome is 33/23 x 100% or 143 cM. Long chromosomes, such as chromosome 1, measure more than 300 cM. Also, genetic maps based on offspring of females are longer than those of males because crossovers are more frequent in females.

The great advantage of genetic mapping is that it does not require any molecular knowledge of the gene of interest. The only requirement is that the gene must exist in two or more alleles because otherwise there would be no phenotypic difference to observe. (For example, if all people were albinos, no underlying gene could be mapped. Indeed, there would be no reason to assume that there are any genes affecting skin pigmentation.) Genes with multiple alleles are called **polymorphic** (Greek for having different forms), and the resulting phenotypes are known as **polymorphisms**.

D. DNA Markers

Genetic mapping of human diseases has often relied on molecular polymorphisms that are neutral to selection, such as the blood antigens mentioned earlier. Also known as **molecular markers**, they have two properties that make them ideal for mapping. First, none of the marker versions is selected against, and therefore the underlying genetic alleles are all common. Second, molecular marker alleles are codominant, meaning that a heterozygous individual expresses both alleles. This makes them much easier to detect than recessive alleles, which may go unnoticed in heterozygotes.

As the HGP was advancing, it became possible to use molecular markers that are not associated with an expressed gene. Now any nucleotide sequence that is *unique* (occurs only once per genome) and *polymorphic* (may vary from one individual to another) could be used as a **DNA marker.** How long does a nucleotide sequence need to be in order to be unique? Because there are four nucleotides, the chance for any particular sequence of twenty nucleotides to occur randomly is one in $4^{20} = 1.1 \times 10^{12}$. So the probability for any such sequence to occur twice, by chance alone, in the human genome of 3.2×10^9 nucleotides is only about 0.003. Thus, most sequences of twenty or more nucleotides are useful as DNA markers.

Several types of DNA markers are used for genetic mapping, in courts of law, and for other purposes. The simplest and most abundant markers are **single nucleotide polymorphisms (SNPs)**. These are unique DNA sequences with alternative bases that are present in at least 1 percent of a human population. Thus, an extended DNA sequence may be exactly the same in all humans except for one position, which may read "G" in most individuals but "T" or "A" in significant numbers of others (Figure 9.6). SNPs result from point mutations that have occurred once and hence have been passed on through the generations. The closer related two individuals are, the more SNP alleles they share.

Some SNPs are located in the coding regions of genes, where they may affect the amino acid sequence and hence the biological function of the encoded protein. Examples discussed previously include mutations in the *FOXP2* and *ASPM* genes, which seem to have played a major role in the evolution of the human language and brain (see Chapters 3 and 7). Most SNPs, however, are located in the vast regions of "junk" DNA that have no known function. More than a million of such SNPs have

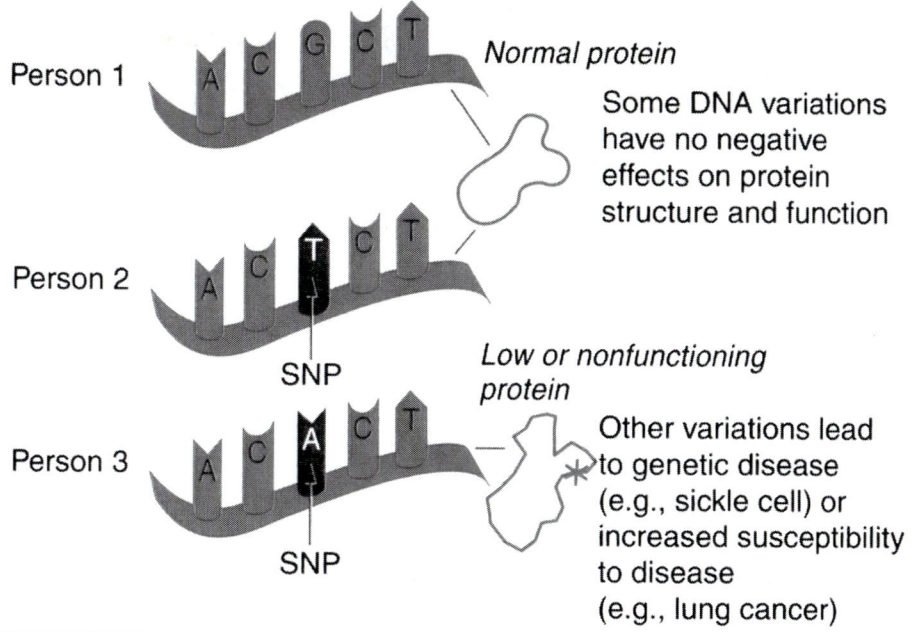

FIGURE 9.6:
Single nucleotide polymorphisms (SNPs) are unique genomic sequences in which a single nucleotide occurs with two or more alternative bases. Most variants have no effect on protein structure and function while some cause genetic disorders.

been identified so far; they will be of great interest for studying human evolution and migration.

DNA markers of a different type, now commonly used in forensic applications, are known as **variable number tandem repeats (VNTRs)**. They consist of repetitive units of two or more nucleotides, clustered together and oriented in the same direction (Figure 9.7a). VNTRs originate from errors in DNA recombination or replication. The location of each VNTR in the genome is constant, but the *number of repeats* at a given location varies among individuals, usually between 5 and 40. The *number of nucleotides per repeat* ranges from 2 to 50 or more. VNTRs with short repeats, usually four or less, are also known as *short tandem repeats (STRs)* or *microsatellites.*

The unique sequence surrounding a VNTR can be used to selectively amplify the DNA segment containing the VNTR. The *length of* the amplified segment then depends on the number of repeats, which can be determined by gel electrophoresis (Figure 9.7b). Due to the variety of alleles (numbers of repeats) for each VNTR, most people inherit two different alleles from their parents. Thus, if one compares any two people for *multiple* VNTRs, the chances of finding identical alleles for all of them is vanishingly small. Currently, a standard set of 13 VNTRs is used by the U.S. government to identify a person with enough certainty to convict a criminal or to exonerate an innocent person.

E. Sequencing and Annotating Genomic DNA

The discovery of DNA markers has greatly facilitated *genetic* mapping as discussed earlier. A DNA marker can also be mapped *physically* by matching it to the sequence

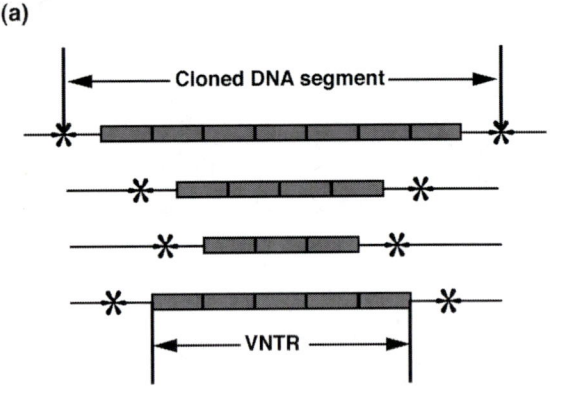

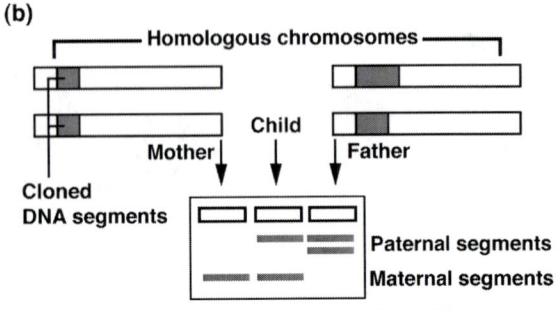

FIGURE 9.7:
Variable number tandem repeats (VNTRs). a: Each VNTR is at a specific chromosomal location and consists of a variable number of repetitive nucleotide sequences (boxes) surrounded by unique sequences (lines). Recognition sites (crosses) in the latter can be used for selective amplification of the VNTR. **b:** Amplified segments are separated by length using gel electrophoresis. Results are shown for one VNTR of a couple and their child. The mother is homozygous for her $(CGGT)_8$ allele, while the father is heterozygous for $(CGGT)_{25}/(CGGT)_{30}$. The child inherits the maternal allele and one of the two paternal alleles. The length of the VNTR segment relative to the entire chromosome is exaggerated.

of a particular clone of a *contig* (see Figure 9.3). Thus, DNA markers make it possible to *superimpose* physical and genetic maps. For every human chromosome, and for any chromosome region of particular interest, there are now detailed maps showing all known genes and DNA markers. The detailed chromosome maps, and the exact nucleotide sequence for each of them, were the main goals of the HGP. These goals have been accomplished, on budget and ahead of schedule. Supplemented with the complete genomic sequences of more than a hundred other organisms, this is a treasure trove of freely available information that is revolutionizing the life sciences.

Much of the remaining work is referred to as **annotation** of the human genomic sequence. It includes finding all human genes and exploring their functions. If one looks at raw genomic sequence, it is not obvious where genes begin and end, or what their functions may be. Fortunately, the regulatory sequences that signal transcription start points are fairly conserved in evolution. The same holds for the *splicing signals* that separate *exons* (parts of a gene that are eventually translated into protein) from intervening introns. Exons must be *open reading frames (ORFs)*, that is, free of the stop codons that terminate the translation process. Thus, running raw DNA sequence through computer programs that recognize transcription start points, splice signals, and ORFs is usually the first step of annotating genomic sequence.

Next, the genetic code is used to translate ORFs into **predicted amino acid sequences**, which are compared to extensive databases containing the amino acid sequences of proteins with known functions, such as transcription factors, cell adhesion molecules, or kinases (enzyme adding phosphate groups). Each of these has characteristic domains functioning as DNA-binding sites or as docking sites for phosphate donor molecules or other reactants. Thus, by comparing a predicted amino acid sequence to the amino acid sequences of known proteins, one can usually come up with a reasonable hypothesis on the function of the newly identified gene product.

F. What Have We Learned from the Human Genome Project?

While much of the annotation work is still going on, a few results have already emerged. Less than 5 percent of the human genome is occupied by genes, including their regulatory regions. Most other genomic sequences are often called "junk DNA" although previously unknown functions are still emerging, such as the encoding of small regulatory RNA molecules. While not every human gene may have been found yet, the total number is likely to be around 25,000. This number came as a surprise, since the going estimate before the HGP had been around 100,000. Considering that the roundworm *Caenorhabditis elegans* has more than 18,000 genes and the fruit fly *Drosophila melanogaster* more than 13,000, it seems that the complexity of an organism is based not so much on the number of its genes but on the **networks** of transcription factors and RNA processing signals that control how genes are utilized. As an analogy, one may realize that good and mediocre chess players all use the same set of chessmen and rules but differ in the combinations of moves they can make.

The annotation of the human genome is confirming on a grand scale an experience that many researchers have made before on single genes: The functional domains of proteins, and often entire proteins, show an astounding degree of **evolutionary conservation**. Thus, the predicted amino acid sequence from an unknown human gene may have an almost exact match in a database of known mouse or fruit fly proteins! It seems that a core of a few thousand genes has been used over and over in animals large and small. Many of these core genes are of medical interest: Genes involved in human cancers, senescence, and neurological diseases usually have counterparts not only in mammals but also in the fish *Danio rerio*, the fruit fly *Drosophila melanogaster*, and the roundworm *Caenorhabditis elegans*. Because these smaller animals are easier to keep in the laboratory, and because experimenting with them is ethically less problematic, they are often used as **model organisms** for research on human diseases.

Of particular interest is the comparison of the human genome with that of the chimpanzee (*Pan troglodytes*). The two genomes differ only by 1.2 percent in terms of single nucleotide changes, in fair agreement with earlier estimates based on the decrease in melting temperature of chimp/human hybrid DNA (see Chapter 3). Most of these differences have no functional significance because they occur in the "junk" regions of genomic DNA. However, about two thousand genetic differences occur in actual genes, including the *FOXP2*[+] and *ASPM*[+] genes discussed in Chapters 3 and 5. The human alleles of these genes, which seem to have been positively selected in hominin evolution, will now receive particular attention.

G. Identification of Disease Genes

One of the strongest motivations for launching the HGP has been the desire to identify human **disease genes**, genes that cause human diseases if they function abnormally. Many of these diseases are "simple" (controlled for the most part by single genes), as opposed to "complex" (controlled by several genes). The genetic defects underlying **simple diseases** are now being identified using data from the HGP.

Suppose you want to characterize a disease gene, D, which is unknown in molecular terms. In many cases, there will be some preliminary information, including an assignment of D to a particular chromosome region and references to human families showing the disease. So one can ask the living members of those families to provide samples of blood or saliva, from which genomic DNA can be prepared. Using established methods, one determines each family member's alleles for DNA markers along the chromosomal region that is suspected to contain D. From pedigrees showing the disease status and marker alleles of family members, D is *mapped genetically* between two DNA markers, labeled c and d in Figure 9.8.

Next, the *contig* covering the chromosomal DNA between c and d is scanned for genes that have already been annotated and for any additional open reading frames that may have been overlooked. Do any of the predicted amino acid sequences make

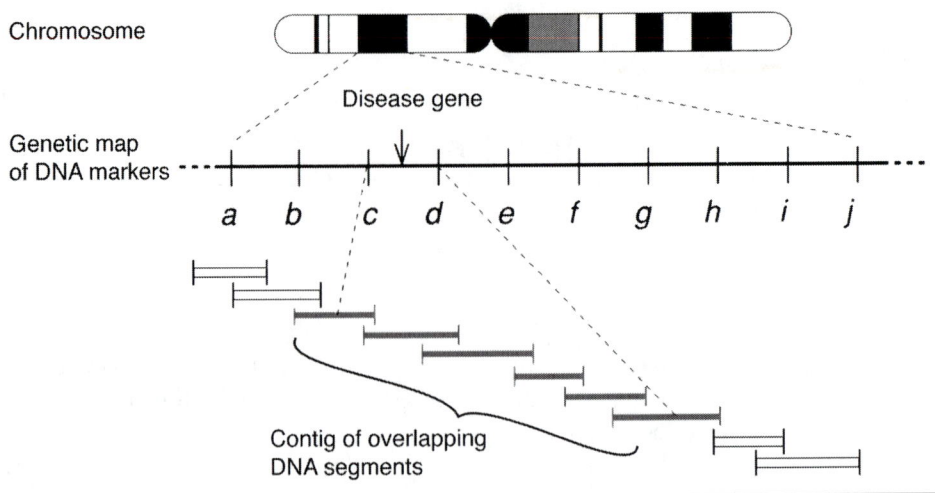

FIGURE 9.8:

Finding a disease gene (D) by a combination of genetic mapping between two DNA markers (c, d) and physical mapping using a contig.
Redrawn from *Los Alamos Science* No. 20 (1992) p. 99

biological sense in the context of the disease symptoms? If so, these candidate genes from all family members are sequenced to see whether they are mutated in all diseased individuals but not in their healthy relatives. If these sequence comparisons reveal a disease-correlated mutation, then the last step needs to be repeated with at least one other family. If one finds another disease-correlated mutation in the same gene, then the gene is probably *causing* the disease, because it is very unlikely that two independent mutations in the same gene are just coincidental with the disease.

From here, one would proceed with creating a strains of model organisms in which the gene of interest has been modified using procedures to be discussed in Chapter 11. If the genetically modified animals show symptoms that are comparable to the human disease, then the first step has been taken in elucidating the molecular basis of the disease and in developing a therapy. Genetic analyses along these and similar lines have helped to characterize more than 1,500 clinical disorders in molecular terms (Figure S9.d). Related information is freely available on the Internet at www.ncbi.nlm.nih.gov/omim/

Beyond the map-based discovery of genes causing simple diseases, it may be possible in the future to identify genes involved in **complex diseases**, which are controlled by several genes. Such diseases, like macular degeneration or schizophrenia, affect many people. Related strategies would rely on massive searches for *associations* between the diseases and single nucleotide polymorphisms (SNPs). These searches take advantage of the fact that SNPs can be discovered by DNA microarrays, which can test tens of thousands of DNA sequences simultaneously. Such microarrays are also the key technique for a new kind of individualized medicine.

H. Use of Genomics for Individualized Medicine

The availability of complete DNA sequences has spawned a new discipline, known as **genomics**, which analyses entire genomes rather than single genes. Human genomics is now becoming the basis for **individualized medicine**, which gauges medical treatment to the individual patient's genetic predisposition.

Box 9.1: DNA Microarrays

Key tools in genomics are **DNA microarrays**, or *gene chips* (Figure 9.9). They are glass slides with tiny dots of DNA in a gridlike pattern. Each dot contains many copies of a single-stranded DNA sequence linked to the glass surface. Typically, these sequences are chosen to represent *every gene* of *a species*. Other microarrays represent genes of particular interest in their normal alleles and in common mutant variants.

To find out, for example, how a particular type of human cancer changes the overall pattern of gene activities in a certain type of tissue, microarrays are used to compare the mRNAs synthesized in cancer cells with those made in normal control cells. To this end, mRNAs extracted from primary tumor cells are tagged with a red fluorescent dye, whereas mRNAs from normal cells are labeled with a green fluorescent dye. The red mRNA mix is then applied to a microarray representing all human genes in a known order. Any mRNA in the mix that is sequence-complementary to a gene on the array will hybridize. Perfectly matched mRNAs will stick, while others are washed away.

The array is then scanned by a laser beam that registers the intensity of red fluorescence at each DNA dot. A signal indicates that labeled mRNA has stuck there, which means that the corresponding gene was transcribed. The overall result is a pattern of spots with different degrees of red fluorescence, showing which of *all* genes in the microarray have been actively or not so actively expressed. This pattern is stored in a computer before all red probe is washed off the microarray, and the whole procedure is repeated with the green mRNAs from the normal control cells. A computer then compares the two fluorescence patterns and calculates how strongly each gene was up- or down-regulated in the tumor cells relative to the control cells. To the human eye, superimposition of the two patterns creates yellow dots for genes that are expressed in both cell types, while black fields represent genes transcribed in neither.

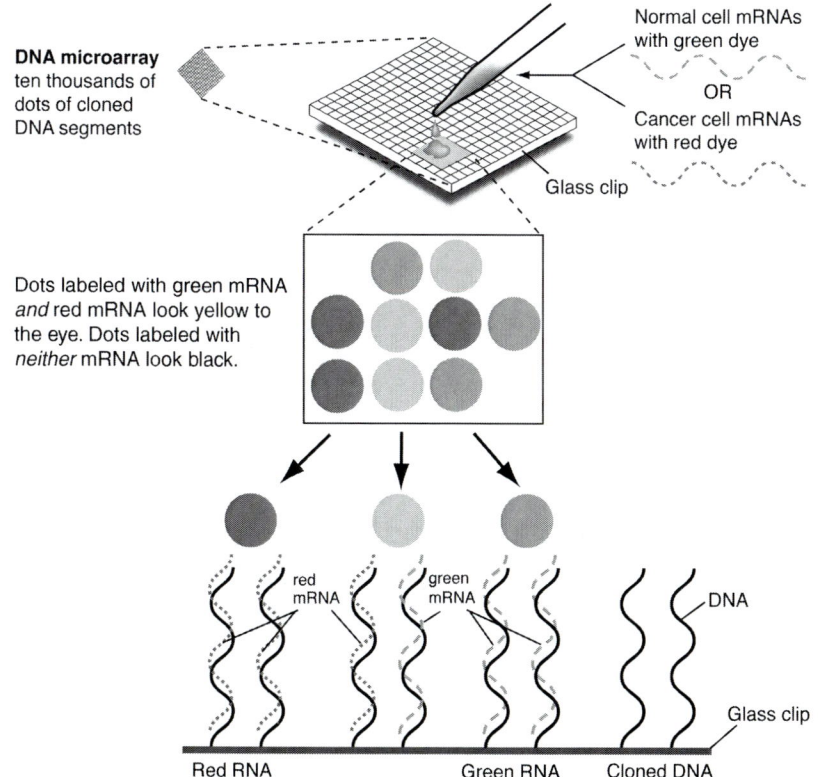

FIGURE 9.9:

A **DNA microarray** is prepared on a glass chip, which is subdivided into 50,000 to 500,000 fields. Each field has attached to it multiple copies of a unique DNA sequence. The DNA dots are hybridized with fluorescent probes representing mRNAs from different tissues.
Redrawn from Krogh D., *Biology* (3rd ed. 2005)

Some clinics now use DNA microarrays to distinguish cancer patients with different risks. Tumors detected early in a screening program are sometimes treated by surgery and radiation therapy alone, without chemotherapy, because the latter carries its own risks and is very draining. However, some patients so treated do have recurrences, with cancer cells spreading to other organs. With the benefit of hindsight, they should have received chemotherapy early on. So the goal is to distinguish patients with a good prognosis (recurrence very unlikely) from patients with a poor prognosis. The traditional microscopic inspection of the primary tumor has been of limited utility in making this decision. Scientists have therefore been looking for better ways of predicting the risk of recurrence for a given primary tumor.

Cancers originate through a series of mutations that disrupt genetic control circuits for cell division, cell adhesion, and other cellular interactions. The disruption may be caused by lack of function in some genes, or overactivity of other genes, or both. How the disruption cascades into a cancer depends on the patient's genetic predisposition and many intercellular signals that affect gene activities. Thus, by studying the genes that are up-regulated or down-regulated in primary tumors, one should be able to discern **signature changes in gene activity** associated with a good prognosis from patterns that come with a poor prognosis. The assigned reading at the end of this chapter illustrates this effort. The same rationale will apply to other diseases, such as asthma or schizophrenia, in which similar symptoms can be caused by different disturbances in genetic control circuits.

Human genomics will promote individualized medicine in many other ways. Knowing a patient's genetic predispositions will often help a physician to gage the dose of a medication that is best for this person. Because people differ in the ways they metabolize drugs, a dose that is too low to have any therapeutic effect for one patient may kill another. Indeed, adverse reactions to drugs are a leading cause of hospitalization and death in the United States. For the same reason, many potentially useful drugs could not be marketed in the past, but improved **pharmacogenomics** may change that.

In other cases, patients were found to use different enzymes for the same physiological effect. For example, a popular class of blood pressure–lowering drugs interfere with *angiotensin-converting enzyme (ACE)*, which causes smooth muscle cells in blood vessels to contract. However, the first ACE inhibitors were ineffective in about 20 percent of all patients. A search for genes encoding proteins similar to ACE led to the discovery of another gene encoding ACE2, an alternate enzyme used by those patients who are refractory to ACE1 inhibitors. The development of drugs inhibiting ACE2 is now underway.

I. Ethical, Legal, and Social Implications

The scientists who started the HGP were keenly aware of its **ethical, legal, and social implications (ELSI)**. From the beginning they have set aside 3 percent of their yearly budgets for studying ELSI through research grants, workshops, symposia, and public forums. The overall goal has been, and still is, to help frame legislation that will make the benefits of genomic information available to citizens while protecting them from possible misuse.

Genomic information is already available commercially from direct-to-consumer services. Given the medical importance of such data, their collection, accuracy standards, and storage **need to be regulated**.

The most important issue is the **privacy** of genomic information because potential employers and insurers stand to gain by discriminating against carriers of unfavorable alleles. Clearly it should be illegal for clinics or researchers to share genetic information in any way that identifies the individual carrier.

The U.S. government has passed the **Genetic Information Nondiscrimination Act (GINA)** in 2008. It protects individuals against discrimination in health insurance and employment. (The military is exempt, as it is from covering the costs of treating many conventionally diagnosed diseases.) One motivation for passing GINA was to encourage Americans to take advantage of genetic testing as part of their medical care.

A related question is under which conditions a judge can admit the use of DNA-based evidence in court, or should be able to compel a person to provide a blood/saliva sample for genetic testing. In each case, it must be regulated which parts of the obtained genomic information (personal identifiers, disease alleles, etc.) may be used and how the misuse of the remaining information is to be prevented.

In some cases, knowledge of a genetic predisposition to disease can help mitigate the disease by early detection, medication, and lifestyle choices. In other cases, patients who are diagnosed with a high-risk allele are left with anxiety and no help because further information on how to manage or treat the disease is not yet available. Persons who feel uncertain about how much of their genomic information they want to know may seek advice from a genetic counselor who can help clients to decide whether to seek complete genomic information, to reveal only selected data, or to blot out specific data. Such counselors can also inform clients about the likelihood of passing on certain predispositions to disease to any present or future children.

EXERCISE PAGE FOR CHAPTER 9

Student Last Name _____

Student First Name _____

Discussion Section _____

On the course web site, use the link on the **syllabus** to find a pdf file of your **assigned reading** for this chapter. The same reading is referenced below. Use the bulleted list of questions to test your knowledge of this reading.

van't Veer et al. (2002). Gene expression profiling predicts clinical outcome of breast cancer. *Nature* **415:** 530–536.

- What is the key technique employed by the authors, and how does it work in principle?
- What tissue samples were analyzed, and at which time were they removed from the patients? How did the authors group their patients?
- What primary data did the authors collect, and how were they processed for the presentation shown in their Figure 1?
- In Figure 2b, left panel, how were the 70 genes selected, and how were the 78 tumors selected and then rank-ordered? How does the rank-ordering of the primary tumors correlate with the clinical outcome after 5 years?
- What are the data tallied in Table 1, and what is the argument made by the authors based on these data?

Feedback

On a scale from 1 to 10, the being the best, please rank the above reading for

1. Interest and Relevance to the topic (10 being most interesting/relevant) _____
2. Readability (10 being most clearly written, easy to understand) _____

In the remaining space, enter any comments that you may have on this reading.

CHAPTER TEN

STEM CELLS AND CLONING

In the course of normal development, most cells proceed from an **embryonic state**, which is unspecialized, to a **differentiated state**, in which they carry out specialized functions. For example, red blood cells are donut-shaped (see Figure 8.4) and filled with hemoglobin, which allows them to squeeze through the finest blood vessels and to ferry oxygen and carbon dioxide efficiently throughout the body. Nerve cells, or neurons, have long extensions by which they send signals to muscles and other neurons. The transition of cells to the differentiated state is called **cell differentiation**.

What does cell differentiation mean in terms of gene activity? As a rule, all cells retain a full complement of genetic information. But like people who buy a complete copy of a newspaper but read it selectively, cells use their genetic information selectively. All cells express a group of **housekeeping genes**, which encode proteins required for basic cell functions, such as generating chemical energy or controlling the cell cycle. In addition, each type of differentiated cell expresses a set of **special genes**, which support the characteristic features of these cells. This general phenomenon is known as **differential gene expression**.

The molecular mechanisms of cell differentiation have recently attracted new attention in the context of stem cells. There are **tissue stem cells**, which are reserve cells used by the organism to replace worn-out differentiated cells. And then there are **embryonic stem cells**, which can be prepared from early embryos and are of medical interest because they might be used for cell replacement therapies. Interest in cell differentiation has also been rekindled by the experience that animals and human cells seem to find it difficult to switch from one differentiated state to another, or to return to the embryonic state. This has become apparent from the small success rate of attempts to **clone** animals, that is, to generate genetically identical offspring from a single individual, as in the case of Dolly the sheep.

In this chapter, we will explore recent stem cell and cloning experiments, including some of their medical and ethical implications, against the background of the fundamental question of how cells can use the same complement of genetic information and yet become so different.

A. Embryonic Cells Divide Actively, Are Undifferentiated, and Can Regulate

When an animal egg has been fertilized, it will begin a series of cell divisions. In humans, fertilization normally occurs in the upper oviduct (Figure 10.1). The fertilized egg, or **zygote**, then travels down to the uterus, while undergoing a series of cell divisions. An embryo consisting of 10 to 30 cells is called a **morula**, which is Latin for mulberry. About four days after fertilization, a fluid-filled cavity appears in the embryo, which is then called a **blastocyst**. During its travel down the oviduct, which takes 4 to 5 days, the embryo remains surrounded by a sturdy, transparent eggshell.

After arrival in the uterus, the embryo hatches from the eggshell and implants itself into the inner uterine layer. The hatched blastocyst consists of two groups of cells with different functions. The outer layer of cells, called the **trophoblast**, will contribute to the placenta. The inner group of cells, known as the **inner cell mass (ICM)**, forms a small mound eccentrically attached to the inside of the trophoblast. The ICM will give rise to the embryo proper.

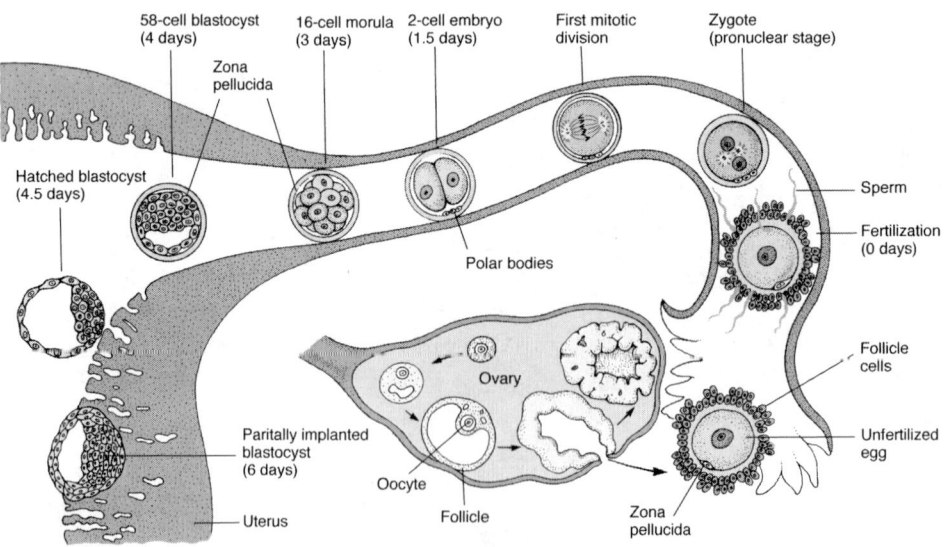

FIGURE 10.1:

First week of human development
From *Analysis of Biological Development* by Klaus Kalthoff. Copyright © 1996 by McGraw-Hill. Reproduced with permission of The McGraw-Hill Companies.

Embryonic cells can be characterized by three features: They divide actively, are undifferentiated, and can regulate. The first two features are almost self-explanatory. Typical embryonic cells stay in the cell cycle, in which mitotic divisions alternate with periods of DNA replication, RNA synthesis, and at later stages, growth. Embryonic cells are also undifferentiated, with a roundish or cuboidal shape and lacking specialized organelles or excessive amounts of any particular protein. The third characteristic of embryonic cells, being able to regulate, needs more explanation.

To developmental biologists, the **fate** of cells is what becomes of their descendants in the course of normal development. In many animals, embryonic cells can depart from their fate, a process called **regulation**. In humans, regulation occurs naturally in the development of *monozygotic twins,* that is, twins developing from one zygote. At any time between the 2-cell stage and an advanced blastocyst, an embryo may divide into two halves, each of which develops into a normal individual (Figure 10.2). In this

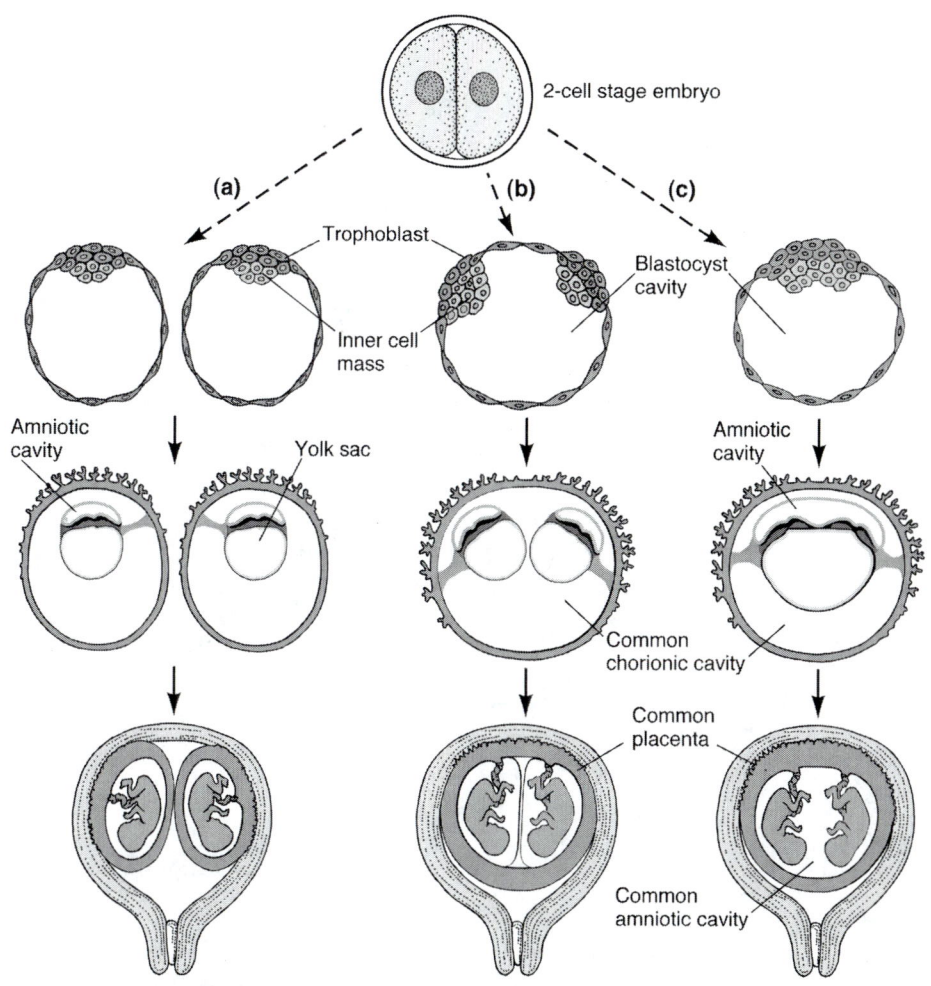

FIGURE 10.2:

Monozygotic twin development in humans. The fertilized egg divides spontaneously at the 2-cell stage (a) or later (b,, c). Each half regulates to form a complete individual. From *Analysis of Biological Development* by Klaus Kalthoff. Copyright © 1996 by McGraw-Hill. Reproduced with permission of The McGraw-Hill Companies.

kind of regulation, the fate of each embryonic half *increases* from half an individual to a whole individual.

Under different circumstances, regulation can also mean that the fate of cells *decreases* from what it normally would have been. In experiments with mice, embryos can be flushed out of the oviducts at the eight-cell stage, released from their eggshells, and stuck together. If the embryos came from different mouse strains, the composite embryo is a **chimera**, that is, an individual composed of genetically different cells. Such a chimera will develop into one oversized blastocyst, which can be implanted into a foster female. The chimera will then grow into a normal-sized mouse that is an integrated mosaic of cells and may have up to eight parents (Figure 10.3).

As demonstrated by the development of twins and chimeras, mammalian (and many other animal) embryos can adjust their growth and development to a reduced or augmented number of constituent cells. This remarkable ability shows that cells are able to exchange signals and adjust their behavior as needed for the development of normal-sized and harmoniously proportioned individuals. Based on the human embryo's ability for regulation is the new service called *pre-implantation genetic diagnosis*, mentioned earlier. In this procedure, eggs fertilized *in vitro* are kept in culture until the 8-cell stage, when one cell is removed for genetic testing. So far it seems that the removal of one cell does not diminish an embryo's ability to develop into a healthy baby.

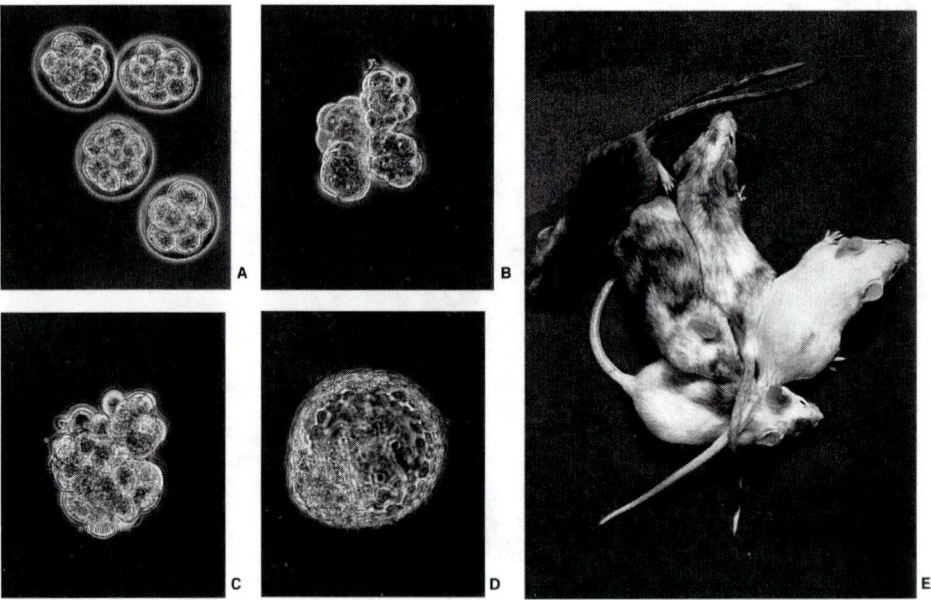

FIGURE 10.3:
Regulation in mouse development. (a-d) When four embryos are removed from their egg shells and aggregated they form one blastocyst. **(e)** When such a chimera is implanted into a female it develops into a normal mouse. It will show fur patches of different color if the component blastocysts came from different mouse strains. From *Analysis of Biological Development* by Klaus Kalthoff. Copyright © 1996 by McGraw-Hill. Reproduced with permission of The McGraw-Hill Companies. Photos courtesy of V.E. Papaioannou, from Figures 9 & 17, Papaioannou, V.E. and Dieterlen-Lievre, F. 1984.

B. Differentiated Cells Express Special Sets of Genes

Cell differentiation initially occurs during fetal development, which in humans begins with the third month of gestation. However, cell differentiation continues throughout the life of an organism as worn-out cells are replaced by new ones. Cell differentiation has three outstanding properties: It produces a limited number of cell types, is generally irreversible, and is based on the expression of special sets of genes.

Humans and other large vertebrates are made up of trillions of cells, but a much smaller number of cell types. Each **cell type** is defined by its morphology and by characteristic molecules that can be detected with antibodies or other specific probes. Familiar examples include nerve cells, blood cells, and muscle cells. Each of these cell types is subdivided further. Muscle cells, for example, include skeletal muscle fibers, cardiac muscle cells, and smooth muscle cells. Skeletal muscle fibers, in turn, are subdivided into red fibers, white fibers, intermediate fibers, muscle spindles, and satellite cells. Some cells take many different shapes. In particular, neurons (nerve cells) have processes of different lengths and branching patterns. While one can quibble about the exact number of cell types in the human, many professionals agree on about 200.

The differentiated state of a cell is generally irreversible. Most differentiated cells do not revert to the embryonic state or transform into other cell types as part of normal development. Bone cells do not normally become muscle cells, and muscle cells do not turn into neurons. Under unusual circumstances, one can observe dramatic exceptions to this rule. For instance, salamanders are well known for their ability to regenerate lost limbs. Even more spectacularly, an experimentally removed lens of a salamander's eye can be regenerated from the iris of the eye, even though iris cells and lens cells have entirely different shapes and synthesize different sets of proteins.

Each cell type has a distinctive set of structural and functional characteristics, which are based on the synthesis of particular proteins, such as hemoglobin in red blood cells or actin and myosin in muscle. How then does a cell determine which of their genes are expressed and which remain silent? One key to differential gene expression is a class of proteins known as **transcription factors**, which bind specifically to matching DNA recognition sequences associated with the transcribed regions of genes. Each cell type has a specific combination of transcription factors that activate the expression of certain sets of target genes while inhibiting the expression of others. The synthesis of such transcription factors often signals the beginning of cell differentiation before it becomes apparent morphologically.

Control of gene expression by transcription factors requires that their target recognition sequences be available for binding. Here is where another mechanism of cell differentiation comes in. Chromosomal DNA is associated with *histones* and other proteins, and the tightness of the association changes with the cell cycle. During mitosis, DNA is coiled, supercoiled, and looped so that chromosomes are visible as individual units. During interphase, chromosomes are loosened up so that they form a diffuse meshwork. In this phase, some chromosome regions are sufficiently open for transcription to occur (Figure 10.4a). Other chromosome regions remain closed so that transcription factors cannot bind to their DNA recognition sequences (Figure 10.4b). Regional differences of chromosome condensation depend on *DNA methylation* and *histone deacetylation*, which will be discussed later in this chapter. Generally, chromosome condensation patterns are passed on from each cell to its descendants.

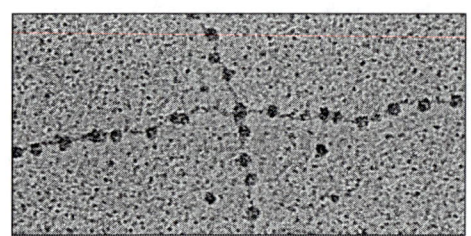
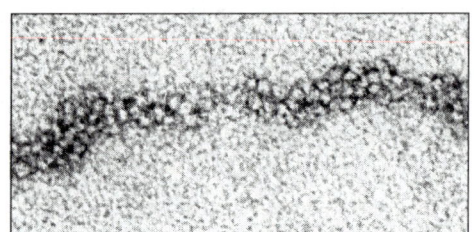
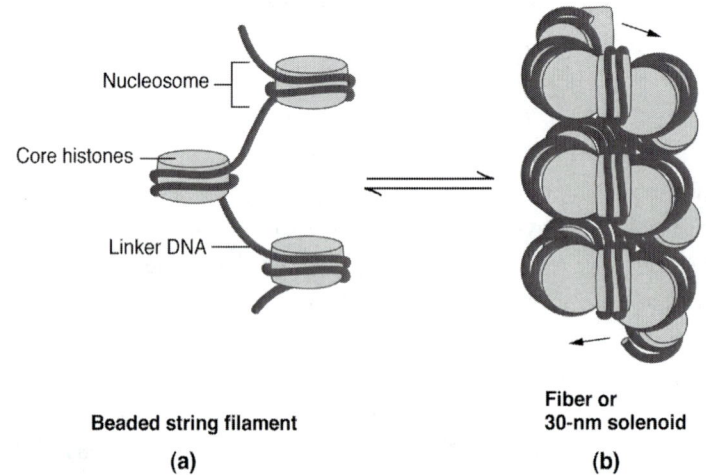

FIGURE 10.4:

Chromatin condensation. In eukaryotic chromosomes, DNA is packaged with histones and other proteins. **(a)** decondensed and **(b)** condensed conformation. From *Analysis of Biological Development* by Klaus Kalthoff. Copyright © 1996 by McGraw-Hill. Reproduced with permission of The McGraw-Hill Companies.

While transcription factors act on individual genes, chromosome condensation affects larger segments of genomic DNA including multiple genes.

C. Tissues Renew Themselves from Tissue Stem Cells

Some differentiated cells, including liver cells, retain the ability to divide. However, most cell types have lost this ability. In these cell lineages, worn-out cells are replaced from small pools of reserve cells, known as **stem cells**. These extraordinary cells are defined by three criteria (Figure 10.5).

- They are *undifferentiated*.
- They have an *unlimited capacity to divide*.
- They *give rise to two kinds of daughter cells: more stem cells* and **progenitor cells**. The latter are still undifferentiated but already committed to develop into certain types of differentiated cells.

Stem cells were first discovered in adult tissues that have a rapid turnover of cells, such as epidermis and hematopoietic (blood cell forming) tissue. We will call them **tissue stem cells** if we want to distinguish them from *embryonic stem cells*, which will be discussed later in this chapter. Tissue stem cells reside in protected microenvironments called **niches**, and signals from the niche keep them in their state of "stemness." As needed, they proliferate and leave the niche to form progenitor cells. This process of **mobilization** is controlled by intrinsic genetic programs and by extracellular signals.

The generation of progenitor cells and more stem cells may follow the *local asymmetry* model (Figure 10.5a) or the *serial asymmetry* model (Figure 10.5b). In either case, progenitor cells go through **amplifying divisions** before they become differentiated cells. Also independently of local or serial asymmetry, a given population of stem cells may be **unipotent** (giving rise to one type of progenitor cell) or **pluripotent** (giving rise to two or more different types of progenitor cells). Examples of pluripotent stem cells are those that form the progenitors for red blood cells and various types of white blood cells.

Renewal from stem cells is observed in most cell populations. The rate of renewal, or **turnover**, varies greatly from one cell type to another. Probably the fastest turnover occurs among the cells that line the inside of the intestine. These cells are exposed to mechanical stress from passing food and chemical attacks from digestive enzymes and acids, so that their average lifetime is about one day. Indeed, a significant portion of feces consists of dead cells from the intestinal lining. At the other end of the spectrum, the turnover of neurons is so slow that for a long time it was believed that humans older than a few years would not form any new neurons.

Developmental biologists have assumed that the potency of stem cells would be limited to those cells that are formed naturally where the stem cells reside. Thus, blood stem cells were expected to form only blood cells, epidermal stem cells only epidermis, and so forth. However, experiments to test this assumption have revealed the ability of at least some tissue stem cells to form cells beyond their normal repertoire. In the most spectacular case observed so far, neural stem cells from mouse brain were injected into the bloodstream of mice whose own blood stem cells had been destroyed by X-irradiation. Surprisingly, the injected mice formed new blood cells that carried the genetic markers of the neural stem cell donors.

D. Embryonic Stem Cells Can Form Any Type of Differentiated Cell

Different lines of experimentation have led to the discovery of **embryonic stem cells (ES cells)**, so called because they are derived from embryos and meet the same criteria that were originally used to define tissue

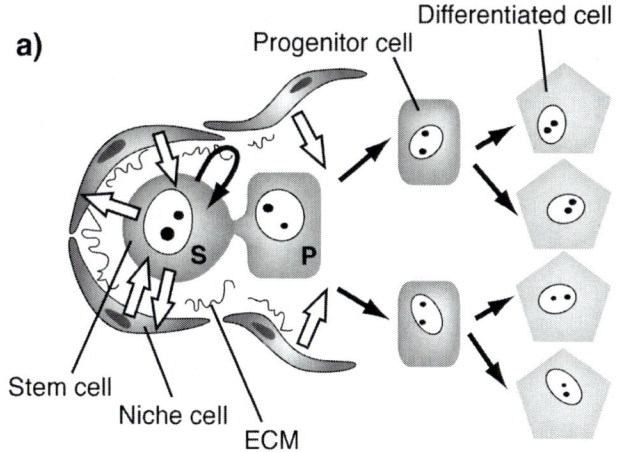

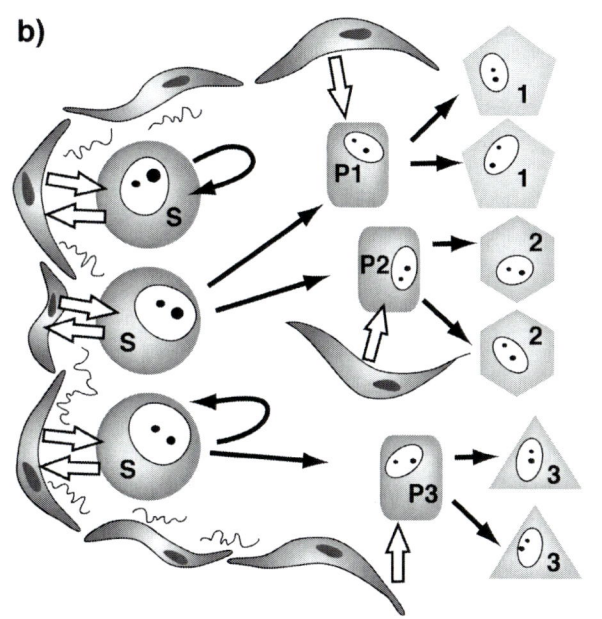

FIGURE 10.5:

Tissue stem cells reside in niches of cells and extracellular material (ECM). Signals (open arrows) between niche and stem cells maintain the stem cells in their state. **A: *Local asymmetry model:*** Stem cell divides asymmetrically, giving rise to two different daughter cells. One daughter is again a stem cell (S), and the other daughter is a progenitor cell (P). P undergoes amplifying divisions before progeny form a particular type of differentiated cell. **B: *Serial asymmetry model:*** S divides, and sometimes both daughter cells are again stem cells. Other times both daughters are progenitors. In either model, there may be one type of progenitor or a few types of different progenitors (P1, P2, and P3). Redrawn after Watt F.M. and Hogan B.L.M. (2000) Out of Eden: Stem cells and their niches. *Science* **287:** 1427–1430

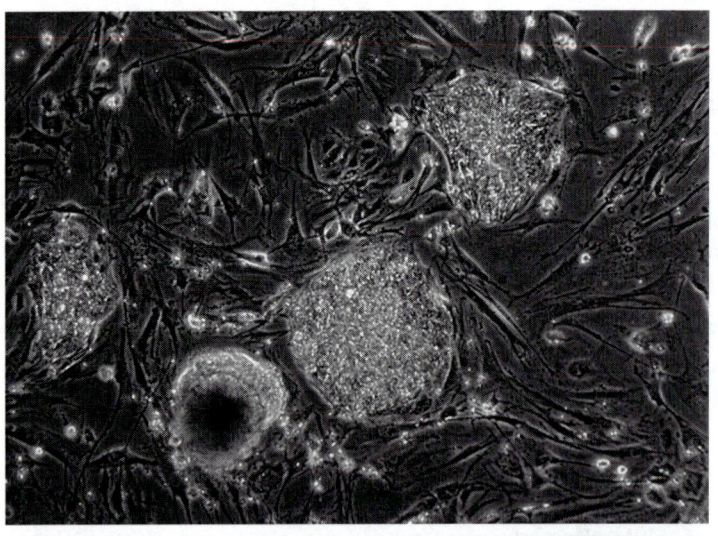

FIGURE 10.6:
Embryonic stem cells growing in culture. Courtesy of the University of Wisconsin Board of Regents.

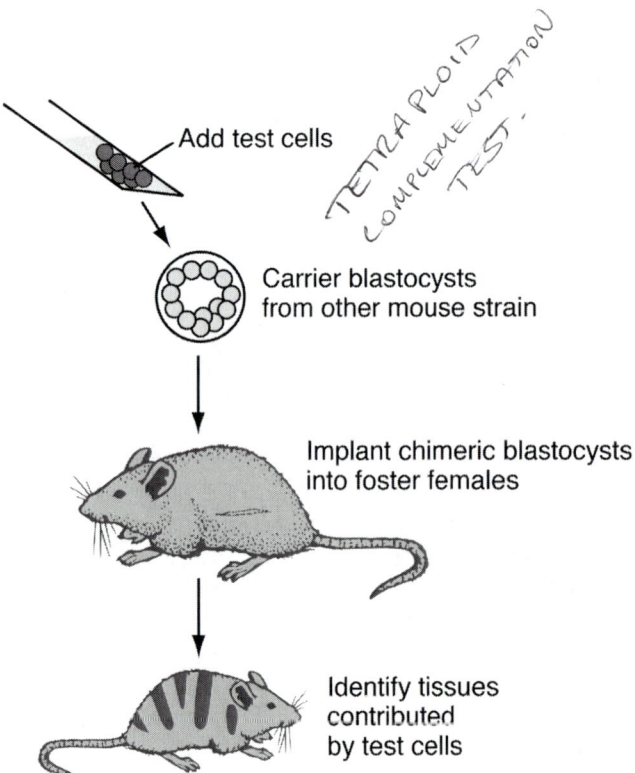

FIGURE 10.7:
Complementation test. Test cells (black) are injected into carrier blastocysts, which are implanted into foster females. Genetic markers allow to determine which tissues of developing pups are derived from the test cells. From *Analysis of Biological Development* by Klaus Kalthoff. Copyright © 1996 by McGraw-Hill. Reproduced with permission of The McGraw-Hill Companies.

stem cells: ES cells are undifferentiated, divide indefinitely, and can give rise to progenitor cells. Their salient feature, however, is their **pluripotency**, defined as the ability to form a wide range of differentiated cells.

The standard way of preparing ES cells starts from embryos at the blastocyst stage (see Figures 10.1 and S10.a). The inner cell mass (ICM) is isolated and cultured in media containing growth factors, that is, signal proteins that stimulate cells to divide. In some of these cultures, mounds of cells (Figure 10.6) will develop that look undifferentiated, divide actively, and synthesize marker proteins also known from tissue stem cells. Such cells are selected for their ability to grow in culture for many passages without change in properties. Selected cell lines are then tested for pluripotency. This can be done *in vitro* by adding to the culture medium growth factors known to initiate specific pathways of differentiation. Alternatively, cells can be tested *in vivo* by injecting them into the skin of mice. The injected cells then form tumors called *teratomas*. These often contain many types of differentiated cells, demonstrating pluripotency of the injected cells.

Mouse ES cells are not only pluripotent but in fact **totipotent**, that is, capable of forming completely normal, fertile individuals. This was shown by a special form of **complementation test** (Figure 10.7), which takes advantage of the astounding ability of mammalian embryos to *regulate*, as discussed earlier. Briefly, the cells to be tested are injected into carrier blastocysts, which are brought to term by foster females. The test cells and carrier blastocysts are taken from mouse strains with different genetic markers, so that tissues derived from each can be distinguished later on. Test cells from embryos or ES cell lines always "join in" with the carrier ICM and contribute substantially to all tissues, including gametes, of the developing pups.

An advanced version of this test is known as **tetraploid complementation**. Here, the carrier blastocyst is raised from an embryo in which the first two embryonic cells have been fused. The resulting single cell develops into a blastocyst that is *tetraploid*, meaning each cell has *four* sets of chromosomes instead of the *two* sets present in a normal *diploid* cell. In a tetraploid blastocyst, the trophoblast portion develops whereas the ICM does not. However, if diploid ES cells are injected into a tetraploid carrier blastocyst, then the added cells can take over the role of the *entire* ICM and give rise to a normal, fertile pup. By this stringent test, mouse ES cells are indeed *totipotent*.

ES cell technology was developed using mice, but the basic properties of mouse ES cells seem to be shared by human ES cells. In particular, human ES cells have been shown to be pluripotent *in vivo* and *in vitro*. Whether human ES cells are totipotent is not known because a tetraploid complementation test with such cells would be unethical.

ES cell lines are highly prized by developmental biologists and medical researchers alike. In basic science, ES cells are ideal objects for studying the complete series of signals that govern the development of cells from an embryonic state through a stepwise series of commitments to a differentiated cell. For medicine, human ES cells hold the promise of **cell replacement therapies** for heart disease, early onset diabetes, Parkinson disease, and other disorders caused by the malfunction of a particular cell type. The ideal scenario is to transfer the equivalent of committed progenitor cells that can replace the defective cells in the patient. Obviously, both basic and applied researchers will have made much progress when they have learned to mimic the development of, for example, pancreatic beta cells (the ones missing in type 1 diabetes) in cell culture.

A concern about using ES cells as a source of replacement cells is that ES cell are very actively dividing, undifferentiated cells—as are tumor cells. Thus, a relapse of any of the transferred progenitor cells into the original ES cell state could be perilous for the patient. However, most medical treatments come with risks that must be weighed against the potential benefits, and it may be possible to eliminate this particular risk by suitable genetic engineering (see Chapter 11) of the replacement cells.

E. Research with Human ES Cells Is Politically Controversial

In the United States, federal grant support is critical to biomedical research, but work on human ES cells has been politically controversial and treated differently by different administrations. Opponents of ES cell work consider human life sacred at any stage of development and therefore reject the standard way of generating human ES cells, which involves the disassembly of human blastocysts.

Proponents of research on human ES cells point to the great scientific and medical value of these cells. They also stress that the blastocysts used to make human ES cell lines are "spares" from fertility clinics, which routinely fertilize more eggs *in vitro* than are implanted into a woman's uterus. The spare blastocysts are deep frozen for later use in case the first implantation has not resulted in a successful pregnancy. Unused spares are kept frozen indefinitely and may eventually have to be discarded.

The critical question is at which stage of development a human acquires the dignity and the legal protection of a person. The answer to this question obviously depends on one's definition of a person. If one holds that personhood begins at fertilization, then a blastocyst is a person (Figure S10.b). If one agrees that a person needs at least a rudimentary nervous system, which makes it possible to feel pain or to respond to external signals, then clearly a blastocyst is not a person.

The British parliament has passed a law including a "14-day rule." Up to this time limit properly proposed and approved research on human embryos is now legal in Britain. Currently, there is also a financial incentive for British couples who donate spare blastocysts for research: part of the costs for their IVF treatment are defrayed by the government. The 14-day rule has been incorporated into the guidelines for human embryonic stem cell research published by the International Society for Stem Cell Research, which were written by scientists, ethicists, and legal experts from 14 countries including the United States.

It may be possible to side-step the controversy about human ES cells by generating equivalent cells using methods that do not require the sacrifice of blastocysts. Several research teams have reprogrammed adult cells from mice as well as humans into *induced pluripotent stem cells (iPS cells)* by adding to the cells extra copies of genes encoding transcription factors that are characteristic for ES cells (see Chapter 11).

F. The Ease of Reproductive Cloning Depends on Species and Stage of Development

Even in the hands of the most skillful experimenters, only one in about ten cell cultures prepared from an inner cell mass yields an ES cell line. Although ICM cells have undergone little development, coaching them into a state of unlimited division and complete pluripotency is not easy. The same experience has been made in attempts at **reproductive cloning**, that is, making genetically identical replicas of an organism.

Plants and simple animals do the equivalent of reproductive cloning naturally in the form of **asexual reproduction**. Plants can multiply through modified roots or stems, such as potatoes or the "runners" of strawberries. Indeed, one can take a potato leaf, digest its cell walls to liberate individual cells, and grow each cell back into a whole fertile plant (Figure S10.c). Similar experiments have been done with cells from carrot roots and tobacco stem pith. This means differentiated plant cells are easily returned to mitotic division and totipotency. Likewise, polyps and jellyfishes multiply by budding off small individuals. However, higher animals do not have such asexual reproduction. Even with salamanders, which are famous for their powers of regeneration, it has not been possible to regenerate a whole individual from a limb, let alone a single cell.

Why is it so difficult to return a differentiated animal cell to a rapidly dividing, totipotent state? The first explanation proposed was that the cytoplasm of animal cells may contain factors that restrict cells to their differentiated states. To test this hypothesis, researchers developed a technique known as **nuclear transfer**, which brings the nucleus of a differentiated cell back into an enucleated egg from the same species. The rationale was that the cytoplasm of an egg, which is known to support the development of a whole organism, should be able to dilute or overcome any factors that had restricted the potency of the nucleus in the differentiated donor cell.

The nuclear transfer technique was first developed for amphibians. The recipient egg was pricked with a glass needle, and the egg nucleus was pushed out through the puncture. The donor cell of the nucleus to be transferred was sucked into a fine glass pipette with a bore small enough to break the cell while leaving the nucleus intact. The nucleus with a bit of surrounding cell cytoplasm was then injected into the enucleated host egg. The re-nucleated eggs were monitored under the microscope. Those that did not develop properly were discarded. Embryos that reached the *blastula* stage, which is comparable to the mammalian blastocyst, were scored for their ability to develop into tadpoles. This latter stage was selected as an endpoint because most organ systems are functioning by then.

The percentage of nuclear transfers that result in normal development decreases steeply with the age of the nuclear donor cells. While nuclear transfers from blastula stages were mostly successful, the success rate dropped to about 1 percent with nuclei taken from tadpoles, regardless of the cell type used as a donor. Similar results have been obtained with all animal species tested.

G. Cloning by Nuclear Transfer Seems to Require Chromosome Remodeling

A modified nuclear transfer technique designed for mammals was first developed with sheep (Figure 10.8). Through a slit cut into the eggshell, the researchers removed the egg nucleus and then placed the donor cell for the new nucleus next to the enucleated egg. The adjacent plasma membranes were fused by means of an electric shock to create a re-nucleated egg that also contained the donor cell cytoplasm. Such eggs were cultured *in vitro* until they reached the blastocyst stage. Well-developed blastocysts were implanted into foster ewes (female sheep), which were then monitored for the progress of their pregnancy.

It was critical in these experiments to ensure that any lambs born did not result from host eggs that had not been enucleated properly or from unnoticed matings of the foster ewes. Eggs and foster ewes were therefore taken from a sheep race (Scottish Blackface) that differed from the race (Finn Dorset) used to provide the donor cells for nuclei. Even though the racial differences were obvious, DNA tests were done to prove independently that the lambs born showed the genetic markers of the nuclear donor cells.

These experiments proved to be frustrating because nearly all re-nucleated eggs died at various stages of development. From 277 enucleated eggs fused with cells derived from the mammary gland of a ewe, a single live lamb was born and named "Dolly" (Figure 10.9). This experiment became known as "cloning" because an organism that develops from a re-nucleated egg is a genetically identical to the organism that donated the nucleus. Similar experiments have been carried out successfully with mice, pigs, goats, cattle, horses, rhesus monkeys, cats, and dogs, using nuclei from different types of cultured cells. In all these experiments, the success rate has been extremely low, with a hundred or more transfers required for each successful case.

The high failure rates of nuclear transfers may be ascribed to technical difficulties, but this is unlikely to be the major cause. In a controlled series of experiments with mice transferring nuclei from fertilized eggs, 2-cell embryos, 4-cell embryos, and differentiated cells, the success rate plummeted rapidly from 34 percent to 3 percent, even though all other experimental parameters were kept constant. These results suggest that the embryonic development of mammals—and probably other

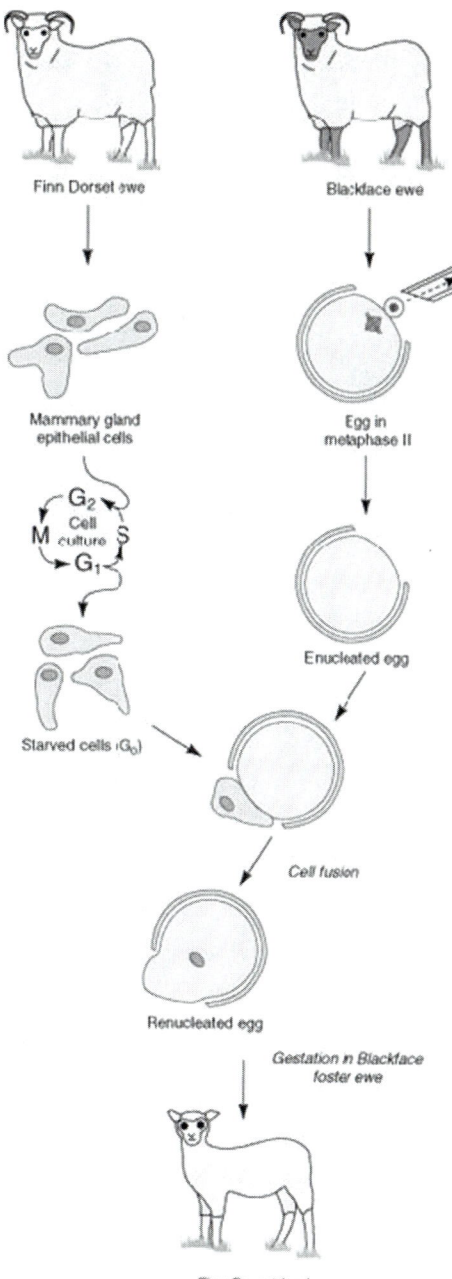

FIGURE 10.8:
Cloning Dolly the sheep. Cultured mammary epithelial cells from a Finn Dorset ewe were used as nuclear donors. They were fused with enucleated eggs obtained from Blackface ewes. The re-nucleated eggs were cultured to the blastocyst stage and then implanted into Blackface foster ewes. A total of 277 nuclear transplantations resulted in a single live-born lamb, named "Dolly"; she showed the racial characteristics and DNA markers of Finn Dorset sheep (see Figure 10.9).
From *Analysis of Biological Development* by Klaus Kalthoff. Copyright © 1996 by McGraw-Hill. Reproduced with permission of The McGraw-Hill Companies.

FIGURE 10.9:
Dolly and the scientist (Dr. Ian Wilmut) who created her. Dolly was cloned as shown in Figure 10.8. Image © remi Benali/Corbis.

animals as well—entails some rapid changes in cell nuclei that render them less and less suited for supporting the development of a whole new organism. The nature of these nuclear changes is not well understood.

Many researchers working in this area think that nuclei from all but the youngest donor cells need to undergo a process of **chromosome remodeling** before they can support the rapid cell divisions and the expression of virtually all genes, both of which are required for the development of an entire new organism. This view is supported by data on the *sequence complexity* of cellular RNA, which is a measure of the number of genes expressed. The results show that the sequence complexity decreases during embryonic development, indicating that genes are transcribed more selectively.

Another observation related to chromosome remodeling is known as **genomic imprinting**. It is best described as the *reversible and sex-specific silencing* of certain genes in the life cycle of mammals. Genetic studies on mice have shown that about 30 to 40 genes are imprinted in developing sperm, while a similar number of different genes is imprinted in developing eggs (Figure 10.10). The imprinted genes remain silent in most cells of the developing organism, while the homologous genes, which were introduced through the gamete from the other sex, are expressed. Only in the offspring's *germ line*, the cells that will form the next generation of gametes, is the imprinting first erased and then reset depending on the offspring's sex.

Imprinting seems to have evolved in mammals as part of a conflict over the amount of maternal nutrients diverted to her offspring via placental circulation or nursing. Males imprint genes that would otherwise restrict the amount of maternal nutrients received by their offspring, whereas females imprint genes that would otherwise divert too many nutrients from themselves.

The mechanisms of imprinting involve **DNA methylation** and **histone deacetylation**, both of which make DNA less accessible for transcription factors. Conceivably, there are more mechanisms of silencing genes in semipermanent ways that happen as part of normal development and would have to be reversed for cloning by nuclear transfer. It appears that this remodeling does not come about as readily in egg cytoplasm as the inventors of nuclear transfer had hoped for, and indeed, the reversal of imprinting in the normal life cycle occurs in primordial germ cells, that is, before the formation of egg or sperm (see Figure 10.10).

H. Reproductive Cloning of Humans Should Not Be Attempted, but Therapeutic Cloning May Be Medically Useful

When the news of Dolly the cloned sheep reached the general media, the inevitable question was: "Will humans be cloned in the future?" Nuclear transfer technology may be appealing to couples with reproductive handicaps that cannot be overcome by other methods. For instance, couples with one infertile partner may wish to have a child based on nuclear transfer or on in vitro fertilization using a clone-derived gamete. Parents who lost a child in an accident or crime may want to have the victim cloned from salvaged cells. Vain people may want to pass on all of their genes in one complete set instead of pooling half of their genes with half of those from a partner.

However, given the horrific failure rates of nuclear transfer experiments with various mammals, attempts to use the same technique with humans would most likely entail large numbers of spontaneous abortions, stillbirths, and babies born with all manner of ailments. Such attempts would therefore be grossly irresponsible at this time.

Even if this or another cloning technique could be made safe for human use, there would still be legal, economical, and psychological problems to resolve. How would the rights and obligations of parenthood be divided among the donor of the nucleated cell, the egg donor, and the foster mother? Who would provide health insurance for a cloned baby? How would a person feel growing up with a parent who is an older version of himself or herself? In view of these problems, the National Academy of Sciences of the United States has recommended that reproductive cloning of humans be banned and that it should be "reconsidered only if ... the procedures are ... safe and effective, and if a broad national dialogue on societal, religious, and ethical issues suggests that reconsideration is warranted."

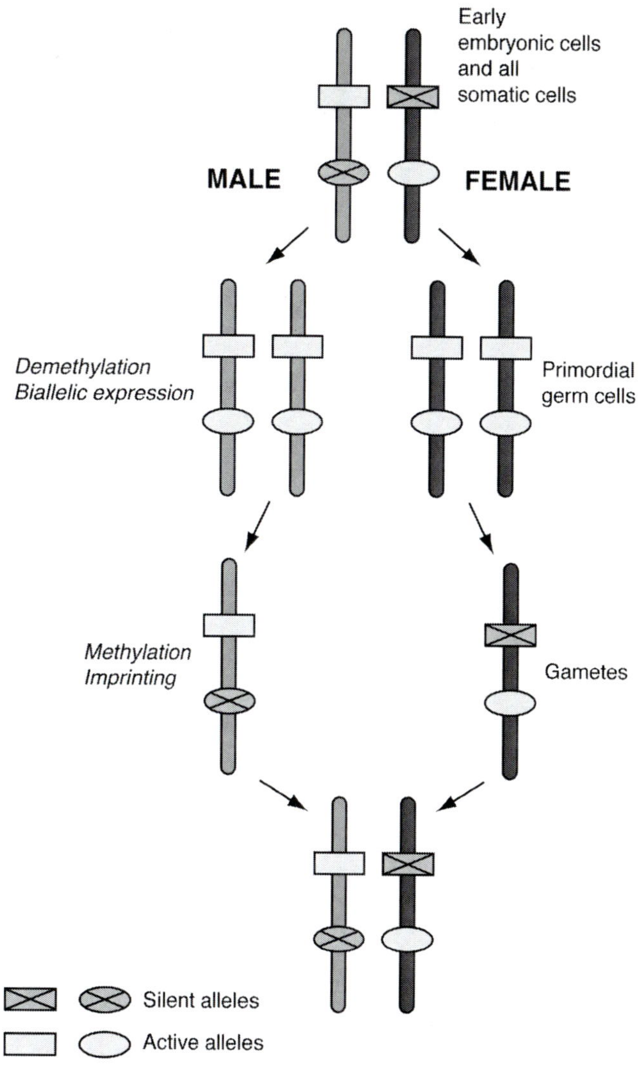

While the **reproductive cloning** of humans, in the sense of making babies, should not be attempted, there is another form of human cloning that could be medically useful and at least ethically defensible. Known as **therapeutic cloning**, it is a form of *cell replacement therapy* that combines nuclear transfer with the use of ES cells.

FIGURE 10.10:

Genomic imprinting in the mammalian life cycle. One set of genes (rectangles) is inactivaed (crossed out) in maternally inherited chromosomes (dark bars) whil another set of genes (ovals) is silenced in paternally inherited chromosomes (light bars). The patterns of inactivation are conserved upon fertilization and development until they are erased in the primordial germ cells. Another round of sex-specific silencing occurs when new gametes are formed.

A problem with using cultured ES cells for cell replacement therapies, as discussed earlier in this chapter, is *immune rejection*. The recipient's immune system usually recognizes transplanted cells or organs as non-self and rejects them. The rejection can be avoided by immune suppression, but this treatment leaves the patient vulnerable to infections and tumors. An elegant solution to this problem would be to use **isogenic** replacement cells, that is, cells carrying the patient's own genome. In some cases, it may be possible to stimulate the patient's own *tissue stem cells* to create the needed type of replacement cells. However, this approach is limited by the slow growth and the restricted potency of tissue stem cells. Alternatively, one may harness the fast growth and pluripotency of ES cells by making them isogenic with the patient.

In a futuristic scenario, a person who suffered a heart attack could be treated as follows (Figure 10.11). A donated human egg is enucleated and fused with a suitable cell taken *from the patient,* and the re-nucleated egg is grown in culture to the blastocyst stage. The inner cell mass is harvested and used to make ES cells. These are cultured with a suitable mix of growth factors to initiate their development as heart muscle cells. These cells could replace the damaged part of the patient's heart and would not be rejected because they carry the patient's own genome. Similar treatments could be devised for treatments of Parkinson's disease, type 1 diabetes, and similar diseases. As

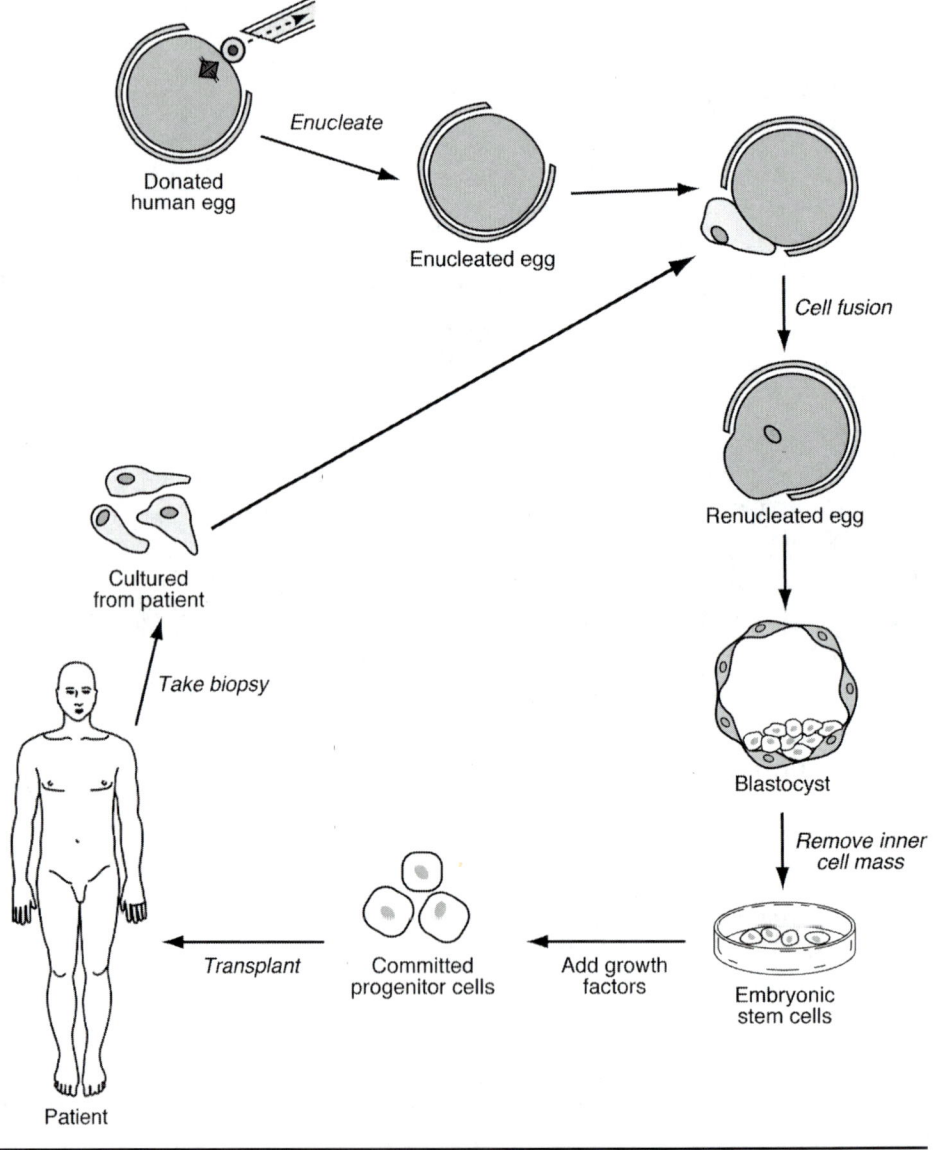

FIGURE 10.11:
Futuristic scenario of cell replacement therapy using the patient's own cells as nuclear donors to create isogenic ES cells.

with any medical treatment, the potential benefit will have to be weighed against the risk of side effects, such as tumorous growth of the replacement cells.

If a disease is caused by a genetic disorder, cell replacement therapy with isogenic cells would not provide a permanent cure because the replacing cells will carry the same genetic liability as the failed original cells. In such cases, ES cells have the additional advantage of being eminently suited for *gene therapy* (see Chapter 11).

Despite its medical potential, work on human ES cells has been hampered by political controversy, as discussed earlier. The National Academy of Sciences of the United States has recommended that, in contrast to *reproductive cloning*, which should be banned, *therapeutic cloning* should be "pursued aggressively." It has been argued that promoting therapeutic cloning while prohibiting reproductive cloning may seem difficult, given their overlap in technology. However, the situation is similar with other tools and techniques that can be used for beneficial as well as for unethical purposes.

EXERCISE PAGE FOR CHAPTER 10

Student Last Name _____

Student First Name _____

Discussion Section _____

On the course web site, use the link on the **syllabus** to find a pdf file of your **assigned reading** for this chapter. The same reading is referenced below. Use the bulleted list of questions to test your knowledge of this reading.

Lanza, R., and Rosenthal, N. (2004). The stem cell challenge. *Scientific American* **June 2004:** 93–99.

- How do the authors define the goal of embryonic stem cell (ES cell) research?
- Which traditional tests do scientists use to determine whether they were successful in establishing an ES cell line? What other test do the authors envision for the future?
- How can the development of ES cells in vitro be manipulated?
- At which stage of development might they be best suited for human cell replacement therapy? Do the authors provide an example?
- As an alternative way of making ES cells, the authors mention "parthenotes." Which advantages of parthenotes do they mention? Can you think of any disadvantages?
- What needs to happen with a nucleus from a differentiated cell when it is transferred back into an enucleated egg for cloning? Which results indicate that this process is difficult and incompletely understood?
- What do the authors say about the prospects of using human adult stem cells for cell replacement therapies?

Feedback

On a scale from 1 to 10, the being the best, please rank the above reading for

1. Interest and Relevance to the topic (10 being most interesting/relevant) _____
2. Readability (10 being most clearly written, easy to understand) _____

In the remaining space, enter any comments that you may have on this reading.

CHAPTER ELEVEN

GENE THERAPY

At the heart of more than a thousand human diseases lies a single mutated gene. Patients may be treated either by restoring normal gene function or by targeting the mutated cells for destruction. Differentiated cells also have been returned to embryonic appearance and pluripotency by adding extra copies of genes expressed strongly in embryonic stem cells. Plants and animals have been provided with exogenous genes to make them more useful for research, agriculture, or pharmacy. However, such innovative gene therapy would be unethical with humans.

A. Gene Therapy Introduces a Foreign Gene into a Recipient Cell

All gene therapies introduce an exogenous gene, or **transgene**, into a host cell (Figure 11.1). In many situations, the transgene should be expressed for the lifetime of the host cell, and potentially its progeny. In this case, the transgene should be integrated into the genome of the host cell, which will then treat the transgene like an endogenous gene. Such a host cell is called genetically **transformed**. In other situations, one wants the transgene to function only for a limited period of time. In this case, the transgene should typically not be integrated into the genome of the host cell, which is then called a **transfected** cell.

Transgenes are supplied to cells for different purposes.

- **Restorative gene therapy** is designed to restore the normal (wild-type) function to host cells that are defective because a genetic disorder.

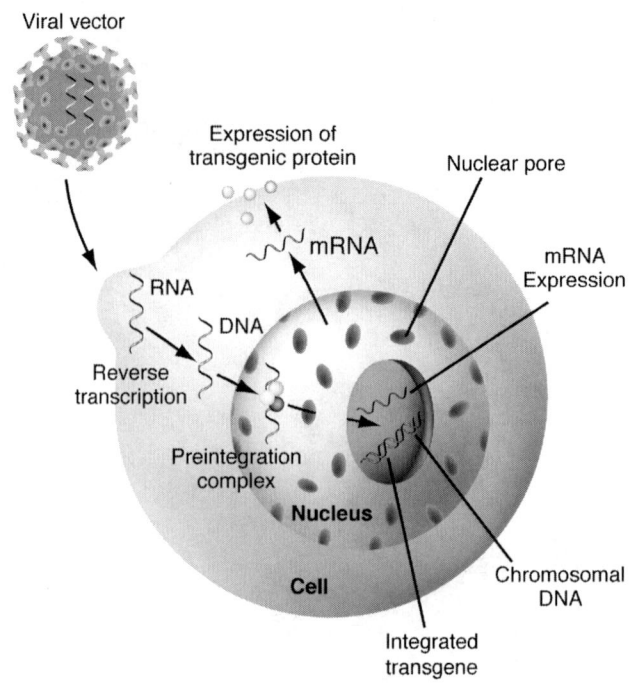

FIGURE 11.1:

Gene therapy. In most current gene therapies, a viral vector is used to insert a foreign gene ("transgene") into the genome of the targeted cell. Ideally, the target cell will express the transgene like a resident gene.

- **Innovative gene therapy** aims to "improve on mother nature" by providing host cells with additional copies of their own genes, or with wild-type genes from other species, or with experimentally modified genes.
- In **targeted destruction gene therapy**, cancer cells are supplied selectively with genes that render them vulnerable to a chemotherapeutic drug or to attack by the patient's own immune system.

While restorative and targeted destruction therapies are designed mostly for human medicine, innovative gene therapy has been used widely in research, agriculture, and pharmacy. The use of innovative gene therapy for humans would be very problematic, as we will discuss towards the end of this chapter.

B. Many Current Gene Therapies Use Viral Vectors

In most gene therapies today, the transgene is simply *added* to the host cell's genome without removing any part of the host's genomic DNA (Figure 11.2). The method of *replacing* a specific segment of the host cell's genomic DNA with an engineered transgene is cleaner, but more demanding, and will be discussed later in this chapter.

The most commonly used methods of gene addition enlist the help of *viruses* that insert their DNA into the chromosomal DNA of their host cells as part of their normal life cycle. This method takes advantage of the natural ability of viruses to breach the plasma membrane of their host cells, and in some cases, to insert their viral genes into the host's chromosomal DNA. Many viruses enter only specific types of cells as their

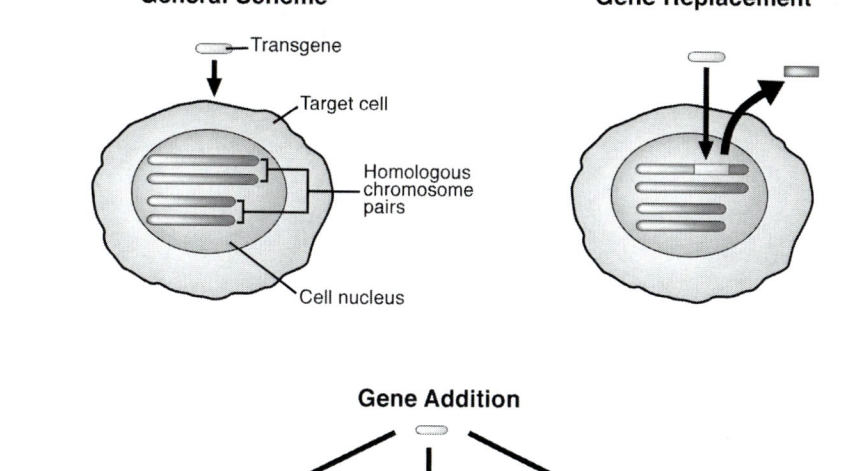
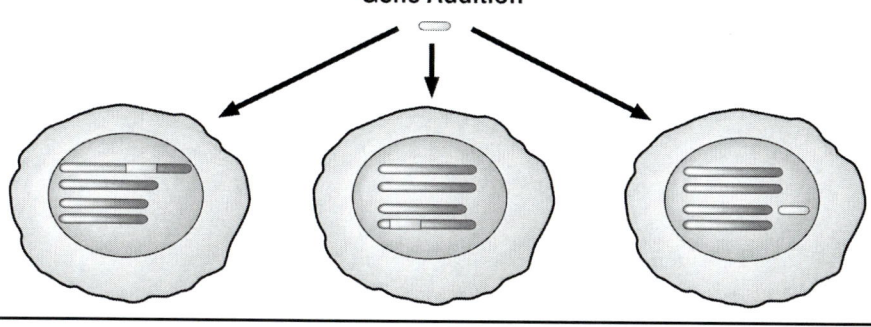

FIGURE 11.2:

Gene Addition vs. Gene Replacement. While gene addition is random, gene replacement cleanly inserts the transgene into the native positive of the replaced gene.

host cells while ignoring other cells. This selectivity allows scientists to choose vectors that enter the target cells of a gene therapy specifically, or at least preferentially. Such viruses are used as **vectors**, or carriers, for transgenes.

A first step in the engineering of a viral vector is the removal of certain viral genes, which eliminates the virus's ability to damage or kill its host cells. The removal of pathogenic genes also makes room for the transgene, the "payload" that the viral vector is designed to carry.

Many viral vectors used in gene therapy are known as **retroviruses**, which use RNA as their genetic material. Once in the host cell cyctoplasm, they reverse-transcribe their RNA genome into DNA (see Figure 11.1). The DNA transcript then enters the host cell nucleus, where it inserts itself into the chromosomal DNA. While this is the ultimate goal of many gene therapies, the insertion into the host genome occurs *randomly* and may wreak havoc on nearby resident genes. Specifically, insertion of the viral DNA may deregulate neighboring genes that control the cell cycle, with the result of tumorous growth. Conversely, random insertion of a viral carrier may inactivate a critically important resident gene, such as a tumor suppressor gene.

Retroviruses include the lentiviruses, the murine leukemia virus (MLV), and the HIV virus. MLV gains access to chromatin only in dividing cells, making it suitable for assisted suicide of tumors embedded in tissues with little mitotic activity, such as the brain. In contrast, lentiviruses can enter intact cell nuclei and are therefore suited for nondividing target cells.

Another group of viruses used in gene therapy are **adenoviruses**, which cause the common cold in humans. They have DNA as their genetic material, but the viral

DNA replicates in the host cell nucleus without inserting itself into the host genome. The use of adenovirus-derived vectors therefore avoids the hazards of random insertion into the host genome, but other problems have been encountered. The adenovirus elicits allergic reactions in humans, the intensity of which differs from one patient to another. Infected cells are also attacked by the host's immune system, further compounding the difficulty of finding a dosage of loaded vector that is large enough to have a therapeutic effect but still safe for most patients.

The **gene transfer efficiency**, that is, the fraction of target cells that become transgenic, is still low with viral carriers. The critical step of delivering the transgene to the targets cells is generally more feasible outside the patient's body ("***ex vivo,***" as opposed to "*in vivo*"). The target cells are removed from the patient, cultured to make them receptive to the transgene, exposed to the transgene, and then returned to the patient. Blood cells are eminently suited for *ex vivo* therapy.

A futuristic use of viral vectors is the "magic bullet" strategy, in which the vector would be modified so that it will stick selectively to its target cells. The viral vectors could then be simply injected into a patient's bloodstream and allowed to diffuse all over the body, knowing that they will deliver their genetic payload selectively to their targets. Non-viral delivery systems, including liposomes (tiny membraneous vesicles) and gold particles delivered by "gene guns" are being developed but are not widely used yet.

C. First Human Gene Therapy: Severe Combined Immunodeficiency (SCID)

Figure 11.3 shows the design of the first successful gene therapy, which was begun by a research team in the United States in 1990. The patients, two girls aged 4 and 9, were suffering from **severe combined immunodeficiency (SCID)**, an inability to fight off infections that usually result in early death. In these two cases, SCID was caused by the lack of a functioning gene for **adenosine deaminase (ADA)**, an enzyme that metabolizes the A nucleotide of DNA, deoxyadenosine phosphate. For some reason, the defect is especially damaging for white blood cells, including B- and T-lymphocytes. One traditional method of treating ADA is transplantation of bone marrow, the tissue that produces lymphocytes, but a compatible donor may not be available. Another method is the regular injection of the ADA protein, which is extremely expensive.

The ADA^+ gene was a great candidate for gene therapy: The gene seems to be the only one controlling the amount of ADA enzyme produced. Also, the activity of the gene is regulated in a simple "always-on" fashion, not by a complex regulatory network as in many other genes. In addition, the gene's level of activity is not critical as long as a minimum amount of ADA enzyme is made. Thus, there was a good chance that giving the patients lymphocytes with a single working ADA^+ gene would cure the disease.

In the case of the two girls suffering from SCID, a normal human ADA^+ gene was inserted in a vector made from the MLV retrovirus. Lymphocytes prepared from each patient's blood were mixed *ex vivo* with MLV vector carrying its ADA^+ payload. After time for infection and testing, the lymphocytes were returned to the patient's bloodstream. One of the patients, Ashanthi de Silva, whose treatment had begun at age 4, continued to improve. Three years after treatment, more than half of her circulating lymphocytes still harbored the transgene and produced ADA. She has been able to attend public schools

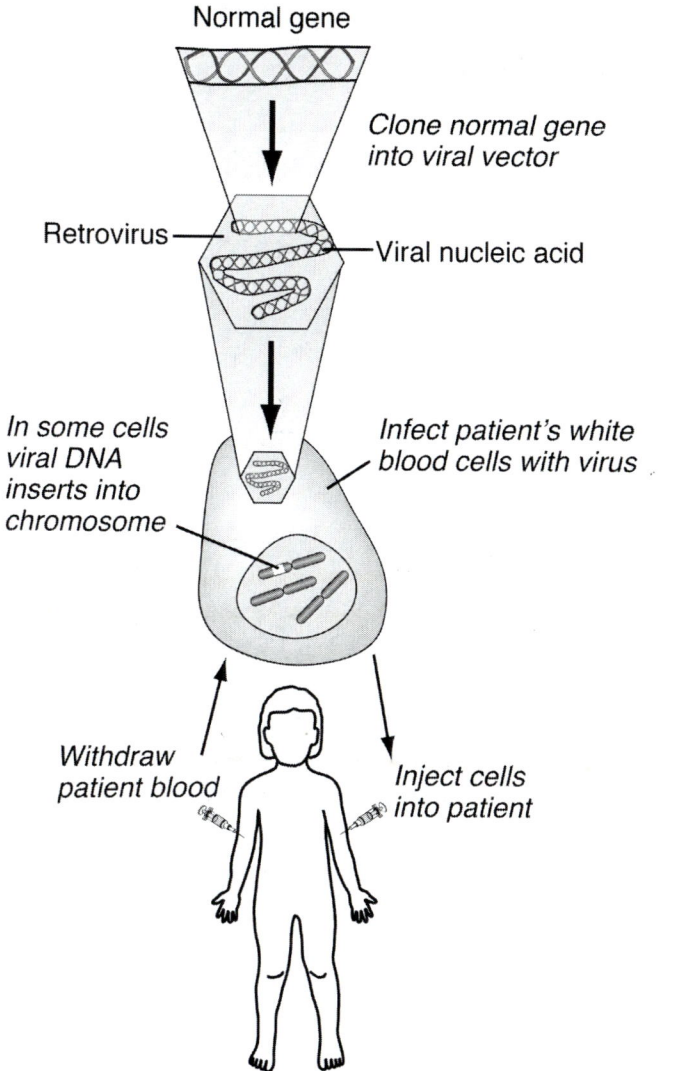

Figure 11.3:
Design of the first successful gene therapy. The curative gene was cloned into a retroviral vector. Lymphocytes prepared from a blood sample of the patient were infected with the loaded vector "ex vivo" and then returned to the patient's blood stream.

and is leading a healthy, active life (Figure S11.a). The second patient did not fare so well. She developed an immune response to her virus-infected blood cells, and less than 1 percent of her circulating lymphocytes contained the ADA gene.

The fact that Ashanthi's lymphocytes expressed both vector and ADA^+ genes beyond the normal lifetime of differentiated lymphocytes suggests that perhaps the cell cultures prepared from her blood unexpectedly had included some circulating *stem cells*. Subsequent gene therapies of ADA-deficient children therefore focused on tissues known to contain blood *stem cells*, such as bone marrow and umbilical cord blood. Cell cultures prepared from these tissues were infected with retroviral vectors containing the ADA^+ gene and then reinjected into the patients. As expected, consecutive blood samples taken from the patients contained increasing numbers of lymphocytes descended from the genetically transformed cells.

D. Some Gene Therapies Have Been Successful but Others Were Disastrous

Another form of SCID is caused by a defective X-linked gene (*SCID-X1*) for a receptor protein that is essential for lymphocyte differentiation. A clinical trial in France began in 1990 with the treatment of two children aged 8 and 11 months. Bone marrow cells from the patients were infected *ex vivo* with a retroviral vector containing a functioning *SCID-X1* gene, and the infected cells were transferred back into the patients. They improved to the point where their lymphocyte development was near normal, and they could return to their homes to lead regular lives. In a follow-up trial, nine out of ten children were treated, apparently successfully. A similar trial in England was also successful.

However, excitement turned into worry when the **first casualties** occurred. Three children enrolled in the French trial developed T-cell leukemia more than two years after the treatment. One of the children died while the other two recovered. Analysis of their lymphocytes showed that the retroviral vector had inserted itself in the regulatory region of a gene promoting cell division, causing deregulated growth.

Another setback occurred in 2000 when an 18-year-old volunteer, Jesse Gelsinger, died in a clinical trial for gene therapy of a genetic disorder causing an inability to form ornithin transcarbamylase, an enzyme required for the safe breakdown of dietary protein. Jesse's death was apparently caused by a massive immune response to the adenoviral vector used to deliver the transgene. The grief about the loss of his life was aggravated by at least the appearance of a conflict of interest: One of the lead investigators had a financial interest in a company that stood to profit if the trial was successful. In a related lawsuit, the U.S. Department of Justice alleged that the trial should have been halted earlier and that the lead investigators misrepresented earlier findings to the National Institutes of Health (NIH) and the Food and Drug Administration (FDA), which were overseeing the study. The lawsuit was settled out of court.

The deaths of two patients in clinical trials have left scientists, physicians, and administrators wondering whether the risks of current gene therapies are outweighed by the promise of a cure. On the one hand, it seems more responsible to halt ongoing trials and focus on basic research until the reactions entailed by the use of viral vectors are better understood, or until safer methods of delivering transgenes to their target cells have been found. On the other hand, there are patients for whom all conventional treatments have failed, and for whom gene therapy in its present form is the only hope. The patients' ardent desire for a cure, the complexity of viral interactions with their human hosts, and conflicts of interest may combine in unfortunate ways that are hard to foresee and regulate.

The current trend is to proceed cautiously with a reduced number of clinical trials, with very restrictive criteria for the patients who may be enrolled, and with tightened rules of oversight. One of the ongoing trials is for X-linked adrenoleukodystrophy (ALD), a severe brain-demyelinating disease in boys caused by the deficiency of a transporter protein; the conventional treatment is by transplantation of hematopoietic (blood forming) cells. Two boys for whom no suitable donor was available received gene therapy of their own hematopoietic cells *ex vivo* with the normal ALD gene in a lentiviral vector. The transformed cells proliferated in the boys, and the progressive demyelination of their brains was halted.

E. Induced Pluripotent Stem Cells

In a pioneering set of new experiments, two Japanese researchers—Kazutoshi Takahashi and Shinya Yamanaka—have transformed differentiated cells from mice as well as humans into **induced pluripotent stem cells (iPS cells)**, which resemble embryonic stem cells (ES cells, see Chapter 10). It seems virtually certain that the discovery of iPS cells will be rewarded with the Nobel Prize for Medicine or Physiology (Figure S11.b).

Using retroviral vectors, researchers added to the genomes of target cells extra copies of genes that are regularly expressed in ES cells (Figure S11.c). In the original experiment, Takahashi and Yamanaka tested 24 candidate genes on skin cells (fibroblasts) from mice. The transgenes were added to the cultured cells in retroviral vectors. A combination of four genes encoding different *transcription factors*, namely *Oct 3/4, Sox 2, c-Myc,* and *Klf4*, was successful in conferring "stemness," that is, unlimited division, embryonic morphology, and the ability to produce any type of differentiated cell (pluripotency). Mouse iPS cells were later shown to be *totipotent* (able to give rise to complete, fertile mice) by the *tetraploid complementation assay* (see Chapter 10).

Follow-up experiments were aimed at avoiding the oncogenic effects of some of the stemness-inducing transcription factors and the disruptive effects of viral insertion into the host cell genome. In particular, researchers tried to avoid the use of *c-Myc*, which is notorious as a proto-oncogene. Others built on the observation that the stemness-inducing transcription factors need to act only temporarily. Using a known inducible recombination enzyme, they seamlessly removed the transgenes after they had done their stemness-inducing work.

In medicine, iPS cells hold much promise for several reasons. Patient-specific iPS cell cultures can be obtained from small biopsies, even from elderly patients (Figure S11.d). Such iPS cell cultures are used as models for studying human diseases and as screens for therapeutic drugs. Also, iPS cells hold great potential for cell replacement therapy because they are *isogenic* with the patient, so that there will be no immune rejection.

If the need for cell replacement is due to a genetic disorder, patient-derived iPS cells can additionally be transformed with a disease-correcting gene. A proof-of-concept experiment was done with cells from patients with *Fanconi anaemia* (FA), a chromosomal instability disorder that leads to malignancies, especially of blood-forming cells. Cells from FA patients were transformed into FA-corrected, patient-specific iPS cells that were able to form hematopoietic progenitor cells.

F. Gene Replacement Is Safer and More Accurate than Gene Addition

Many of the hazards encountered in current gene therapies result from the random insertion of the transgenic DNA, regardless of the genomic neighborhood in which the insertion occurs. Such problems could be avoided if the transgene were targeted to a known site in the resident genome where it could be expected to function but do no harm. For the purposes of restorative gene therapy, it would be best if the defective resident gene were cleanly *replaced* by its normal allele (see Figure 11.2).

Gene replacement takes advantage of cellular enzymes that unwind, cut, and re-ligate DNA. Together, these steps can lead to the crosswise rejoining of double-stranded DNA molecules that are *aligned in parallel* and have *identical nucleotide sequences.* This molecular mechanism, known as **homologous recombination**, causes the phenomenon of crossing over in meiotic cells (see Figure 9.5). However, because the hallmark of meiosis, the pairing of homologous DNA sequences, is a rare event in other cells, the few cells in which recombination has happened must be isolated from tens of thousands of other cells in which it has not.

A powerful method to select the few cells in which homologous recombination has occurred is shown in Figure 11.4. Cultured cells are incubated with a transgene, which has been cloned in a circular DNA known as a **plasmid**. To facilitate its uptake, the plasma membrane of the cells is made porous by electric shock or other means. In *restorative gene therapy,* the plasmid includes the curative normal allele that is to be exchanged for its defective counterpart in the cellular genome. In addition, the plasmid includes another gene that acts as a **selectable marker**. It confers on transformed cells the ability to grow in the presence of a toxic drug that kills all cells that do not

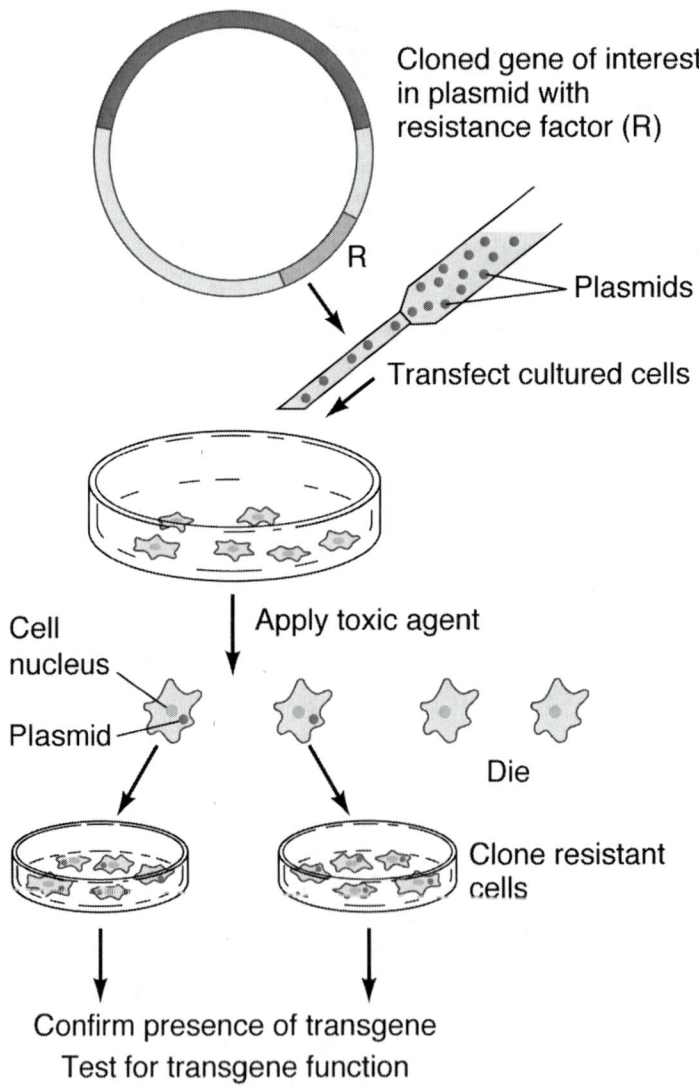

FIGURE 11.4:
Homologous recombination traced with a Selectable marker. From *Analysis of Biological Development* by Klaus Kalthoff. Copyright © 1996 by McGraw-Hill. Reproduced with permission of The McGraw-Hill Companies.

have the selectable marker. Thus, only those few cells that have successfully incorporated the transgenic plasmid will survive. A commonly used selectable marker is the Neo^R gene, which allows cells to grow in the presence of neomycin and related compounds.

To allow homologous recombination, identical DNA sequences must be aligned in parallel. To this end, the plasmid also contains **site-specific sequences** copied from sequences in the resident genome that bracket the DNA segment to be replaced by homologous recombination (Figure 11.5). If recombination occurs between plasmid DNA and residential DNA in both "brackets," then the host DNA integrates both the transgene and the selectable marker, both of which will then be replicated along with the host cell's genome in future cell cycles. In the case of a restorative gene therapy, the cell has now replaced its defective gene with a functional one. The same technique has been used to **"knock out"** genes in order to study how a cell or organism behaves in the absence of the gene. For this purpose, the transforming plasmid contains no curative gene, and the site-specific sequences are selected from *within* the target gene, which is then inactivated by inserting the selectable marker into it. Similarly, one can **"knock in"** an additional gene to any selected part of a host cell's genome.

The clean replacement of a defective resident gene with a functioning transgene has been demonstrated in a "proof of principle" experiment with mice (Figure 11.6). These mice had no functioning immune system because they were deficient for the *Rag2* gene, which is required for the DNA recombinations that generate the diversity of lymphocytes. The first steps in their therapy were the same as envisioned for human therapeutic cloning (see Chapter 10). Adult mouse cells were used as donors of nuclei that were transferred to enucleated eggs, and developing blastocysts were sacrificed to make embryonic stem cells (ES cells). These were transformed with a plasmid containing the normal $Rag2^+$

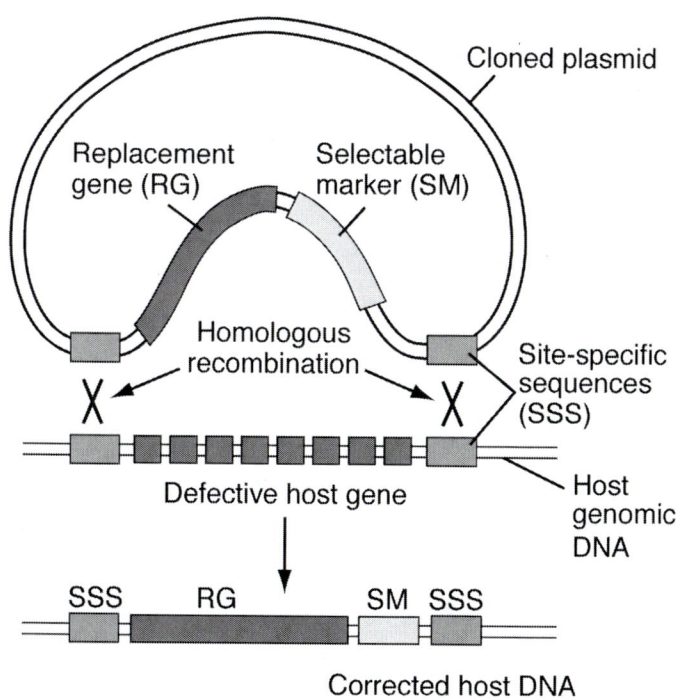

FIGURE 11.5:
Gene replacement by homologous recombination

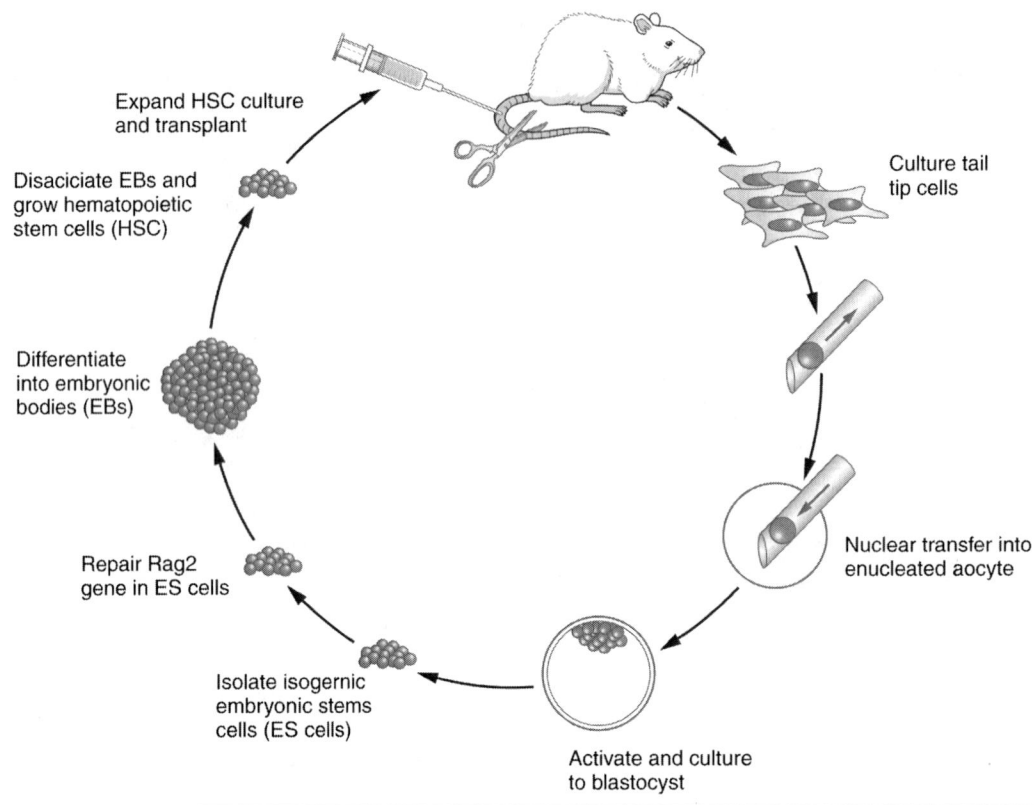

FIGURE 11.6:
Combine Nuclear Transfer and Gene Repair for Cell Replacement Therapy. The treated mice were deficient for the recombination activating gene 2 *(Rag2)*. Redrawn after Rideout W.M. et al. (2002) *Cell* **109:** 17-27

gene, a selectable marker, and site-specific sequences directing the construct to the defective resident *Rag2⁻* gene. The genetically repaired ES cells were cultured with the appropriate growth factors to promote their development into committed blood cell progenitors. These cells were injected into the bloodstream of the immunodeficient mice. The transferred cells were genetically identical to the recipient mice, except that the cells had a wild-type allele of the *Rag2⁺* gene. Four weeks later, the treated mice had formed functional lymphocytes, and antibody proteins were detected in their blood.

The results show that therapeutic cloning can be combined with homologous recombination for an effective gene therapy based on gene replacement rather than gene addition. Can the same set of techniques be used for human cell replacement? There would be several advantages to this strategy. The greatest one, as discussed earlier, would be avoiding the risks associated random transgene insertion into the host genome. Also, replacing the defective resident gene with its normal allele would leave the latter embedded in the network of interactions with neighboring genes in which it has evolved. Moreover, while gene addition can cure only recessive disorders, gene replacement would also be effective against dominant disorders, which are caused by a deregulated allele in the patient's genome.

In regard to patient safety, the greatest risk of cell replacement therapies may be a potential relapse of transferred committed progenitor cells into a state on uncontrolled division, which would give rise to a tumor. Building a genetic "suicide switch" into the transferred cells may eliminate this risk. Making isogenic replacement cells by nuclear transfer involves using human eggs for research and medicine. This ethical and political problem would be avoided by making replacement cells from iPS cells.

G. From Transgenic Cells to Transgenic Organisms

So far we have discussed gene therapy for *cells*. Of particular interest is gene therapy of *stem cells,* as mentioned earlier, because of their longevity. Beyond this level, there is much use for **transgenic organisms**, which carry a specific transgene in *all* of their cells, not only in a particular cell type or lineage. Methods for generating transgenic organisms are also known as **germ line therapies** because they include the *germ line*, that is, the lineage of cells that produces gametes (see Figure 12.2). Transgenic organisms therefore pass on their transgenes to their offspring. Germ line therapies have been carried out routinely with plants and animals. However, there are currently no approved clinical trials for any human germ line therapy. This is because the inherent risks of gene therapy would be multiplied by the transgene's presence in cells where it is not needed. This concern is compounded by the prospect that any errors or unforeseen side effects would be passed on to future generations.

There are many ways of creating transgenic plants and animals. We will focus here on methods used with mammals in research, agriculture, and pharmacy. One way of making a transgenic mammal is to inject the cloned gene into a pronucleus of a fertilized egg (Figure 11.7). This method was first used to add a rat growth hormone gene to mice, which then grew to twice the normal size (photo at bottom of Figure 11.7). Spurred by this initial success, the method has been used with other mammalian species. However, it has not been possible to control the chromosomal site where a transgene becomes integrated or even the number of transgene copies inserted into the host genome. Sometimes transgenes are integrated and then lost again. The level of transgene expression can therefore not be controlled, and there is a risk of silencing a host gene by the random insertion of a transgene.

Better ways of making transgenic mammals rely on targeted insertion of the transgene using homologous recombination (see Figure 11.5). Such experiments have been carried out extensively with mouse embryonic stem cells (ES cells) because their active division facilitates the use of selectable markers (see Figure 11.4). Since transformed ES cells are readily cloned, they can be tested extensively *in vitro* for proper insertion and activity of the transgene. ES cell clones that have passed all relevant tests are then turned into complete transgenic mice by *tetraploid complementation,* as described in the previous chapter (see Figure 10.7). These mice are typically heterozygous for the transgene, but they can be bred to homozygocity if desired.

Another way of turning a transgenic cell into a transgenic animal relies on the *nuclear transfer* technique used to generate Dolly the sheep (see Figure 10.8). This technique can of course be used with cloned *transgenic* cells as nuclear donors. This method has been plagued by low success rates, but with hundreds of nuclear transfers, a few healthy transgenic mammals are usually obtained.

The opportunities to make complete, fertile transgenic mammals have met with great interest in pharmacy, agriculture, and basic science. The following examples will illustrate the wide range of applications.

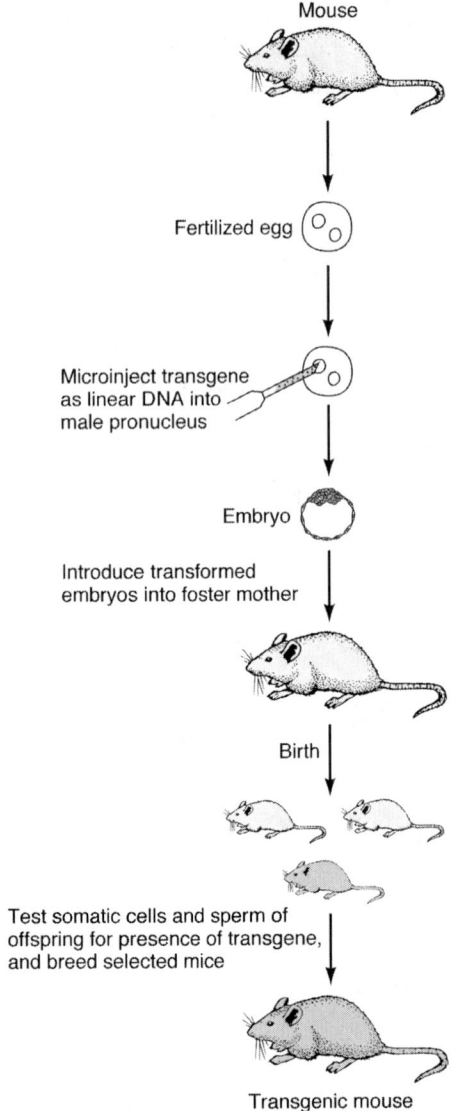

FIGURE 11.7:
Generating supersized mice by microinjecting a cloned transgene for growth hormone into one pronucleus of a fertilized egg. From *Analysis of Biological Development* by Klaus Kalthoff. Copyright © 1996 by McGraw-Hill. Reproduced with permission of The McGraw-Hill Companies.

H. How to Make Pharm Animals, Leaner Pigs, and Smarter Mice

Mammalian milk contains everything a young mammal needs to thrive, including a diverse mixture of proteins. These proteins are encoded by genes that are expressed specifically in mammary glands. Key elements in this specificity are the regulatory regions of these genes. Indeed, any gene with the regulatory region of a sheep milk protein gene will be expressed in the mammary glands of a sheep. This rule holds for **fusion genes** in which the *regulatory* region of the sheep milk protein gene is spliced in front of the *transcribed* region of a human pharmaceutical protein. Therefore, a female sheep that has been made transgenic for such a fusion gene will be "fooled" into producing the human protein as part of her milk.

This type of experiment was done in the same institute that previously cloned Dolly the sheep. This time, they set out to make sheep that were transgenic for a fusion gene combining the regulatory region of sheep ß-lactoglobulin (BLG, the most abundant milk protein) with the transcribed region of a human factor IX (a blood-clotting protein). The sheep BLG/human factor IX fusion gene was spliced to a selectable marker and used to transform cultured embryonic sheep cells as shown in Figure 11.8. Selected cells were cloned and tested to make sure the fusion gene was integrated into the cellular genome. Cloned cells were used as donors for nuclear transfer to enucleated eggs. Two female lambs, named Molly and Polly, developed and carried the transgene in all their body cells. They were expected to produce human factor IX in their milk as adults. Similar attempts to create "pharm animals" for the production of human insulin, proteins that dissolve blood clots to treat acute heart attacks, and iron transport protein to be used as an infant formula additive, are under way.

Another series of experiments was aimed at producing leaner pigs. Eggs were made transgenic for bovine growth hormone (bGH) gene by microinjection into a pronucleus, as shown for mice in Figure 11.7. Developing pigs grew faster and yielded pork with less fat as intended (Figure S11.e). However, they also suffered from a host of maladies including stomach ulcers, painful joint inflammation, heart enlargement, skin irritation, and kidney disease. A new attempt was made using a transgene for insulin-like growth factor-I (IGF-I). The IGF-I transgenic pigs have some of the same consumer-friendly meat characteristics as the pigs made transgenic for bGH, without showing the health problems of the latter.

Some of the most spectacular transgenic animals created for basic research are the "*Doogie* Mice," so named after the TV character *Doogie Howser*, who became a brilliant medical doctor while he was still a teenager (Figure S11.f). The creators of the *Doogie* mice, Joe Tsien and colleagues at Princeton University, were interested in the molecular basis of learning and memory. Their focus was on the so-called NMDA receptors, which are embedded in the plasma membranes of neurons and strengthen their synapses if the connected neurons are active at the same time (see assigned reading at the end of this chapter).

The ability to make transgenic organisms, in which specific genes have been added, modified, or deleted, has revolutionized the study of gene function. This achievement was recognized by the award of the 2007 Nobel Prize in Medicine or Physiology to Mario Capecchi, Oliver Smithies, and Sir Martin Evans.

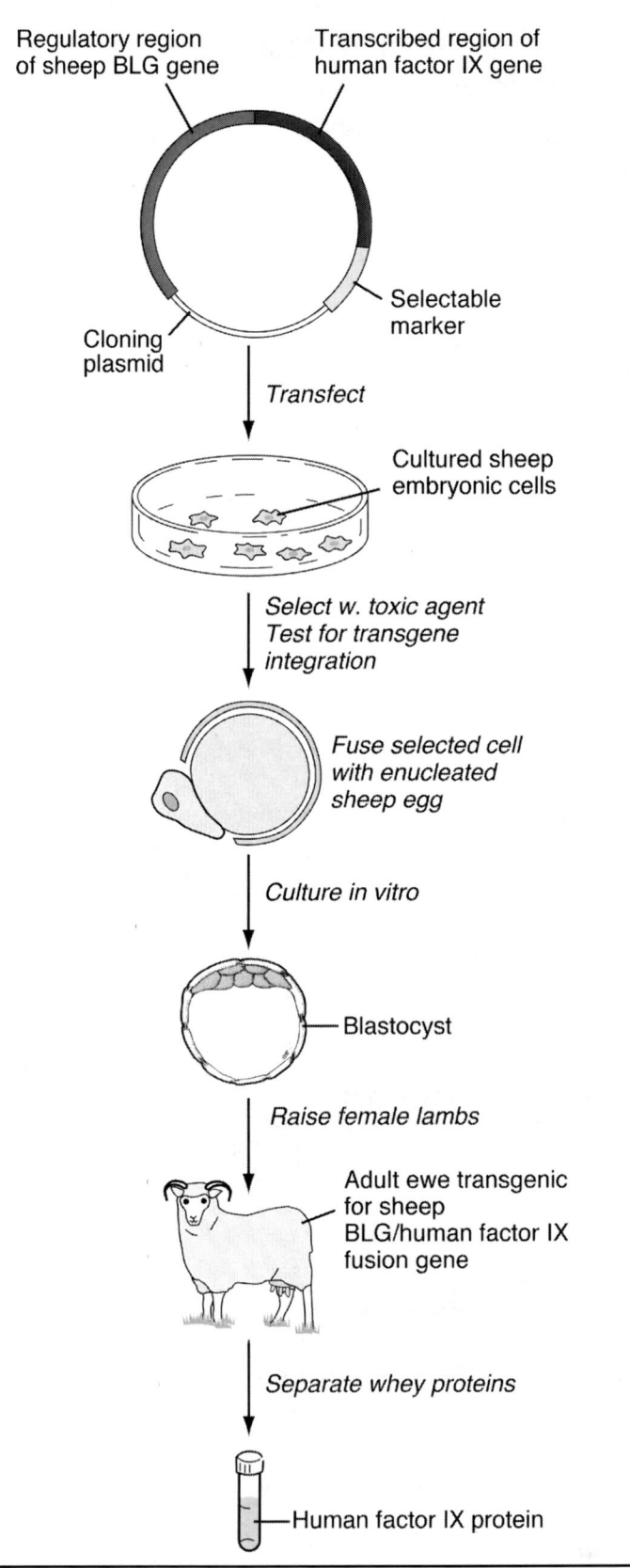

FIGURE 11.8:
Making human pharmaceutical proteins in "pharm" amimals

I. Restorative versus Innovative Germ Line Therapy for Humans

Applying any of the currently available methods for germ line gene therapy to humans would be grossly unethical. This picture may change, though. Within a decade or two, it may be possible to derive *transgenic eggs or sperm* in laboratory culture from ES cells or iPS cells that have incorporated a transgene by homologous recombination. Under these conditions, human germ line therapy would be an extended form of in vitro fertilization, with gametes grown from genetically modified cells of one partner or both. Methods for deriving eggs and sperm from mouse ES cells and iPS cells are well under way, and efforts to adapt these methods for humans will be pursued vigorously because of their potential for circumventing human fertility problems. Thus, in a not too distant future, we may reach the point where human germ line therapy could be deemed safe and effective. Would it also be ethically defensible?

At this point, it is important to distinguish between *restorative* and *innovative* gene therapy. So far, all human gene therapies approved for clinical trials have been restorative, because they were designed to correct a well-defined genetic disorder, thereby restoring a normal genome as it had evolved naturally. However, some parents may want to have children with extra genes for being taller, more athletic, more intelligent, or more beautiful. The required germ line therapies would be innovative, because they would create novel genomes that may not have been tested in evolution, or if they were, they have been selected away.

Attempts to improve on natural genomes may be successful, as apparently was the case with the *Doogie* mice. However, these mice did not have to compete in the wild. Quite possibly, their enhanced memory and learning abilities were associated with a loss in some other skill that is required for a mouse to survive in the wild but not in a laboratory, such as a keen sense of smell or a robust circadian rhythm. The health problems of the pigs made transgenic for bovine growth hormone should serve as a warning. Most genes in complex organisms, especially in the human, are *pleiotropic*, meaning that they affect more than one trait (see Figure 3.9). Genes that contribute to traits like stature or intelligence are likely to affect other traits as well, so that attempts at "tuning up" one trait may compromise other traits. Because such "side effects" are hard to predict, and because they cannot be tested conclusively in laboratory animals, innovative germ line therapies for humans should not be attempted.

Shinya Yamanaka, the principal investigator in the original iPS cell study, has expressed concern about the prospect of deriving human gametes from iPS cells. At his behest, the Japanese science ministry has notified all universities and research agencies not to produce gametes from iPS cells and not to implant embryos made with iPS cells into human or animal wombs. Patent applications that would make the unauthorized use of iPS cell technology illegal are also under way.

EXERCISE PAGE FOR CHAPTER 11

Student Last Name _____

Student First Name _____

Discussion Section _____

On the course web site, use the link on the **syllabus** to find a pdf file of your **assigned reading** for this chapter. The same reading is referenced below. Use the bulleted list of questions to test your knowledge of this reading.

Tsien, J. T. (2000). Building a brainier mouse. *Scient. Amer.* **April 2000:** 62–68.

- What is long-term potentiation (LTP), and how is it thought to work in memory formation and learning? *Long term potentiation, long-lasting stimulation strengthens synaptic connection between neurons via NMDA receptor*
- Outline the structure and function of the NMDA receptor, and its presumed biological role in LTP. *Tetramere 2 NR1's + 2 NR2's pore that allows influx of Ca^{2+}. recognize glutamate + Mg^{2+} efflux.*
- How does the NR2 subunit of the NMDA receptor change with age, and why may Tsien have chosen the NR2B gene to make *Doogie* mice?
- How did Tsien use the LoxP/Cre system to make *conditional* knockout mice? *2 mice — 1: flanked NR1 by LoxP "cutsites" 2: Cre enzyme attached to on switch only works in brain*
- How do mice with a loss of NMDA function in the brain behave? *abnormal spatial*
- How did Tsien make his *conditional* gain-of-function mice? [p. 66B] *& a str...*
- How do mice with enhanced NR2B synthesis in the brain behave?

Feedback

On a scale from 1 to 10, the being the best, please rank the above reading for

1. Interest and Relevance to the topic (10 being most interesting/relevant) _____
2. Readability (10 being most clearly written, easy to understand) _____

In the remaining space, enter any comments that you may have on this reading.

↓ offspring has both genes

In hypocampus CA1 region, Cre splices out NR1 gene.

↓ young NR2B ——→ old NR2A
stay open longer

CHAPTER TWELVE

AGING AND SENESCENCE

When people say a person is "getting old," they may just mean the chronological progress of adding one birthday after another. We will refer to this process as **aging**. However, "getting old" also comes with connotations of becoming infirm and getting closer to death (Figure 12.1). We will call the latter process **senescence**.

While aging is inevitable and occurs to all objects, senescence is a unique phenomenon of multicellular organisms. They die and leave dead bodies. Only their eggs or sperm have a chance to live on as their offspring. The lineage of cells that give rise to gametes is defined as the **germ line** (Figure 12.2). Germ line cells are potentially immortal and form unbroken chains of cellular continuity between generations (Figure S12.a). All other cells of a multicellular organism are called **somatic cells**; they are bound to die and may be viewed as the temporary caretakers of the germ line. In humans and many other animals, germ line and soma are separated early during embryonic development.

Why do we grow old and die? We will first consider some ultimate causes of senescence and then consider two of the many proximate causes that have been proposed: oxidative damage and the shortening of telomeres.

A. What Is Senescence?

There are many physiological parameters that could be used to measure senescence: number of push-ups, time for a 100 m dash, ability to memorize telephone numbers, and many more. A problem with these measurements is that they do not progress in synchrony. A person who is getting old in terms of push-ups may still be doing well in terms of memorizing numbers, and

FIGURE 12.1:
Oldest verified person on record. This photograph shows Jeanne Calment of Arles in France on her 122nd brithday. She died a few months later, on 4 August 1997.
Image © PARROT PASCAL/CORBIS SYGMA

so on. Is there a way of integrating the various processes of senescence? Yes, there is, and insurance companies have figured it out a long time ago.

A person taking out term life insurance makes a contract with an insurance company. The client pays a premium for the company's promise to pay a benefit to a designated beneficiary in case the client dies within a year after closing the contract. Any potential client studying the company's price charts will notice quickly that, for a given benefit, the premium to be paid increases with age: steeply beyond age 40, and even more steeply beyond age 80. Are the insurance companies just ripping off old people, or is there a statistical reality behind their tables?

To answer this question, one needs to study **survival curves**, which can be drawn from two types of data.

- **life tables** showing how many individuals of a **cohort**, such as all men born in the United States in 1894, were still alive in 1895, 1896, and so forth
- **census data**, indicating how many individuals in a present population are 1 year or older, 2 years or older, and so forth Figure 12.3a shows survival curves for females in the United States based on census data collected for three different years.

Life table data are more difficult to collect. Census data may be confounded by migration. For our purposes, either data are generally useful.

Survival curves for humans vary depending on sex as well as time and place where the data were collected. Survival curves for females, based on census data collected at different times in the United States, are shown in Figure 12.3a. For example, the curve for 1960 shows some decline during the first few years of human life ("child mortality"), followed by a shallow decline until around age 40,

FIGURE 12.2:
Ther germ line is a continuous lineage of cells that are capable of forming eggs or sperm. Germ line cells are potentially immortal, connecting individuals of successive generations in an uninterrupted chain. The other cells–collectively called soma–will die, having served as temporary caretakers of the germ line.

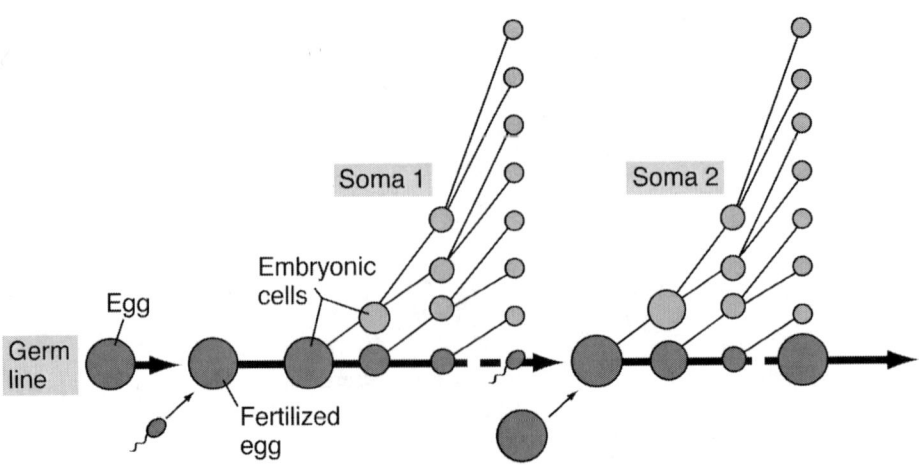

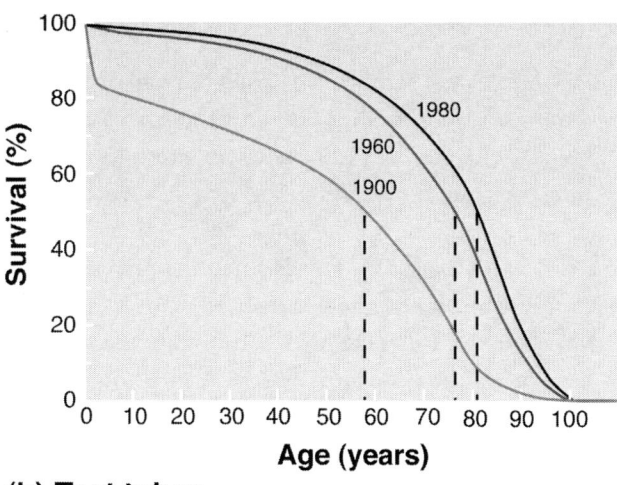

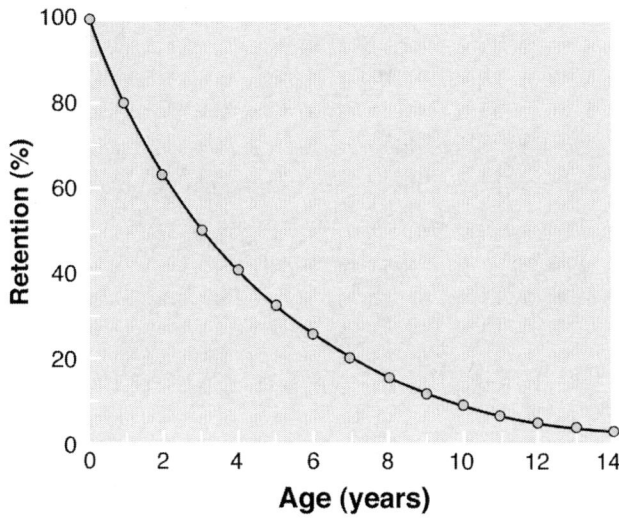

FIGURE 12.3:
Survival of living organisms versus retention of inanimate objects. (a) Survival curves for human females in the U.S.A., based on census data for 1900, 1960, and 1980. Each curve traces how many females were *at least* 1 year old, at least 2 years old, etc. The vertical lines indicate the age to which 50% of the population had survived in each census. **(b)** Retention curve for a cohort of test tubes placed into service at a given time in a laboratory. Tubes disappeared due to theft, misplacement, or accidental breakage.
From *Analysis of Biological Development* by Klaus Kalthoff. Copyright © 1996 by McGraw-Hill. Reproduced with permission of The McGraw-Hill Companies.

and then an increasingly steep decline with progressing age before a slower terminal decline.

Survival curves of organisms look different from the retention curves of nonliving objects, such as a cohort of glass test tubes taken into service in a laboratory on a certain date (Figure 12.3b). The percentage of test tubes retained after 1, 2, 3, and more years decreased exponentially, due to misplacement, theft, accidental breakage, and so on. This means the *fraction* of glasses lost each year was constant (in this case, 20 percent per year). Such a curve is expected if the probability for a glass tube to be lost *does not change with age*. In other words, this type of curve characterizes a potentially immortal population that is decimated only by random accidents or losses.

The difference between survival of living organisms versus retention of inanimate objects becomes even clearer if one plots the *rate of change* versus age (Figure S12.b). The **mortality rate** is defined by the following equation.

$$m(t) = \frac{\text{Number of individuals who die while they are } t \text{ years old}}{\text{Number of individuals who were alive at the beginning of year } t}$$

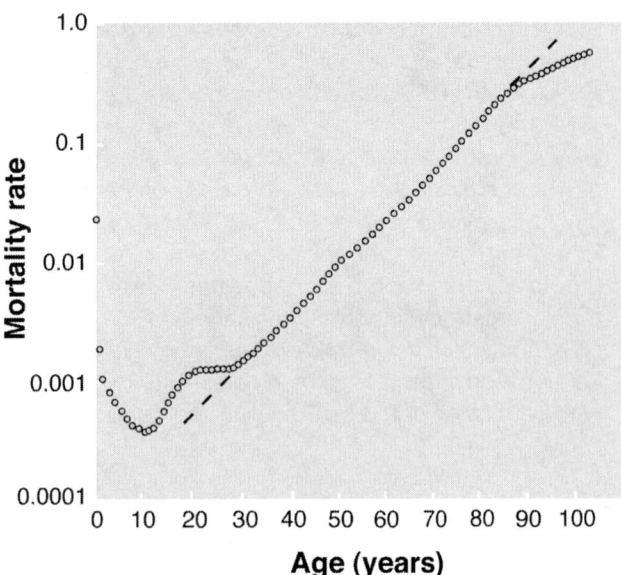

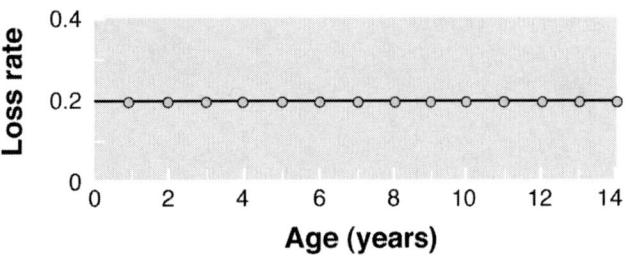

FIGURE 12.4:
Mortality rate of organisms versus loss rate of inanimate objects. (a) Human mortality rate based on the 1960 census data show in Figure 12.3a. The mortality rate was high in children, due to childhood diseases, and then lowest around age 10. It increased again for teenagers, mostly due to accidents. Between 30 and 85 years of age, mortality increased exponentially, resulting in a straight line in this semi-logarithmic plot. **(b)** Loss rate of the test tubes shown in Figure 12.3b. The loss rate was constant at 20% each year.
From *Analysis of Biological Development* by Klaus Kalthoff. Copyright © 1996 by McGraw-Hill. Reproduced with permission of The McGraw-Hill Companies.

The loss rate for inanimate objects is defined correspondingly. Figure 12.4b is a plot of the loss rate of the batch of test tubes shown in Figure 12.3b. The loss rate for test tubes is a straight horizontal line. A plot of the mortality rate of a human population looks fundamentally different (Figure 12.4a). The rate is appreciable during the first few years, then very low around age 10, followed by an *exponential increase* after age 30. (This is when the premiums for term life insurance go up steeply.)

The rapid increase of the mortality rate toward the end of the lifespan is observed in all organisms. We will adopt this statistical phenomenon as our definition of *senescence*. It avoids the incongruence of different physiological criteria mentioned earlier and instead integrates over all causes of death.

It is important to distinguish senescence from *aging*, which simply means growing chronologically older. Unfortunately, some authors use the terms "aging" or "ageing" when they actually mean senescence. While aging is inevitable, it is not immediately obvious why there is senescence. It seems wasteful to grow big bodies only to let them die and grow new ones. Why don't we keep ourselves in good repair and live indefinitely?

B. Bad Explanations of Senescence

Seemingly plausible explanations of senescence often turn out to be circular arguments. One such "explanation" argues that humans in particular show senescence because they have post-reproductive life spans, during which they are no longer subject to natural selection. The problem is that most animal species—and some human males—reproduce throughout life, and menopause seems to be specific to human females. Even so, in primitive societies, most people die before reaching post-reproductive age. Thus, if a decrease of reproduction in old age occurs at all, it would be part of general senescence. Ascribing senescence to loss of reproduction then becomes a circular argument.

According to another "explanation," senescence and death have evolved to make room for the younger, more productive, members of a population. Again, it is implied from the beginning that oldsters are less fit, although this is the contention to be proven. Another fallacious assumption in this argument is that, in a population without senescence, there would be lots of Methusalahs. While this would be the case in developed human societies, there is only a small fraction of old individuals in nondeveloped human societies or in animal communities, where predation, diseases, accidents, and other random events take a steady toll. This statistical effect is recognized in the following, more rational, explanations of senescence.

C. Mutation Accumulation Hypothesis of Senescence

In a natural population that is decimated by random accidents, there are many more youngsters than oldsters (Figure 12.5). This would be the case even in a hypothetical population of potentially immortal individuals as long as the death rate from disease, predation, and such is high. In such populations, even if young and old *individuals* had the same reproductive fitness, say, one offspring every five years, the youngsters *as a group* would have many more offspring simply because there are more youngsters.

Under these circumstances, any mutations that reduce fitness have a greater effect, and will therefore be selected against more strongly, if they are expressed *early in life*. In contrast, mutations reducing fitness *late in life* will be selected against only weakly, just because most members of the population have already died from random causes before the mutations take effect.

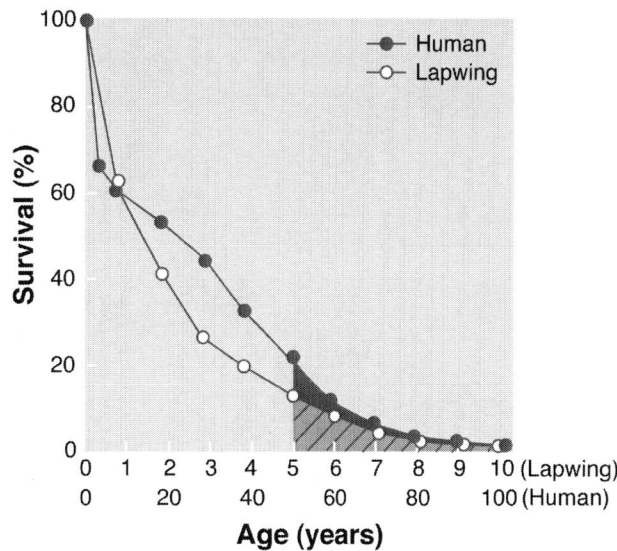

FIGURE 12.5:
Few oldsters. Survival curves are plotted for a bird (English lapwing) and for a human population (British India, 1921–1930) living under primitive conditions. Individuals in the 2nd half of their maximum life span (shaded areas) constitute only about 10% of the entire populations.
From *Analysis of Biological Development* by Klaus Kalthoff. Copyright © 1996 by McGraw-Hill. Reproduced with permission of The McGraw-Hill Companies.

According to the **mutation accumulation hypothesis**, senescence is caused by deleterious mutations acting late in life, which have accumulated *in the germ line over many generations*. The accumulation has been due to *decreasing force of natural selection* in the older and therefore *smaller* segment of a population.

This explanation of senescence is free of circular arguments and based only on two observations:

- Older individuals are less frequent in a population simply because they have been exposed to random hazards for longer periods of time
- There are random mutations, and most of them reduce fitness. Mutations are caused by radiation, chemicals, and the limited accuracy of the DNA repair processes that remove damaged or misreplicated DNA segments.

Empirically, there are many examples for genetic variants reducing fitness *late in life,* including osteoporosis, type 2 diabetes, cataracts, most cancers, Alzheimer's disease, arthritis, and atherosclerosis.

Predictions derived from the mutation accumulation hypothesis have been tested by **forced selection experiments** with laboratory animals.

Mimicking early accidental death in a laboratory population should decrease longevity because mutations reducing vigor in midlife and later are no longer selected against. This prediction was confirmed in experiments with the flour beetle, *Tribolium castaneum*. Adults were kept in age cohorts and killed after producing their first offspring. After forty generations the median longevity of the population had decreased, in accord with the prediction derived from the mutation accumulation hypothesis.

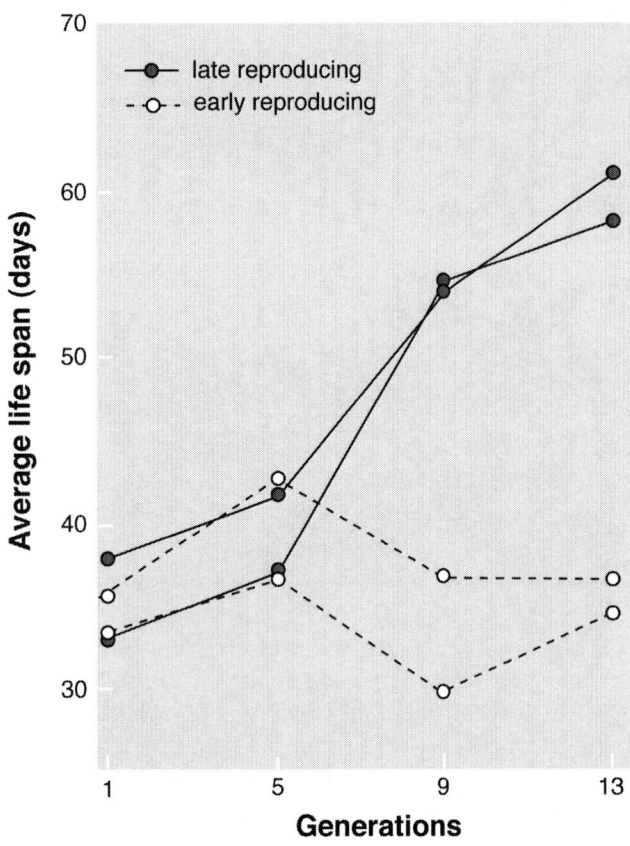

FIGURE 12.6:
Age at reproduction and life span. Two laboratory strains of fruitflies were propagated using only eggs from older females. Two control lines were propagated using only eggs from young females. After just a few generations, the late reproducing strains showed a much longer life span.
From *Analysis of Biological Development* by Klaus Kalthoff. Copyright © 1996 by McGraw-Hill. Reproduced with permission of The McGraw-Hill Companies.

Conversely, if the age of first reproduction is postponed, senescence should be postponed as well, because mutations reducing vigor in mid-life are strongly selected against. This prediction has also been confirmed by forced selection experiments with genetically heterogeneous (outbred) *Drosophila melanogaster* cultures. In strains propagated from old adults—which had lived through selection during midlife—both longevity and fertility at old age increased (Figure 12.6).

In developed human societies, birth rates for women over 30 have increased, while teenage pregnancies have been declining. Assuming that these trends will continue, what would you expect for the average human life span in these societies?

D. Genes with Antagonistic Pleiotropy

An interesting variant of the mutation accumulation hypothesis is based on genes with **antagonistic pleiotropy:** Some alleles of these genes enhance fitness early in life at the cost of decreased fitness later in life, while other alleles of the same genes have the converse effect. As an example, consider genes encoding the enzymes involved in the testosterone metabolism of men. Some alleles of these genes will cause an elevated blood level of testosterone, which would make for stronger bones, more muscle, more aggressive behavior, stronger sex drive, and more sperm production. All of these should translate into enhanced reproductive success, but also into higher rates of prostate cancer late in life. In populations where most men do not grow old enough to get prostate cancer, the high testosterone alleles should be positively selected for. However, the same alleles will contribute to the senescence of men who grow old enough.

Like the mutation accumulation model, the antagonistic pleiotropy model is based on the decreasing frequency of older individuals in a population. But instead of relying on the accumulation of deleterious mutations that are just weakly selected against, the antagonistic pleiotropy model focuses on genetic alleles that provide increased fitness early in life, while imposing a price of reduced fitness later in life. Again like the mutation accumulation model, the antagonistic pleiotropy model deals with *genetic alleles that accumulate in the germ line over many generations.*

In new forced selection experiments with *Drosophila*, strains propagated from old adults showed delayed mortality in comparison to strains propagated from young adults—as had been observed in the earlier experiments shown in Figure 12.6. However, it was also observed that the late-reproducing strains laid fewer eggs early in life. Moreover, the difference in mortality between the early-reproducing and late-reproducing strains *disappeared* for females that were incapable of laying eggs after X-irradiation or due to a mutation. Thus, there seemed to be a trade-off between longevity and early **fecundity** (eggs laid per female per day), and the shorter life span of the early-reproducing flies was better explained by the cost of making eggs than by accumulation of random deleterious mutations.

Genes with antagonistic pleiotropy have also been found in the roundworm *Caenorhabditis elegans*. For instance, recessive mutant alleles in the *age-1* locus show a 65 percent increase in mean life span at the price of reduced fecundity, at least under conditions of periodic starving. Interestingly, the price of reduced fecundity was not paid by the *age-1* mutants under conditions of unrestricted feeding.

In mice, over-expression of the *Klotho* gene extends the life span while reducing fecundity (pups per female per year). The Klotho protein acts as a bloodborne signal that inhibits cellular senescence (see Section G) and programmed cell death.

E. Disposable Soma Hypothesis

A third hypothesis on the ultimate cause of senescence has been called the **disposable soma hypothesis**. It focuses on the separation between germ line and soma (see Figure 12.2), taking the perspective that every individual is the temporary caretaker of his/her germ line. According to the disposable soma hypothesis, this carrier function has evolved to be as economical as possible, especially in regard to the repair of *damage that accumulates in the soma during the life of an individual.*

It may help to see the maintenance of the germ line by successive generations of somatic carriers in analogy to car ownership. The conventional wisdom is that it is best to buy a car new (or almost new) and maintain it well until the costs of repair (in time and money spent) become higher than making the payments for a new car. At that point, buying a new car becomes more economical.

Just as it costs an owner time and/or money to maintain a car, it costs an organism energy to replace or repair its cells and molecules. In both situations, maintenance costs tend to increase with age. In an organism, the principal agents of repair are proteins. Carbonyl (C=O) groups and other damaging modifications of proteins accumulate in various body parts as an individual ages (Figure 12.7). As the proteins that carry out molecular repair functions become damaged themselves, they are bound to work less efficiently, requiring more energy to maintain a given rate of repair. Additionally, genetic mutations in somatic tissues accumulate with age, and again the encoded proteins will work less efficiently. At some point, keeping an organism repaired becomes more expensive than replacing an old soma with a new one.

At least some species gauge the effort spent on soma maintenance to the risk of accidental loss. (Again the analogy of car ownership may be useful: You will tend to invest less in maintaining your car

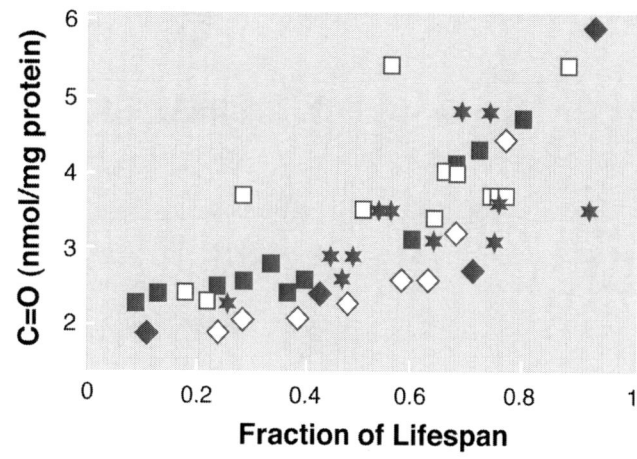

FIGURE 12.7:

Oxidative damage accumulates with age. Carbonyl (C=O) content of protein from [■] human dermal fibroblasts in culture, [★] human eye lens, [□] human brain at autopsy, [◆] rat liver, and [◇] whole fly.
From *Analysis of Biological Development* by Klaus Kalthoff. Copyright © 1996 by McGraw-Hill. Reproduced with permission of The McGraw-Hill Companies.

if the risk of losing it to theft or an accident is high.) In a study on the Virginia opossum, *Didelphis virginiana,* two populations were compared. One population lived on the mainland in South Carolina, where it was decimated by birds of prey and by mammalian predators. The other population lived off shore on an island, which was too small to support large predators. Generally, predators take out weak individuals first and thereby help to maintain the health of their prey populations. However, based on the disposable soma hypothesis, one would expect that the opossums with less predation would invest more into maintenance. The results indicate that in this case the disposable soma effect was stronger: Island females were better able to nurse their young and showed less cross-linking of their collagen fibers.

In summary, there are three evolutionary explanations of senescence:

- According to the *mutation accumulation hypothesis*, mutations reducing fitness late in life accumulate over the generations in the germ line because selection against these mutations is less stringent than the selection against mutations reducing fitness early in life.
- According to the *antagonistic pleiotropy hypothesis*, alleles that increase early fitness at the expense of later fitness are positively selected for and therefore accumulate in the germ line over the generations.
- According to the *disposable soma model,* the most economical way of maintaining a germ line is through periodic renewal of the somatic carrier, because the cost of maintaining the carrier increases as the carrier ages.

Are these explanations mutually exclusive?

All evolutionary theories of senescence imply that the longevity of each species has evolved as part of many trade-offs that optimize reproduction under various ecological conditions, including food availability, likelihood of predation or accidental death, cost of reproduction, and cost of cellular repair processes. This means that human longevity may change as any of these parameters change in developing societies.

F. Oxidative Damage and Organismic Senescence

Researchers interested in *proximate causes* have not found any particular genetic control mechanism that would ensure the timely onset of senescence similar to puberty or menopause. Depending on species, senescence seems linked to a wide range of physiological processes. We will focus on two of them, namely, *oxidative damage* and *loss of telomerase activity* in mammals.

In a process known as **oxidative phosphorylation**, eukaryotic cells generate chemical energy in the form of ATP by reducing oxygen to water. Each oxygen molecule (O_2) is turned into two molecules of water (H_2O) by gaining four electrons (e^-) and four protons (H^+). Oxygen reduction occurs in the inner mitochondrial membrane. The four-step process (Figure 12.8) generates three intermediates:

- superoxide radical ($\cdot O_2^-$)
- hydrogen peroxide (H_2O_2)
- and hydroxyl radical ($\cdot OH$)

All three intermediates are highly reactive **oxidants** (electron acceptors), while two of them are also **radicals** (having unpaired electrons). Failure to convert these intermediates efficiently leaves them to react with nucleic acids, lipids, and proteins, impairing their normal functions.

Superoxide radical and hydrogen peroxide are the first two intermediates in the reduction of oxygen. Cells have mitochondrial and cytosolic enzymes that break down excess amounts of these intermediates into harmless water and oxygen, thus avoiding the formation of the third and most damaging intermediate, the hydroxyl radical (see Figure 12.8).

- superoxide dismutases convert superoxide radical into hydrogen peroxide and oxygen
- catalases convert hydrogen peroxide into water and oxygen
- peroxidases reduce hydrogen peroxide to water by oxidating other molecules, such as reduced glutathione and other so-called antioxidants (electron donors)

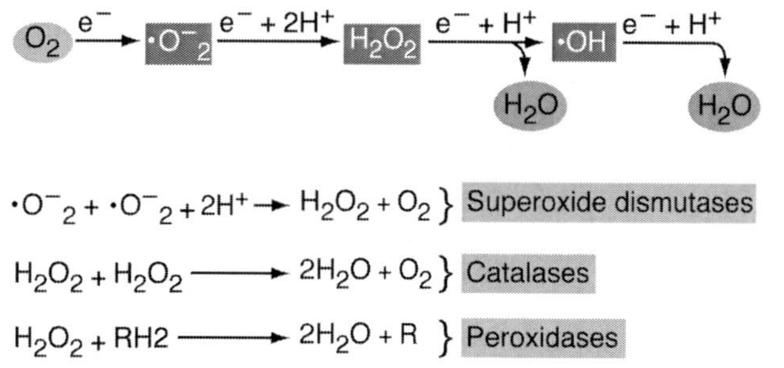

$$O_2 \xrightarrow{e^-} \cdot O^-_2 \xrightarrow{e^- + 2H^+} H_2O_2 \xrightarrow{e^- + H^+} \cdot OH \xrightarrow{e^- + H^+}$$
$$\qquad\qquad\qquad\qquad\qquad\qquad\qquad\quad H_2O \qquad\quad H_2O$$

$\cdot O^-_2 + \cdot O^-_2 + 2H^+ \rightarrow H_2O_2 + O_2$ } Superoxide dismutases

$H_2O_2 + H_2O_2 \longrightarrow 2H_2O + O_2$ } Catalases

$H_2O_2 + RH_2 \longrightarrow 2H_2O + R$ } Peroxidases

FIGURE 12.8:
Reactive intermediates. During oxidative phosphorylation, oxygen is reduced to water by the stepwise addition of electrons and hydrogen ions. Highly reactive oxidants generated as intermediates include superoxide radical [$\cdot O_2^-$], hydrogen peroxide [H_2O_2], and hydroxyl radical [$\cdot OH$]. In eukaryotic cells, several enzymes (names behind brackets) convert these oxidants into less harmful molecules.
From *Analysis of Biological Development* by Klaus Kalthoff. Copyright © 1996 by McGraw-Hill. Reproduced with permission of The McGraw-Hill Companies.

Oxidants damage cellular molecules including DNA, lipids, and proteins. In particular, oxidants

- generate about 10^4 modified DNA bases per cell per day. A small fraction of DNA damage escapes repair mechanisms and accumulates with age
- react with lipids to form *lipid peroxides*, that is, lipids containing an -O-O- linkage. Lipid peroxides change the properties of biological membranes.
- react with amino acids to form *carbonyl groups* (C=O double bond), methionine sulfoxide, tyrosine-tyrosine cross-links, and disulfide bonds. The modified proteins lose some or all of their biological activity.

Many correlational data suggest that oxidant damage to cellular molecules hastens the process of senescence.

- Life spans of organisms are negatively correlated with peroxide levels and positively correlated with antioxidant levels.
- It is estimated that 40 to 50 percent of all proteins in old individuals are present in oxidatively damaged form.
- In humans with premature aging diseases (progeria or Werner's syndrome), the carbonyl content of proteins is much higher than from age-matched normal individuals.
- In fresh cells, the carbonyl content of proteins increases with the age of the cell donor. In cultured cells, the carbonyl content of proteins increases with culture time.

An experimental study on Mongolian gerbils demonstrated the beneficial effects of an antioxidant, *N-tert-butyl-alpha-phenylnitrone (PBN)*, on age-related changes in the brain. Gerbils were injected twice daily with PBN dissolved in saline, while control animals were injected with saline only. In old gerbils, the PBN injections caused a marked decrease in carbonyl residues (the tell-tales of oxidized proteins) in brain tissue (Figure 12.9). When

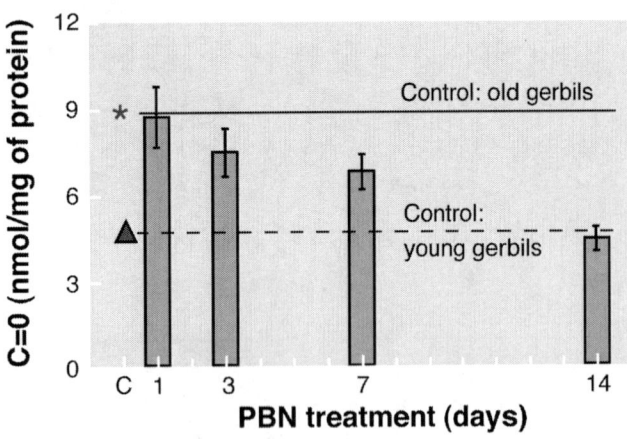

FIGURE 12.9:
Antioxidants reduce protein damage. Daily injections of old gerbils with the antioxidant PBN (N-tert-butyl-alpha-phenyl-nitrone) reduced the carbonyl (C=O) content of their brain proteins to the level found in untreated young gerbils. From *Analysis of Biological Development* by Klaus Kalthoff. Copyright © 1996 by McGraw-Hill. Reproduced with permission of The McGraw-Hill Companies.

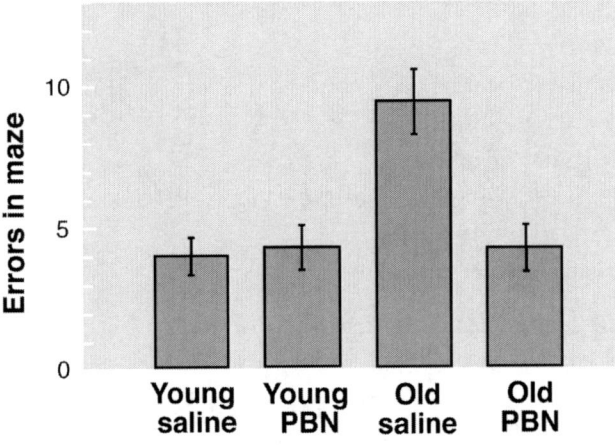

FIGURE 12.10:
Antioxidants help old gerbils. Effects of age and PBN (antioxidant) injections on the patrolling behavior of gerbils. Errors were defined as those arms of a radial maze that were reentered before the gerbil had explored all eight arms of the maze. PBN reduced the number of errors made by old gerbils but did not help young gerbils. From *Analysis of Biological Development* by Klaus Kalthoff. Copyright © 1996 by McGraw-Hill. Reproduced with permission of The McGraw-Hill Companies.

the PBN injections were discontinued, carbonyl residues returned to the levels observed when the experiment began. Parallel to the biochemical indication of less oxidative stress, the spatial and temporal memory of old gerbils also improved as indicated by performance in maze tests (Figure 12.10). In young gerbils, which show less oxidative damage to brain tissue and perform better in maze tests than old gerbils, PBN injections had no significant effect. The results indicate that oxidative damage to proteins may limit brain function at advanced age, and that this process can be reversed by antioxidants.

Because superoxide dismutase (SOD) and catalase play key roles removing oxidants from cells, investigators have tested the effects of increasing the gene dosage for these enzymes. No constant pattern of results has emerged yet.

G. Cellular Senescence and Loss of Telomerase Activity

Normal somatic cells from mammals, when kept in culture, undergo only a limited number of divisions before they arrest. The number of divisions that a line of somatic cells undergoes in culture decreases with the age of the donor of the cells. Human fibroblasts derived from fetal or neonatal tissue can undergo 50 to 80 rounds of cell division. The upper limit of cell divisions cannot be overcome by improved culture conditions. This phenomenon has been called **cellular senescence**, even though its connection to organismic aging and senescence is not clear. Only germ line cells, stem cells, and tumor cells seem to be exempt from cellular senescence and are therefore called **immortal cells**.

Cellular senescence has been linked to the shortening of **telomeres**, distinct structures at the ends of eukaryotic chromosomes, which have been likened to the plastic caps of shoelaces (Figure S12.c). Telomeric DNA consist of hundreds to ten thousands of tandem repeats of a G-rich sequence, which is GGGTTA in humans. The telomeric DNA attracts a protein/RNA complex known as **telomerase** (Figure 12.11). For its elucidation, the Nobel Prize in Physiology or Medicine for 2009 was awarded to Elizabeth Blackburn, Carol Greider, and Jack Szostak. Telomerase has two components:

- telomerase reverse transcriptase (TERT), a protein
- telomerase RNA component (TERC), which serves as a template for TERT

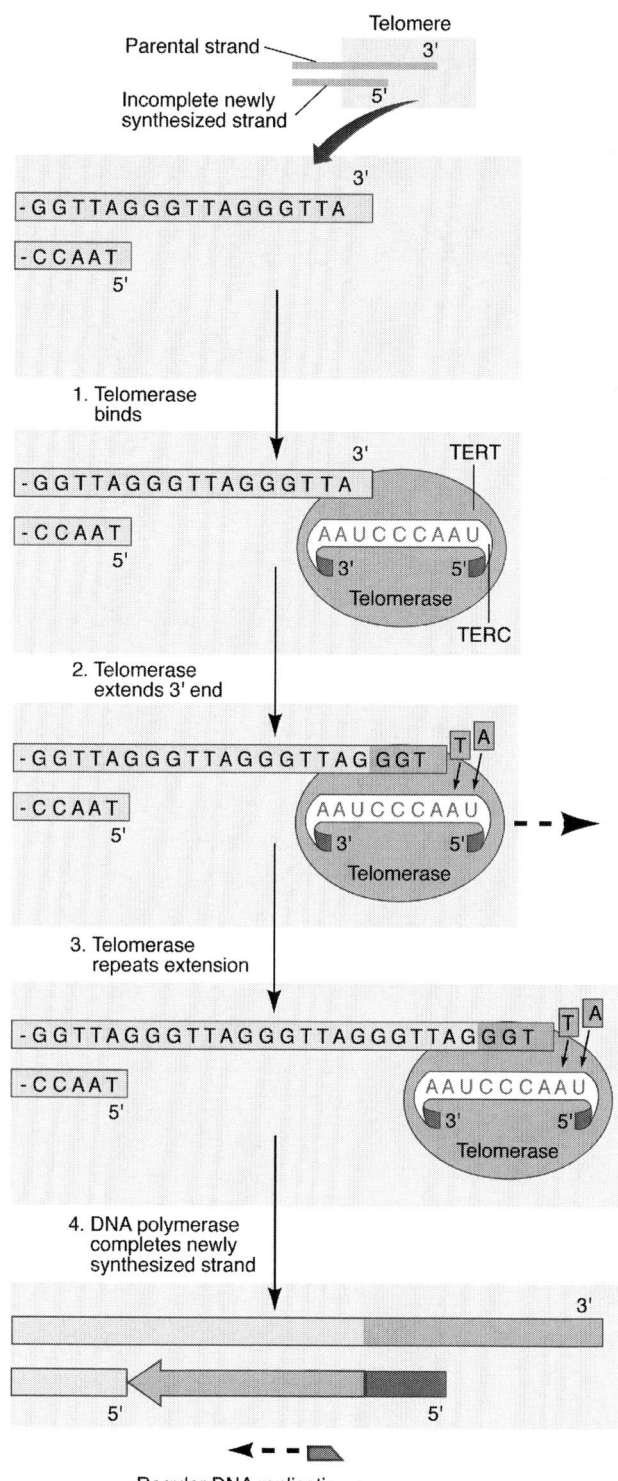

FIGURE 12.11:

Telomerase. For replication of linear DNA, the 3′ end of each parental strand needs to be extended. This is accomplished by telomeric DNA consisting of many GGGTTG or GGGTTA repeats associating with telomerase. The latter consits of a reverse transcriptase (TERT) and an RNA component (TERC) serving as a template for TERT.
From *Analysis of Biological Development* by Klaus Kalthoff. Copyright © 1996 by McGraw-Hill. Reproduced with permission of The McGraw-Hill Companies.

When all telomere components are present, they compensate for the **end-replication problem**, which arises from the fact that DNA polymerase, the enzyme complex that replicates DNA, works only in the 5'-to-3' direction and needs an RNA primer. Therefore, in every round of replication of a linear DNA molecule, the 3' end of each parental strand forms an overhang that cannot be replicated unless it is first extended to sufficient length. This is accomplished by TERC and TERT (see Figure 12.11).

However, TERT is made only in certain mammalian cells, including the germ line, adult stem cells, and most cancer cells. In most normal somatic cells, the genes encoding TERT are switched off. Exceptions include adult stem cells and lymphocytes, which retain the ability to divide throughout the life of an organism and show a low level of TERT synthesis. In cells without TERT activity, each chromosome loses about 100 nucleotides of telomeric DNA in each mitotic cycle. Some cells respond to worn-off telomeres as they do to unrepaired DNA damage: They block the cell cycle at the G1 checkpoint. In other cells, chromosomes without telomeres fuse end to end, forming unstable chromosome configurations that ultimately prevent further mitoses.

Taken together, the data indicate that most normal somatic cells inhibit the synthesis of TERT, and that most cancer cells have resumed TERT synthesis. The lack of TERT in normal somatic cells seems to function as a safeguard against cancer by limiting the number of divisions that a cell can undergo.

Drugs that inhibit inappropriate TERT activity in somatic cells may become a new tool against cancer. This avenue is actively being pursued by pharmaceutical companies, but success may be limited by the ability of at least some cancer cells to maintain long telomeres in the absence of detectable TERT activity.

Attempts to realize the old dream of a "fountain of youth" by somehow keeping TERT genes active in all cells of the body do not seem to be pursued. Since more than 90 percent of human cancer cell lines show TERT activity, it seems likely that instead of living longer, people would die earlier from cancer if TERT genes were routinely expressed in *all* somatic cells.

However, experiments are underway to restore TERT activity *to specific cell lineages* as part of a cell replacement therapy that can be combined with other somatic gene therapies and does not depend on embryonic stem cells (Hornsby, 2007).

Transgenic mice deficient for TERC and/or over-expressing TERT indicate that telomere length and the level of TERT expression are independent determinants of the mobilization and proliferative capacity of epidermal stem cells.

EXERCISE PAGE FOR CHAPTER 12

Student Last Name _____

Student First Name _____

Discussion Section _____

On the course web site, use the link on the **syllabus** to find a pdf file of your **assigned reading** for this chapter. The same reading is referenced below. Use the bulleted list of questions to test your knowledge of this reading.

Flores, I.; Cayuela, M.; and Blasco, M. A. (2005). Effects of telomerase and telomere length on epidermal stem cell behavior. *Science* **309**: 1253–1256.

For larger photos and supporting material, go online to http://www.sciencemag.org/cgi/content/full/sci;309/5738/1253

Background information (see topic 10): Tissue stem cells can self-renew and generate differentiated cells that replace lost cells. Such stem cells reside in protected microenvironments, or "niches." When needed, they begin to proliferate and leave the niche. This process of "mobilization" is controlled by intrinsic genetic programs and by signals from the niche.

- Characterize the three transgenic mouse strains used by the authors to study the effects of telomerase and telomere length on epidermal stem cell behavior.
- How are "label-retaining cells" (LRCs) made visible in this study? Where do they occur in mouse epidermis, and how is this location interpreted?
- What is the TPA treatment that the authors apply to their mice, and why was it used in this study?
- How did the loss of TERC in *Terc*$^{-/-}$ mice affect telomere length, number of LRCs per hair follicle, as well as the disappearance of LRCs from follicles and hair follicle length after TPA treatment?
- How did the loss of TERC in *Terc*$^{-/-}$ mice affect the proliferation of epidermal stem cells ("clonogenicity") *in vitro*?
- How did over-expression of TERT in K5-mTert affect the same behaviors of epidermal stem cells?
- How did the authors address the question whether the stimulating effect of TERT on clonogenicity depended on TERC? What was the result, and how do the authors interpret it?
- What is the significance of the observation that the stimulating effects of telomerase over-expression were not correlated with increased telomere length?

Feedback

On a scale from 1 to 10, the being the best, please rank the above reading for

1. Interest and Relevance to the topic (10 being most interesting/relevant) _____
2. Readability (10 being most clearly written, easy to understand) _____

In the back of this page, enter any comments that you may have on this reading.

CHAPTER THIRTEEN

NATURE AND NURTURE

So-called "nature-versus-nurture" questions abound in public discourse as well as in science. Is human behavior determined by genes or by culture? Most professors in mathematics are male: Is this sex-specific talent or self-perpetuating stereotype? A few percent of all males and females are homosexual: lifestyle choice or genetic destiny? Careful examination of human and animal behavior indicates that both are intricate combinations of inborn and learned components. Can we quantify the extent to which a trait is determined genetically? Yes, we can, but such numbers must be viewed in proper context.

A. It's Nature *and* Nurture, not Nature *versus* Nurture

Historically, **nature-versus-nurture** conflicts have often had political or religious overtones. Indeed, certain political movements have tried to create societies relying almost exclusively on either genetics or education. Either way, the results have been disastrous. The eugenics movements in England and the United States have focused on human genetics. They tried to improve their societies by favoring individuals perceived as having "good genes," and by sterilizing people deemed genetically inferior. In Germany, this mindset has ultimately led to the holocaust. At the other end of the ideological spectrum, communist regimes generally assumed that human attitudes about property and welfare are all learned and completely malleable. When all reeducation attempts failed, they replaced education with terror, only to fail again.

Even scientists studying animal behavior have in the past belonged to "schools" emphasizing either inborn or learned activities. One group, represented by Tinbergen, von Frisch, and Lorenz in Europe, called themselves ethologists. In their view, animals are born with fixed-action patterns, or **instincts**, that are released by appropriate stimuli. For example, the red dot on a seagull's lower beak will invariably prompt nestlings of the same species to open their mouths, and this signal in turn will prompt adults to feed the young. To the extent that behavior was shown to rely on instincts, it was viewed as genetically "hardwired." A different school of scientists studying animal behavior, founded by A. B. Skinner in the United States, called themselves behaviorists. Teaching animals arbitrary actions, such as pressing levers in the laboratory, they came to view animal behavior as the result of **learning**. Learning was also seen as the basis of all human behaviors, including gender roles and sexual orientation.

Since then, many studies have shown that animal as well as human behaviors have both inborn and learned components. These components can be simply additive or acting together in more intricate ways. So it is nature *and* nurture, not nature *versus* nurture.

B. Neuronal Connections Change During Learning

The importance of both genetics and learning in behavior can be demonstrated in whole animals as well as in specific neurons. The enhanced learning ability of mice transgenic for an NMDA receptor gene, as discussed in the assigned reading of Chapter 11, is a powerful demonstration of genetic control. The same study, in the multiple tests given to transgenic and normal mice, was taking advantage of the ability of mice to learn.

The cellular basis of learning and memory has been studied extensively the sea slug *Aplysia californica*. Because this species has large neurons, with ganglia (clusters of neurons) spread out over different parts of the body, it has been a favorite object of neurobiologists. On their dorsal side, the animals have a tube for aspirating water, the **siphon**, and **retractable gills** (Figure 13.1). Gently touching the skin of *Aplysia*'s siphon prompts the animal to retract its gills.

If this stimulus is repeated several times, the protective reflex of gill retraction becomes progressively weaker as long as the touch of the siphon is not associated with any negative experience. This simple form of short-term memory is known as **habituation**.

The gill withdrawal reflex is mediated by simple neuronal circuits, each of which consists of a *sensory neuron*, an *interneuron*, a *motor neuron*, and the *gill muscle* (Figure 13.2). Under habituation conditions, the sensory neuron releases fewer and fewer synaptic vesicles containing *glutamate* as a neurotransmitter. After multiple training sessions, the number of presynaptic terminals of the sensory neuron onto the motor neuron and the interneuron also decreases.

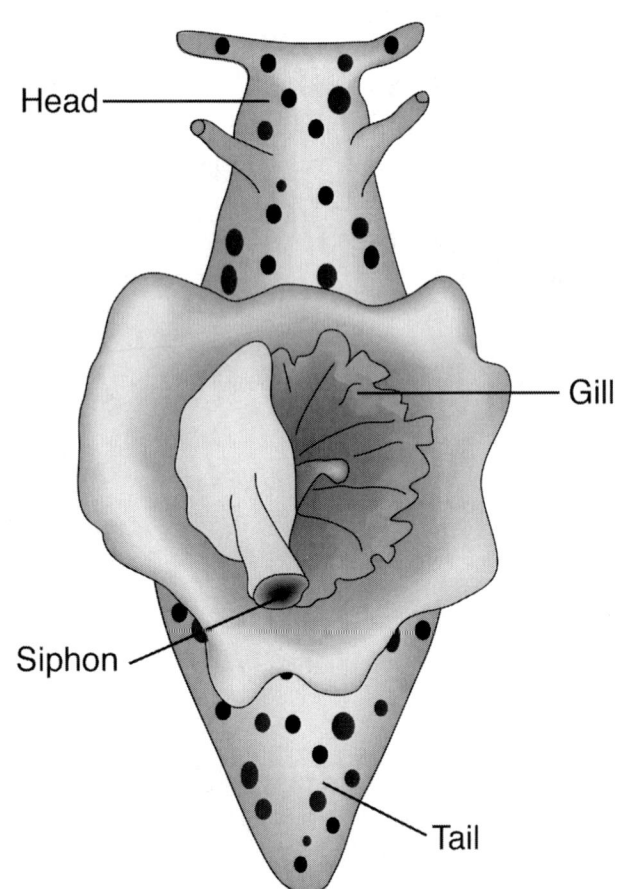

FIGURE 13.1:
The sea slug *Aplysia californica* shown in dorsal view.

A form of learning that reverses habituation is called **sensitization**. If a weak electric shock is delivered to the tail of a habituated *Aplysia* at the same time when the syphon is touched, the gill withdrawal reflex intensifies again. While a few tail shocks sensitize the gill withdrawal reflex only for a short time, repeated shocks cause prolonged sensitization.

The shock-causing sensitization activates a sensory neuron located in the tail (see Figure 13.2). This neuron in turn activates another interneurons, called a *facilitating interneuron,* which releases serotonin onto the terminals of the sensory neurons of the siphon. The serotonin receptors in these terminals lie at the head of the signal transduction chain that generates cyclic adenosine monophosphate (cAMP) from ATP (Figure 13.3).

The cAMP in the siphon's sensory neuron terminal has two effects. First, it phosphorylates potassium (K^+) channels, thereby blocking the release of K^+ ions. This makes action potentials arriving in terminal ends of the sensory neuron last longer. The extended action potentials in turn keep voltage-gated Ca^{2+} channels open longer, with the result of more glutamate being discharged as a neurotransmitter. This kind of sensitization, based on channel modification, lasts for several hours.

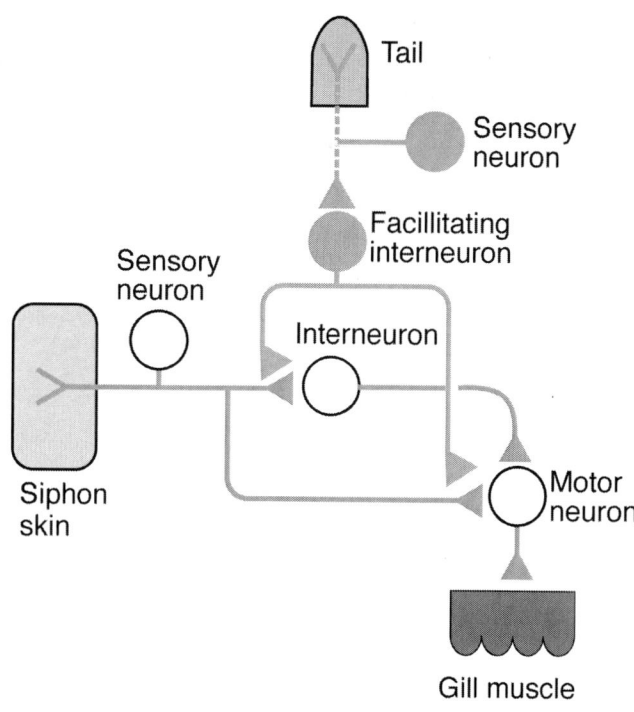

FIGURE 13.2:

Gill withdrawal reflex in *Aplysia*: Habituation (white neurons) and sensitization (gray neurons). The area in the center is detailed in Figure 13.3)

The second effect of cAMP is longer-lasting, acting through the phosphorylation of a transcription factor and consequent gene regulation. One of the morphological results is the formation of additional synaptic terminals between the sensory neuron of the syphon and its target interneuron and motor neuron. This kind of sensitization, based on the formation of additional synapses, lasts for many days or weeks.

In summary, simple neural networks show the hallmarks of both genetic control and learning. They are genetically controlled in the same way that all assemblies of proteins are. At the same time, the key molecules and synaptic connections can change as a result of previous experience.

C. Animals Are Genetically Prepared to Learn

Most animal traits are controlled by genetic and environmental parameters. In simple cases, their effects are just additive. Other traits, in particular behaviors, can be intricate combinations of genetically fixed (instinctual) and learned processes.

Ducks and geese nest on the ground, where their young are vulnerable to predation. A behavior that mitigates this vulnerability is known as (behavioral) **imprinting**, not to be confused with genomic imprinting (see Chapter 10). Young ducks and geese will follow around whatever large moving object they have encountered *during a short period after hatching.* Under natural conditions, this object is almost always a parent, which will defend them against any predator. Under experimental conditions, hatchlings

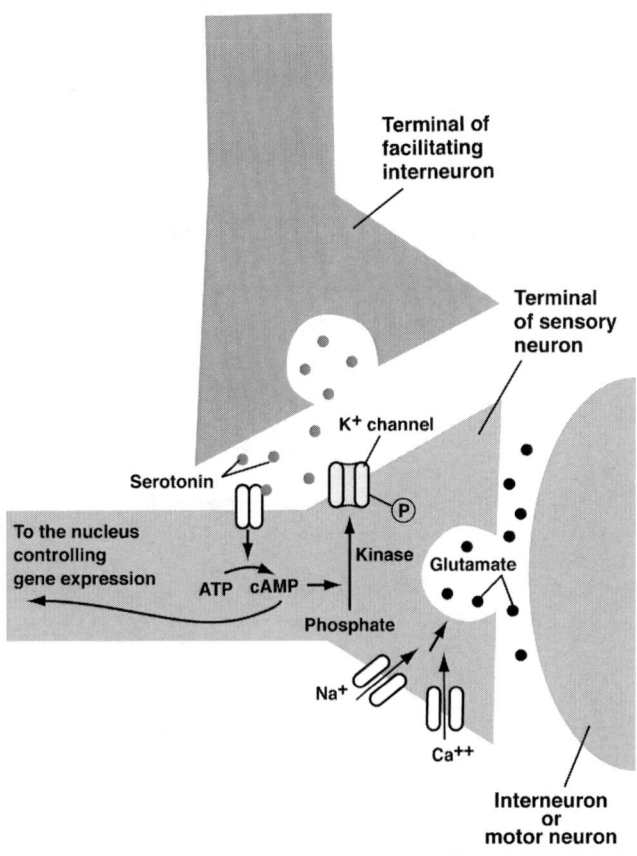

FIGURE 13.3:
Molecular mechanism of sensitization. This diagram shows the central area of Figure 13.2 in more detail. Serotonin released be the facilitating interneuron binds to a receptor on the terminal of the sensory neuron of the siphon. The receptor triggers a signal chain that leads to the formation of cAMP, which activates a kinase that blocks channels releasing K+ ions. As a result, arriving action potentials last longer and Ca++ channels stay open longer. With more Ca++ admitted, more glutamate is released from the sensory nerve terminal

may imprint on a human scientist, as happened when Konrad Lorenz became mother goose (Figure 13.4). In behavioral imprinting, the following around behavior is instinctual, but the identity of the leader is learned.

A similar combination of inborn and learned behaviors is observed in blackbirds, which chase away predators by **mobbing** them and sounding alarm calls. In an experiment, a caged blackbird was shown a stuffed owl (Figure S13.a). As expected, the blackbird mobbed the owl. In a neighboring cage, another blackbird could not see the owl and was instead shown a stuffed honeycreeper, a harmless species. Initially, the second black bird was undisturbed. But when he heard the persistent ruckus that the first blackbird made over the stuffed owl, the second blackbird began to mob and scream at the harmless honeycreeper. Again, the mobbing behavior of blackbirds turned out to be instinctual, while the predator species to be mobbed are learned.

There is a growing awareness that certain genes linking nature and nurture have profound effects on reproductive and social behavior, and are therefore likely to play key roles in evolution. Examples include the genes for two similar neurohormones, vasopressin and oxytocin, as well as their cognate receptors. Both neurohormones are produced in the hypothalamus and released from the posterior pituitaty gland (see Figure 25.1). **Vasopressin** is named after its physiological effects on water retention and blood pressure. **Oxytocin** is named for its hormonal effect of inducing uterine contractions during childbirth. Both neurohormones also have behavioral effects by acting on specific brain region where their receptors are concentrated.

For example, the gene encoding the **vasopressin V1a receptor (V1aR)** plays a prominent part in the social behavior of voles. Male mammals have a surge of vasopressin in the brain after mating. The effects of this surge on the male's reward system and behavior depend on the amount of V1aR in the **ventral pallidum** of their brain (Figure 13.5). This became apparent from experiments with the social and monogamous prairie vole (*Microtus ochrogaster*) and the solitary and polygamous meadow vole (*M. pennsylvanicus*). See assigned reading at the end of this chapter.

The 5' regulatory region of the V1aR gene contains a microsatellite, which is longer in prairie voles than in meadow voles. The length of this microsatellite also varies among prairie voles and is positively correlated with V1aR gene expression. Long-allele males are more inclined to bond with a female partner and to groom their young. Because microsatellites easily change in length during DNA replication, their occurrence in gene-regulatory regions may lead to rapid evolution in gene expression.

FIGURE 13.4:
Behavioral imprinting. Young geese imprinted on ethologist Konrad Lorenz, who happened to be nearby when the geese were hatching. Since then, they followed him around as they normally would do with a parent.
Thomas D. McAvoy/Time & Life Pictures/Getty Images

The V1aR homolog in humans also seems to affect the pair-bonding behavior of men and the marital quality perceived by their wives/long-term partners. Men carrying a particular allele ("334 allele") of the RS3 repeat in the 5' regulatory region score lower on a partner bonding scale, are less likely to be married, and if they are, have lower marital quality reported by their partners.

Oxytocin receptors have a similar but not identical distribution to vasopressin receptors in the brain, and in humans, a surge of oxytocin is correlated with sexual orgasm. In women, oxytocin release also promotes breastfeeding and is stimulated by suckling on her nipples. Several lines of evidence indicate that oxytocin reduces anxiety and stress and promotes pair bonding as well as social interaction.

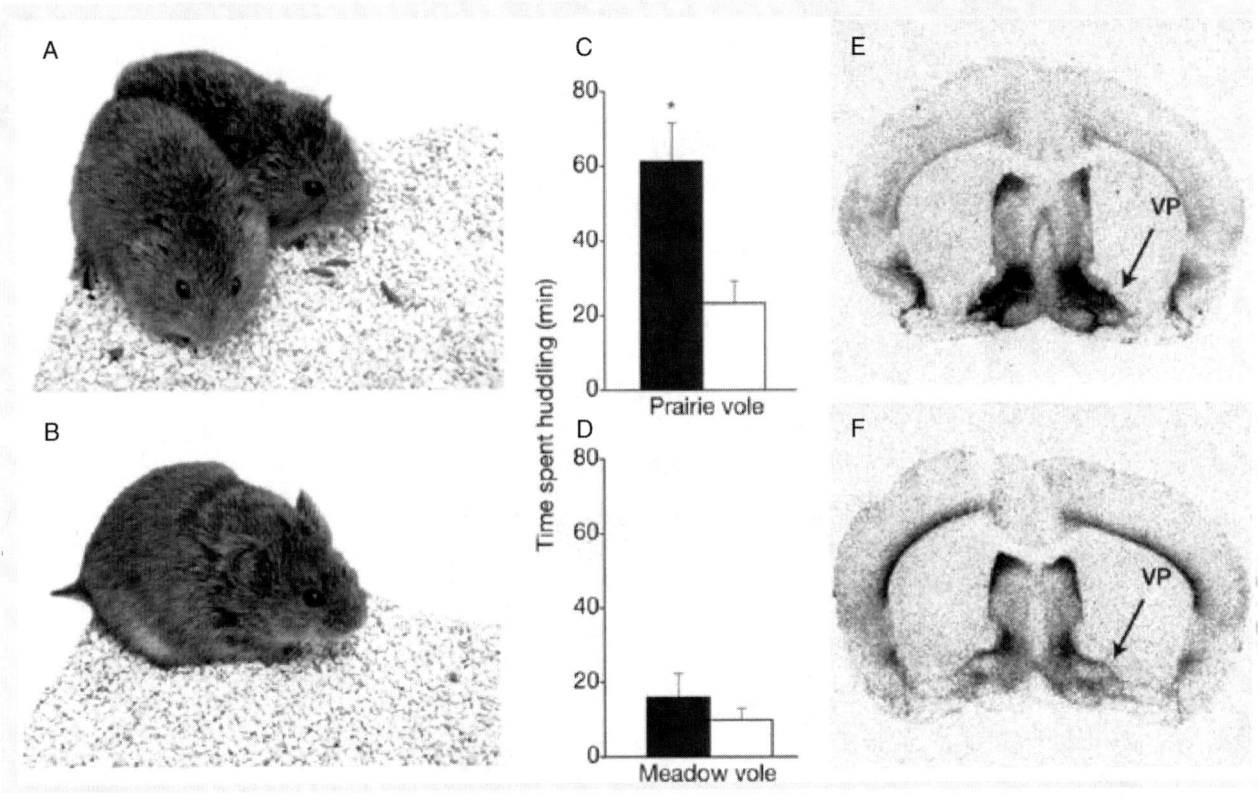

FIGURE 13.5:
Single Gene Driving Behavior and Evolution

A, B: Prairie voles (A) are social and monogamous, liking to 'huddle' with their mate, whereas meadow voles (B) are solitary and polygamous.

C, D: Partner preference test. After mating with a female, a male prairie vole (C) tended to spend significantly more time in contact with his mate (filled columns) than with a new female (open columns) ($P < 0.05$), whereas meadow voles (D) did not form partner preferences and spent little time huddling with either female.

E, F: Autoradiograms of the forebrain showing expression of the gene for the vasopressin 1a receptor *(V1aR)* in the ventral pallidum (VP). *V1aR* expression is stronger in prairie vole VP.

Reprinted by permission from Macmillan Publishers Ltd: *Nature* Volume 429, Copyright © 2004.

D. Some Behaviors Depend on Epigenetic Changes that Are Both Heritable and Reversible.

Recently, there has been a new focus on behaviors that depend on **epigenetic changes in chromatin**, which are passed on during cell division but do not depend on changes in DNA nucleotide sequence. Instead, epigenetic phenomena rely on changes in chromatin structure that make chromosomal DNA more or less accessible to the key proteins required for gene expression, RNA polymerase and transcription factors. There are two known molecular mechanisms underlying epigenetic changes: DNA methylation and histone acetylation.

DNA methylation is the addition of a methyl (CH_3) group to the DNA's cytosine base. This step is catalyzed by several enzymes including DNA (cytosine-5) methyltransferase 1 (Dnmt1). Recognizing half-methylated CpG sequences after DNA replication, Dnmt1 methylates the newly synthesized CpG sequence, in effect propagating the preexisting methylation pattern. Dnmt1 is critical for normal development in mice: *Knock-out mice* (see Chapter 11) lacking a functional *Dnmt1* allele die during gestation. The effect of DNA methylation on transcription is inhibitory. Several studies have shown that the promoter region of genes are methylated when they are inactive and demethylated when they are active.

Histone acetylation occurs by adding an acetyl group ($COCH_3$) to specific lysine residues of chromosomal histone proteins. Again there are enzyme activities, histone acetyltransferase (HAT) and histone deacetylase (HD), catalyzing this step and its reversal. Acetylation removes the positive charge of lysine, thus reducing the affinity between histone and DNA and making the latter more accessible. In effect, then, histone acetylation enhances gene expression.

A particularly intriguing example of an epigenetic effect on behavior has been observed in rats, where females show a heritable variation in **maternal care** (pup licking, grooming, and arched-back nursing). In contrast to pups raised by low-care mothers, pups raised by high-care mothers develop into adults that are more exploratory, less fearful, and showing a healthier stress response characterized by low glucocorticoid levels. Remarkably, the high/low-care mothering styles are passed on to biological as well as to adoptive daughters.

High-care mothering triggers a cascade of molecular events in the hippocampus of pups. They result in greater expression of transcription factor NGFI-A, which activates the gene for the glucocorticoid receptor (GR). Presence of more GR in the hippocampus arguably translates into the neuroendocrine and behavioral characteristics seen in offspring of high-care mothers (Figure 13.6). Recent work has linked the long-term effects and heritability of high-care mothering to the two epigenetic chromatin changes mentioned earlier: *DNA demethylation* and *histone acetylation*.

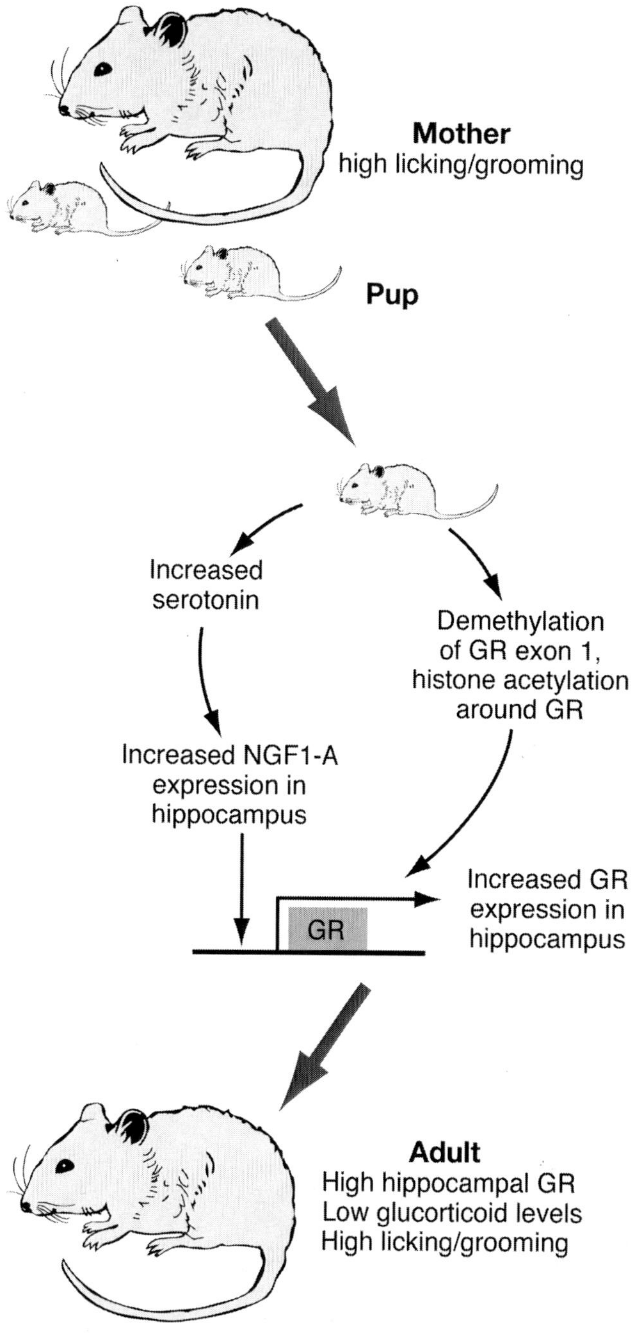

The binding site of the GR gene for the NGFI-A transcriptional activator was found to be demethylated and associated with acetylated histones in rats raised by high-care mothers, in contrast to rats raised by low-care mothers. These differences emerged over the first week of life, were reversed with cross-fostering, and persisted into adulthood. Most importantly, a drug that inhibits histone deacetylase (thereby promoting histone acetylation) changed the behavior of rats raised by low-care mothers towards the behavior normally shown by the offspring of high-care mothers.

FIGURE 13.6:

Epigenetic transmission of mothering style. High-care mothering in rats is inherited epigenetically, by DNA demethylation of the glucocorticoid receptor (GR) gene and acetylation of surrounding histones.
Redrawn from Sapolsky R.M (2004) Mothering style and methylation. *Nature Neurosci.* **7:** 791-792.

This case indicates that environmental effects such as mothering style can cause epigenetic changes that have profound effects on gene expression and behavior. Such epigenetic changes are heritable but reversible.

E. Birds and Apes Have (Some) Culture

Some birds and primates learn by observing and imitating other individuals within the same population. This nongenetic propagation of habits among fellow group members is the minimal definition of **culture** used by anthropologists.

Among songbirds, a rudimentary stage of singing is inborn, but young birds perfect their songs in the company of older birds. In many bird species, the adult songs differ from one population to another.

In a troop of macaques on the Japanese island of Koshima, a two-year-old female began around 1950 to **wash potatoes** in water before eating them. Within a decade, all members of the troop below middle age had adopted the same technique. None of the original troop members are still alive, but the troop is still washing potatoes (Figure 13.7).

Some clear cultural differences were observed in field studies of seven African chimpanzee populations. Each of the sites had a unique culture, or cluster of habitual behaviors. Thirty-nine behavioral patterns were found to be habitual in some communities but absent in others. These patterns occur primarily in grooming, courtship, and tool usage. The behaviors were passed on by imitation and other social interactions.

For example, at Gombe/Tanzania, chimps use 60 cm long sticks to **fish for ants**. They wait until the insects have come halfway up the stick and then withdraw the tool to sweep it off with their free hand, thus gathering a mouthful of ants at a time. At Tai/Ivory Coast, the chimps use sticks half as long, wait only a few seconds, and then use their lips to sweep the ants directly into their mouth. The Tai method nets only one-fourth as many ants per minute, but in 20 years of observation, no chimp at Tai has ever been observed eating ants Gombe-style.

In the Goualougo Triangle/Congo, chimps deliberately fray the tips of their fishing rods with their teeth. The brush-tipped rods attract many more termites, and the chimps learn the new technique from their peers. Other locally diverse behaviors include hammering to open nuts, fanning to keep flies away, and two partners clasping hands over their heads, while grooming each other using the opposite hand.

In summary, chimps have a remarkable ability to invent new customs and technologies, and they pass these on socially rather than genetically. In chimps, then, behavior is not only a composite of inborn and individually learned elements, but also influenced by tradition, or culture, within the population.

FIGURE 13.7:
Culture in non-human primates. Macaques on the Japanese island of Koshima wash potatoes before eating them, a cultural trait started by a female around 1950.
Photograph by Frans de Waal.

In humans, cultural components of behavior are clearly much more prevalent than in chimps or any other animals. Due to superior language and brain capabilities, the process of learning from a fellow group member is much faster and more effective in humans than in chimps.

In the presence of a strong culture in a population, selfish mutants and immigrants may come under stronger social pressure, and *group selection* may become an effective vehicle of evolution (see Chapter 21).

F. Traits May Vary Discontinuously or Continuously

Having convinced ourselves that behaviors are shaped by nature and nurture, can we go further and arrive at some *quantitative measure* telling us how much a *specific trait* is controlled by genetic factors? Such knowledge could be useful for educational and medical purposes. For example, if it were known that obesity is mostly genetic, then one could identify the satiety factors or other signals that are not made properly, and develop drugs that can substitute for them. Conversely, if it were clear that obesity is mostly environmental, then one could focus treatment on diet and exercise, beginning in schools. Quantifying the extent of genetic control, unfortunately, is somewhat technical and often misunderstood. Because related errors play into the hands of chauvinists as well as blind reformers, it is necessary to understand the most commonly applied methods.

The procedures used to estimate the heritability of traits depend on whether the trait varies discontinuously or continuously.

Some traits (of humans as well as other organisms) show **discontinuous, or qualitative, variation**: An individual either shows the trait or does not. Examples include many diseases, such as albinism (Figure S13.b), cystic fibrosis, and sickle-cell anemia. Discontinuously variable traits are often controlled by allelism in a single gene. Such traits are called *simple* or *Mendelian* traits.

Some traits vary discontinuously but are still *polygenic* (depend on allelism of several genes); such traits are called *complex* or *non-Mendelian* traits. Some of these traits are caused by the simultaneous failure of two or more gene functions that are backing up one another. For example, most cancers appear only after several mutations have occurred in the same cell. Other discontinuously variable but non-Mendelian traits are *threshold effects*: The trait may appear whenever an underlying quantitative variable, influenced by multiple genes, exceeds a threshold value.

For up-to-date descriptions of human genetic disorders causing discontinuously variable traits, see OMIM (Online Mendelian Inheritance in Man) at http://www.ncbi.nlm.nih.gov/entrez/query.fcgi?db= OMIM

A different class of traits show **continuous, or quantitative variation**: An individual may express the trait anywhere within a wide range. Examples include height (Figure 13.8), body mass index, skin pigmentation, and intelligence quotient.

Continuously variable traits often show a **normal distribution**, a.k.a. Gaussian distribution, or bell

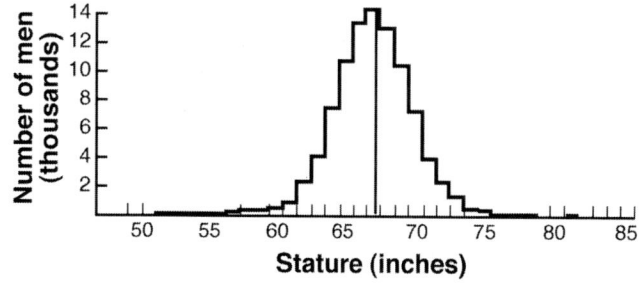

FIGURE 13.8:

Stature (height) as a continuously variable trait. The data were recorded in increments of one inch for Englishman called up for military service 1939. The mean value (vertical line) was 67.5 inches.

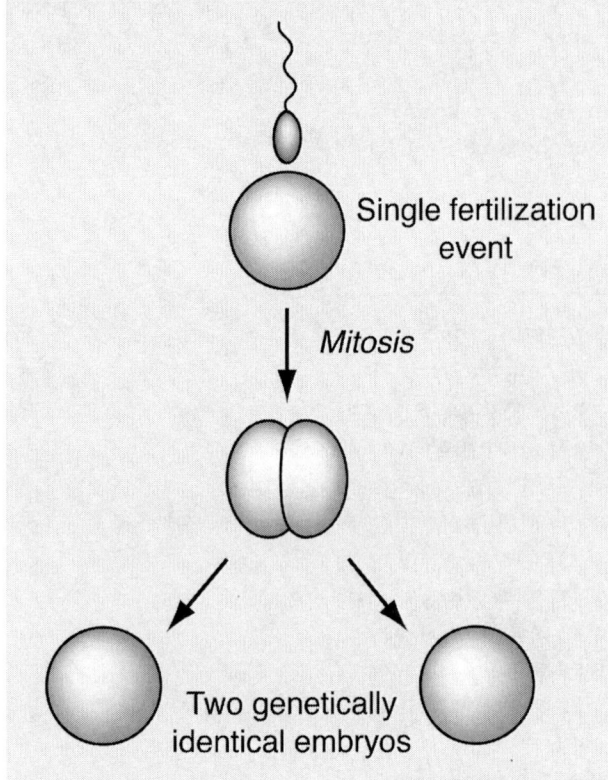

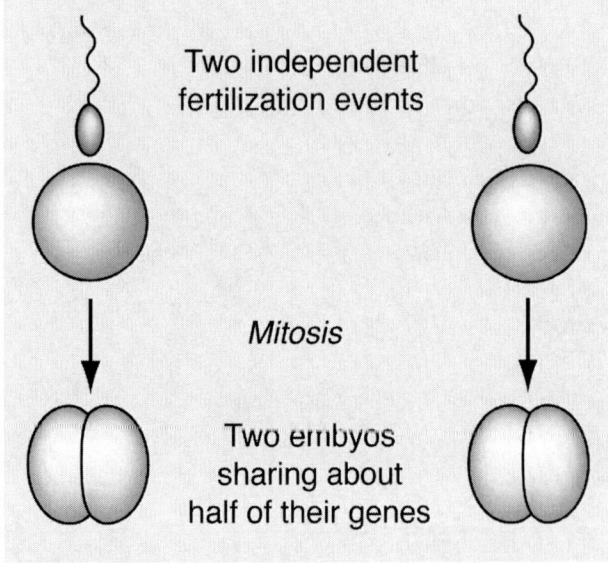

FIGURE 13.9:
Monozygotic and dizygotic twins. (a) Monozygotic ("identical") twins develop from one fertilized egg, or *zygote*, which subsequently splits into two embryos. **(b)** Dizygotic ("fraternal") twins result from two eggs fertilized independently during the same ovulatory cycle.

curve (see Chapter 14). Such a distribution is obtained when a trait is determined by many (genetic and/or environmental) factors, each of which makes a *small, additive* contribution. A good analogy for a continuously variable trait with a normal distribution is a multiple coin toss: You toss, say, 20 coins simultaneously and then count how many of them came to lie heads up. If you do this a thousand times and plot how often you got 0, 1, 2, 3, ... , 20 heads up, you get a bell curve.

The methods used to estimate the heritability of discontinuously and continuously variable traits are fundamentally different.

G. Heritability of a Discontinuously Variable Trait Can Be Estimated by Its Concordance Ratio

The key concept in quantifying the heritability of discontinuously variable traits is **concordance**. Any two individuals who both show the same version of a discontinuously variable trait are called concordant for the trait. For instance, a parent and a child who both suffer from epileptic seizures are *concordant* for epilepsy. If one has epilepsy and the other has not, they are called *discordant*.

The **concordance rate** is defined for a sample of analyzed pairs of individuals of which *at least one* individual has a (relatively rare) trait. The fraction of concordant pairs within a sample of analyzed pairs is called the concordance rate. Thus, if 250 pairs of siblings with at least one epileptic individual were analyzed, and if 31 of these pairs were concordant for epilepsy, then the concordance rate would be 31:250=0.124. In other words, for a pair in which one partner shows a qualitative trait, the concordance rate is the probability that the other partner shows the same trait.

A high concordance rate between relatives may suggest a genetic predisposition for the trait of interest but does not prove the point. The trait might be caused by an infectious agent or dietary factor, which would be shared between members of the same household. A method that avoids this problem is to compare concordance rates for the same trait in *monozygotic versus dizygotic twins*, because any confounding environmental factors should act similarly on both kinds of twins.

Monozygotic twins, a.k.a. "identical twins," result from one fertilized egg, or *zygote*, and therefore have identical sets of genes (Figure 13.9; Table 13.1). The term "monozygotic (MZ)" is preferred because these twins have different fingerprints, and some DNA mutations may occur differently in them after birth.

Dizygotic (DZ) twins, a.k.a. "fraternal twins," result from two eggs fertilized independently during the same reproductive cycle. They share as much genetic information as normal siblings, approximately 50 percent but possibly more depending on parent homozygocity and assortative mating (see Chapter 14). There is a 1:2:1 ratio of boy-boy:boy-girl:girl-girl pairs of DZ twins, while MZ twins are always of the same sex.

The diagnosis of same-sex twins as DZ or MZ has relied traditionally on blood proteins but is now shifting to the use of DNA markers. The frequency of MZ twin births is between 3.5 and 5 per 1,000 live births; the incidence of DZ twin births ranges from about 2 to 50 per 1,000, depending on the population. The recent use of hormonal contraceptives and fertility drugs has increased the frequency of DZ twins and multiple births. More twins are probably conceived and spontaneously aborted. MZ twins are abbreviated as MZT or MZA twins if one wants to specify whether they were reared together or apart. DZT and DZA twins are designated accordingly.

To quantify the heritability of a discontinuously variable trait, one determines the concordance rate for that trait in samples of MZT twins and DZT twins. The concordance rate for MZT twins divided by the concordance rate for DZT twins is called the **concordance ratio**.

If the concordance ratio is near 1.0, then the greater genetic similarity between MZT twins as compared to DZT twins has no significant effect on the concordance rate, and it can be concluded that the genetic control of the trait is small or nonexisting. However, if the concordance ratio is significantly greater than 1.0, then it can be concluded that there is a heritable disposition for the trait.

The greater the concordance ratio, the stronger is the relative effect of genetic factors on the trait (Table 13.2). For example, the concordance ratio for measles, a highly infectious disease, is near 1.0, indicating any genetic predisposition, if it exists, is drowned out by contagiousness. On the other end of the spectrum, the concordance ratios for Down syndrome and cleft lip are very high, indicating a substantial genetic predisposition.

The concordance ratio gives only an overall measure of the genetic control of a trait; it does not reveal the number or identity of the genes are involved. Concordance ratios may also differ from one human population to another.

Table 13.1: Monozygotoc and Dizygotic Twins		
	Monozygotic Twins	**Dizygotic Twins**
Definition	derived from one fertilized egg, or zygote	derived from two independently fertilized eggs
Synonym	identical twins	fraternal twins
Shared Genetic Alleles	100%	approximately 50%
Boy/Girl Pairs	none	50%
Frequency	3.5-5.0 per 1,000 live births	2–50 per 1,000 live births

Table 13.2 Concordance Ratios for Discontinuously Variable Human Traits				
Trait	Mz Conc. Rate (%)	DZ Conc. Rate (%)	Concordance Ratio	Source
Epilepsy	72	15	4.8	Cummings
	59	16	3.6	Sutton
Tuberculosis	56	22	2.6	C
	59	19	3.1	S
Measles			1.0	KK notes
Down syndrome	89	7	12.7	C
Diabetes	65	18	3.6	C
Cleft lip	42	5	8.4	C
Rheumatic fever	25	6	4.5	S
Bronchial asthma	88	31	2.83	S
Peptic ulcer	26	13	2.0	S
Eye color	99	28	4.5	C
Handedness (left or right)	79	77	1.0	C
Male homosexuality	52	22	2.4	Bailey and Pillard
Female homosexuality	50	25	2.0	Bailey and Benishay

EXERCISE PAGE FOR CHAPTER 13

Student Last Name _____

Student First Name _____

Discussion Section _____

On the course web site, use the link on the **syllabus** to find a pdf file of your **assigned reading** for this chapter. The same reading is referenced below. Use the bulleted list of questions to test your knowledge of this reading.

Lim, M. M., et al. (2004). Enhanced partner preference in a promiscuous species by manipulating the expression of a single gene. *Nature* **429:** 754–757.

- What differences did the authors observe between the prairie vole and the meadow vole with regard to the expression pattern of V1aR, and with regard to the dopamine 2 receptor (D2R)?
- What are the correlated differences in behavior?
- What is the basic hypothesis proposed by the authors, and how did they go about testing it?
- What were the three groups of male meadow voles they used in their study, and how did these groups originate?
- How was the "partner preference test" set up, and how did males from the three groups behave in it?
- How did male behavior in the partner preference test change after the animals were pretreated with eticlopride, a *dopamine* receptor antagonist? How did the authors interpret this result?
- What do the authors propose about the "convergence" of an individual recognition pathway and a generic reward pathway in pair bond formation? Which of the two would involve V1aR, and how?

Feedback

On a scale from 1 to 10, the being the best, please rank the above reading for

1. Interest and Relevance to the topic (10 being most interesting/relevant) _____
2. Readability (10 being most clearly written, easy to understand) _____

In the remaining space, enter any comments that you may have on this reading.

CHAPTER FOURTEEN

ESTIMATING THE HERITABILITY OF CONTINUOUSLY VARIABLE TRAITS

We will continue our discussion on the heritability of human traits, that is, the extent to which they are determined by genetic factors. Having dealt with discontinuously variable traits, such as blood groups or diseases, in the previous chapter, we will now focus on continuously variable traits, such as height or IQ.

A. Bell Curves Reveal Genetic and Environmental Control of Continuously Variable Traits

For a continuously variable (or quantitative) trait, a **frequency distribution** shows how frequently each interval of the trait is observed in a population (see Figure 13.8). The frequency distribution is bell-shaped if the trait is controlled by many (genetic and/or environmental) factors, each of which makes a small additive contribution. (Remember the analogy of the multiple coin toss used in Chapter 13.)

A bell curve, or **normal distribution**, is characterized by two parameters, the *mean* and the *standard deviation* (Figure 14.1). The **mean** is the average value of the trait measured. The mean occurs most frequently, and the bell curve is centered over the mean. The **standard deviation (s)** indicates how wide the bell curve is. If one moves one standard deviation to the right and to the left from the mean, then the central part of the curve will

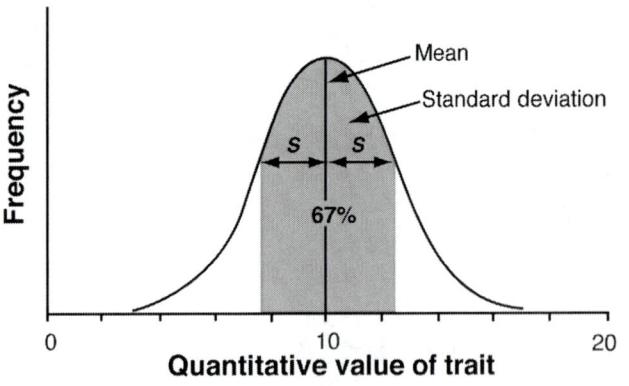

FIGURE 14.1:
Normal Distribution of a Continuously Variable Trait. A normal distribution has the shape of a bell curve, centered over the mean value of the trait. The standard deviation (s) indicates the width of the curve.

include approximately 67 percent of all measurements in the sample. A better definition of s is given on the following page.

The mean and standard deviation of a bell curve change if we manipulate the genetic and environmental factors that contribute a trait's overall variability. For example, assume we measure the body weight of mice sampled from an *outbred* (genetically heterogeneous) population that has been divided into subpopulations kept under various environmental conditions. Assume also that each subpopulation had time to adjust physiologically but not enough time for breeding.

A sample of mice from a subpopulation kept in a heterogeneous environment will show a broad bell curve (Figure 14.2). If instead we would sample two other subpopulations of mice, which are kept under controlled rich or standardized poor environmental conditions, their body weight would show bell curves with higher or lower means. Also, the standard deviations, which now represent only genetic variation, would be smaller than the original standard deviation, which represented both genetic and environmental variation.

Assume the controlled environmental conditions had reduced the standard deviation from 11 grams to 9 grams, then we could ascribe the residual standard deviation to genetic factors, and the ratio of 9:11 would be some indicator of the extent to which the body weight of mice in this population is controlled by genetic factors.

However, there is an important caveat. The fraction of the standard deviation that persisted under controlled environmental conditions is not a property of mouse body weight *per se* but of mouse body weight *in the population we have analyzed.*

To realize the importance of this difference, imagine the results we would have obtained by measuring the body weight of mice from an *inbred* (genetically homogeneous) laboratory strain (Figure 14.3). In this case, the resulting bell curve would still have shown an appreciable standard deviation as long as we had kept the mice in a wide range of environmental conditions. However, under controlled environmental conditions, the standard deviation would have shrunken to nearly zero, a result that could have misled us to assume that the fraction of genetic factors controlling mouse body weight is negligible.

Thus, studying a genetically homogeneous population leads to an underestimate of the relative importance of genetic factors. Likewise, studying a population that lives in a narrow range of environmental conditions leads to an overestimate of genetic factors.

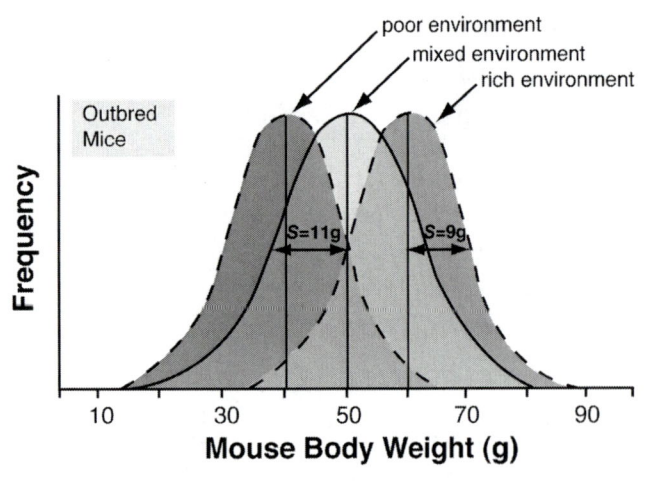

FIGURE 14.2:
Body weight distrubutions of outbred mice kept in different envirionments. Note that the standard deviation is smaller if the environment is standardized rather than heterogeneous.

B. Heritability of Continuously Variable Traits

The extent to which a continuously variable trait is controlled by genetic factors *in a specific population* is described by a numerical value called **heritability**.

The key concept for understanding the heritability of continuously variable traits is **variance**, which is defined as follows.

$$V = \sum_{i=1}^{N} (y_i - m)^2 / N-1 = s^2$$

with:

V = variance
N = sample size
y_i = trait measurements on sample, $i = 1$ through N
m = mean of trait measurements
s = standard deviation

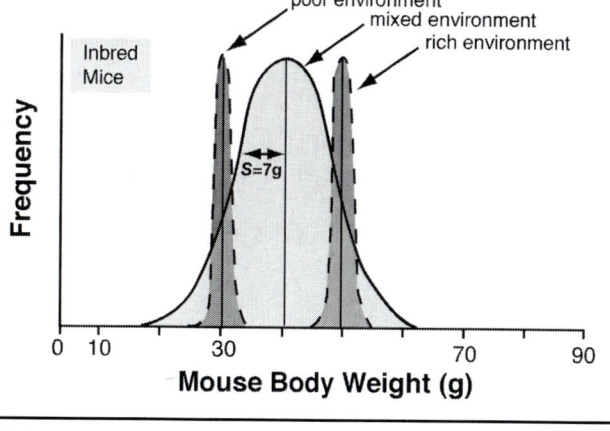

FIGURE 14.3:
Body weight distributions of *inbred, i.e. genetically homogeneous,* mice kept in different environments. Standardized environments reduce the standard deviation (s) to near zero, giving the false impression that body weight may not be affected by genetic factors. However, the results obtained with outbred mice (Figure 14.2) show otherwise.

An advantage of working with variance instead of standard deviation is that the variance can be broken down into additive components.

$$Vt = Vg + Ve + Veg + Vm$$

with:

Vt = total variance
Vg = variance caused by genetic diversity
Ve = variance caused by environmental heterogeneity
Veg = variance caused by environmental/genetic interactions
Vm = variance caused by measurement error.

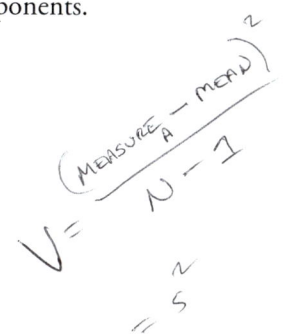

The **heritability** of a continuously variable trait in a population is defined as the genetically caused variance divided by the total variance:

$$H = Vg / Vt$$

Again, it is important to note that any estimate of the heritability of a trait will depend on the measurements made on *a given population*. Results obtained under special circumstances may be misleading. For instance, a trait may be strongly under *genetic* control. Yet, if the sampled population happens to be inbred, then the result will be an underestimate of Vg and hence of H.

Conversely, a trait may be strongly affected by an unknown *environmental* parameter. For example, fluorescent light tends to make toddlers crabby. If investigators who are unaware of this fact would measure the crabbiness of toddlers, and all data were collected in fluorescent light, or all in incandescent light, then the result would be an underestimate of Vt and hence an overestimate of H.

The above equation for heritability can be used directly for outbred populations of nonhuman organisms that can be sampled under variable as well as under constant environmental conditions. In the latter case, Vt can be computed by measuring a sample of individuals, V_m can be determined by repeated measurements of the same individual, and $V_g = V_t - V_m$ because V_e and V_{eg} are zero.

For humans, the definition of H is still the same, but V_g cannot be measured as described above. Instead, H must be estimated based on twin and/or adoption studies. We will focus on twin studies here and include some data from adoption studies in the following chapter.

C. Monozygotic Twins Reared Apart (MZA Twins) Are Remarkably Similar

Monozygotic (MZ) twins carry identical sets of genetic information (if we ignore mutations occurring after fertilization). Of particular interest are MZ twins, who were reared apart (MZA twins) because their parents divorced, gave them up for adoption, and so forth. In contrast to MZT twins, the shared environment of MZA twins is limited to their intrauterine time and a short postnatal period. Many MZA twins grow up without knowing of each other, or not sufficiently motivated to find each other. In the Minnesota Study of Twins Reared Apart, more than 100 MZA pairs have been reunited and given extensive medical and psychological tests (Figure S14.a). In this group, which may not be a perfectly random sample of all MZA pairs, the overwhelming impression was how similar their life histories had played out despite their different postnatal environments.

One of the first MZA pairs to be studied were Jim Lewis and Jim Springer, who had been adopted by separate families in Ohio. When they met in 1979, after 39 years of being apart, their physical appearance, hairstyle, moustache, and body language were astoundingly similar (Figure S14.b). Amid the euphoria over their reunion, astonishing similarities in their lives and behavior became apparent. Both had been named James by their adoptive parents, both had been married twice; first to women named Linda and second to women named Betty. Both had children, including sons named James Allan. Both had at one time owned dogs named Toy. Both chain-smoked, both liked beer, both had woodworking workshops in their garages. Both drove Chevys, both had served as sheriff's deputies in nearby Ohio counties. They had even vacationed on the same beach in the Florida Gulf Coast. Both lived in the only house on their block. Their scores in personality test were often as close as for the same person taking the test twice. Intelligence tests, mental abilities, gestures, voice tones, likes and dislikes were similar as well. So were medical histories: Both had high blood pressure, both had experienced what they thought were heart attacks, both had undergone vasectomies, and both suffered from migraine headaches. They even used the same words to describe these headaches.

Another MZA pair, Gerald Levey and Mark Newman, were separated on day 5 after birth. Although they grew up in different cities, their life histories remained almost parallel. Both became volunteer firefighters, but they did not know each other until they met 31 years later (Figure 14.4). Both men were balding, 6' 6" tall, weighed 250 pounds, sported droopy mustaches, and wore aviator-style glasses and a key ring on the right side of their belt. Physiological and medical tests revealed more similarities. Tests for personality traits, intelligence, and other cognitive abilities also gave similar results. Both men were bachelors attracted to the same type of women, favored the same pastimes, and showed the same mannerisms.

Another pair, Jack Jufe and Oskar Stoehr, had been separated a few months after birth and grown up in very different environments. Jack stayed in Trinidad with their father, a Jewish merchant. Oskar grew up in the all-female household of their German grandmother and attended a Nazi-run school. When the two were reunited as adults in the United States, they both showed the same swaggering gait, a short mustache, rectangular wire-rimmed glasses, and a blue two-pocket shirt with epaulettes. They did not seem to like each other so much as to romanticize their similarities. There were differences in their political leanings: Jack was a self-employed entrepreneur; Oskar was a union man. Oskar was a skier; Jack was a sailor. Jack was divorced; Oskar was married, but his wife was critical of their marriage. In personality tests, Jack and Oskar revealed similar habits and temperaments. Most striking were their shared idiosyncrasies: Both stored rubber bands on their wrists, both read magazines from back to front, both dipped buttered toast in their coffee, and both liked to startle people by sneezing aloud in elevators.

FIGURE 14.4:

MZA twins Mark Newman and Gerald Levey were separated on day 5 after birth and lived apart for 31 years.
Image courtesy of Pete Byron.

D. Heritability Estimates Based on MZA Twin Data

MZA twins have been used extensively for estimating the heritability of continuously variable traits in humans. The key parameter on which these estimates rely is the **intra-pair correlation coefficient (r).** It is defined as the linear regression of pairwise trait measurements as shown in Figure 14.5 for height of MZA twins. By definition, r ranges from 0 (genes are unimportant) to 1 (genetic alleles are everything). Intra-pair correlation coefficients can be determined for any pairs of individuals, such as mothers and their biological children, men and their adopted children, siblings, MZT twins, and so forth.

For intra-pair correlation coefficients of MZA twins, it can be shown mathematically that

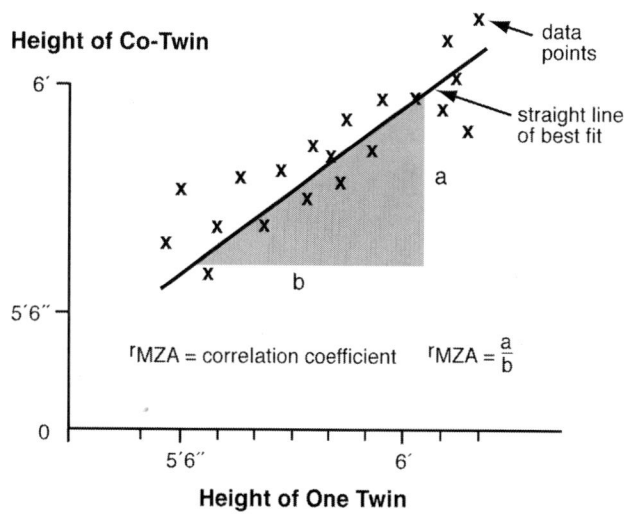

FIGURE 14.5:

Intra-pair correlation coefficient for MZA twins (r_{MZA}). To obtain the r_{MZA} of a trait, such as height, one plots the pairwise measurements as points in a coordinate grid. Here, the x-coordinate represents the height of one twin and the y-coodinate represent the height of his/her co-twin. The incline of a straight line through these points represents r_{MZA}.

$$r_{MZA} = \frac{Vg + Ves}{Vt}$$

with

r_{MZA}	=	intra-pair correlation coefficient for MZA twins
Vt	=	total variance in population
Vg	=	genetic variance of trait in population
Ves	=	variance based on environmental factors shared by the two members of a pair but experienced differently by different pairs

Some authors consider *Ves* to be negligible for MZA twins because of their early separation. Under this assumption, r_{MZA} becomes a measure of heritability

$$r_{MZA} = \frac{Vg}{Vt} = H$$

The assumption that *Ves* is negligible for MZA twins has been criticized on the grounds of

- shared prenatal environment
- shared postnatal environment before separation
- tendency of adoption agencies to choose similar homes for MZA twins

Some studies indicate that the prenatal environment, including maternal diet and drug use, may significantly contribute to variance among pairs of twins. Thus, r_{MZA} may yield an *overestimate* of heritability.

E. Heritability Estimates Based on MZT and DZT Data

A way of avoiding the effects of intrauterine, perinatal, and placement factors on heritability estimates is to measure *intra-pair correlation coefficients* for MZT and DZT twins and multiply the difference (r_{MZT} and r_{DZT}) by a factor of 2 ("**MZ-minus-DZ-times-two**" method).

The intra-pair correlation coefficients for MZT and DZT twins can be expressed as follows.

$$r_{MZT} = (Vg + Ves) / Vt = Vg/Vt + Ves/Vt$$

and

$$r_{DZT} = (0.5 * Vg + Ves) / Vt = 0.5 * Vg/Vt + Ves/Vt$$

The correction factor of 0.5 in the lower equation accounts for the assumption that DZ twins share about half of their genetic alleles. The use of the same symbol (*Ves*) in both equations reflects the assumption that shared environmental factors have nearly identical effects on MZ and DZ twins.

By subtracting the two equations, one obtains

$$r_{MZT} - r_{DZT} = 0.5 * Vg / Vt$$

or

$$2 * (r_{MZT} - r_{DZT}) = Vg/Vt = H$$

The assumption that shared environmental factors (V_{es}) are the same for MZT and DZT twins may not always be warranted. Parents or teachers may approach twins differently, depending on whether they perceive them as "fraternal" or "identical."

Another problem is the assumption that DZ twins (and other siblings) share 50 percent of their genetic alleles on average. Actually, they share *all* alleles for which their parents are homozygous. Also, mate choices may be preferential between partners who share a phenotype, and therefore at least some of the genetic alleles promoting this phenotype. For example, most tall women prefer tall men as marriage partners. This phenomenon is known as **assortative mating**.

Homozygosity and assortative mating do not affect r_{MZT} because MZ twins are genetically as similar as they can be. In contrast, homozygocity and assortative mating inflate r_{DZT} since they cause DZ twins to share more than half of their genetic alleles. Thus, the "MZT-minus-DZT-times-two" method tends to yield *underestimates* of heritability.

In summary, estimates of heritability based on MZA correlations tend to be overestimates, whereas MZ and DZ comparisons tend to underestimate heritability. Therefore, in order to obtain a useful estimate of the heritability of a continuously variable trait, it is desirable to use both the MZA method and the MZT-minus-DZT-times-two method. The true heritability value will lie between the two estimates, and an evaluation of additional evidence may give more weight to one estimate or the other.

F. Heritability of the Intelligence Quotient (IQ) is Politically Controversial

The human behavioral trait that has been measured most extensively is the **intelligence quotient, or IQ**. The commonly used IQ tests measure the proband's (test-taker's) ability to give the meaning of words, to indicate which of several words does not belong with the others, or to decide how a three-dimensional object would look from a different perspective. The scores from different IQ tests are combined and converted to a distribution in which the mean is 100. This *standardization* is done to correct for the effects of age and to eliminate a universal increase in mean test scores over time. Thus, a person's IQ indicates his/her performance relative to his/her age cohort.

Many psychologists feel that the IQ reflects a **general intelligence factor**, which may be described as a person's ability to grasp the essential features of a complex problem so that a solution can be found. IQ scores are fairly accurate in predicting how a child will do in school, and they are somewhat accurate in predicting the socioeconomic status that a person will achieve. People who score above average on IQ tests also seem to be more adept at managing their health problems and tend to have fewer automobile accidents.

Intra-pair correlation coefficients for IQ of adult MZA twins average at 0.74 (Table 14.1). This heritability estimate is probably too high because intrauterine factors—including diet, drugs, and stress—as well as birth complications affect cognitive abilities.

In contrast, IQ correlation coefficients for pairs of MZT and DZT twins of European origin average around 0.85 and 0.59, respectively (see Table 14.1). This would translate into a heritability estimate of (0.85 − 0.59) x 2 = 0.52. This estimate is probably too low because of *assortative mating*, as discussed earlier.

Table 14.1: IQ Correlation Coefficients for Different Pairs of Individuals Reared Together or Apart

Pair Relationship	Reared	IQ Intra-Pair Correlation
Monozygotic Twins	Together	0.85
Monozygotic Twins	Apart	0.74
Dizygotic Twins	Together	0.59
Siblings	Together	0.46
Siblings	Apart	0.24
Midparent/Child	Together	0.50
Single-parent/child	Together	0.41
Single-parent/child	Apart	0.24
Adopting parent/child	Together	0.20

From Devlin, B.; Daniels, M.; and Roeder, K. (1997). The heritability of IQ. *Nature* **388**: 468–471.

Together, more than 200 twin studies indicate that the heritability of IQ scores in populations of European origin is between 0.52 and 0.74.

When IQ measurements different ethnic groups within the United States are compared, Asian Americans score higher than European Americans, who in turn score higher than African Americans on average.

Given that the heritability of the IQ *within each of these populations* is high, the differences *between these populations* have also been interpreted as genetically based. However, this is an unwarranted conclusion, as shown by the experiment diagrammed in Figure 14.2: An outbred mouse population was split, and one subpopulation was kept on a rich diet, another one was fed poorly, and a third one was kept in a heterogeneous environment. After some time sufficient for physiological adjustment but not for breeding, the mean body weights differed significantly between the three subpopulations. Within each subpopulation, the heritability of body weight was high because standardizing the environmental conditions caused only a small reduction in the width of the bell curves. Yet we know that the three subpopulations are genetically indistinguishable because they were derived from the same population, and no further breeding occurred.

The average IQ differential between Asian, White, and Black Americans—historically, 15 points, or one standard deviation—is of a magnitude that could clearly be caused by environmental effects. This is demonstrated by the fact that raw IQ scores (obtained before standardization to $m = 100$) have increased worldwide and by the same magnitude over the past 50 years. This improvement within two generations cannot possibly be genetic. More likely, it is due to improved perinatal care, better nutrition, and a more stimulating lifestyle for more people.

After evaluation of many studies, and after much discussion, members of the Genetics Society of America prepared a resolution, which was endorsed in 1976 by 1,390 scientists. They declared that in their view there was "no convincing evidence as to whether there is or is not an appreciable genetic difference between the races" with regard to IQ. A report published in 1996 by the American Psychological Association has confirmed this assessment.

G. Similarity between MZA Twins May Result from Genetically Driven Niche Choices

The correlation coefficients used for heritability estimates do not account for interactions between genotype and environment. MZA twins—like other people—select congenial environments, or niches, in which they feel they will thrive. Because of their shared genetic alleles, MZA twins will tend to select similar niches, which in turn foster similar tastes, habits, and so on. Thus, the interaction between genetic predispositions and selected environments probably contribute greatly to the uncanny similarities between MZA twins.

MZA pairs demonstrate that two people who share the same genetic information can grow up in very different environments and still wind up being very similar persons and have similar life histories (Figure S14.c). Postulating genetic alleles for wearing similar apparel, wanting to sneeze aloud in elevators, and such seems absurd. Ascribing such idiosyncrasies to environmental factors shared in utero or during a short postnatal period is equally unreasonable.

Bouchard and others have proposed that each person, consciously or unconsciously, chooses the kind of niche in life that fits his/her genetic predisposition. The ability to make such choices would seem to be positively selected for because a good fit between genotype and niche is likely to promote survival and reproductive success. When people are asked how they have made such niche choices, or life decisions, they often refer to "gut feelings," which are controlled by genetic alleles affecting brain chemistry.

Niche selection may well occur in a stepwise fashion. The case of Gerald Levey and Mark Newman (see Figure 14.4), who both became firefighters and sported the same haircut, eyeglasses, and so forth, the process could have played out as follows. Both twins were big and strong, heritable attributes that all fire departments like to see in their recruits. So both twins presumably had a gut feeling that a fire department would be a good place for them to be. Feeling rewarded by an adrenalin rush is largely a matter of brain chemistry and probably linked to certain genetic alleles, which MZA twins would share. Hence, they would both feel rewarded and persuaded to stay with firefighting. Wire-rimmed glasses are more heat-resistant and fit better under protective gear than horn or plastic frames. Firefighters probably have professional journals and annual meetings that promote certain attitudes and appearances, which contribute detail to the firefighter niche.

In summary, the stunning similarities between MZA twins could result from a series of niche choices that optimize the fit between relatively general heritable traits—body size, brain chemistry, and so on—and various available niches. The genetic identity of MZA twins seems to guide them in making very similar niche choices, even in different external environments.

EXERCISE PAGE FOR CHAPTER 14

Student Last Name _____

Student First Name _____

Discussion Section _____

On the course web site, use the link on the **syllabus** to find a pdf file of your **assigned reading** for this chapter. The same reading is referenced below. Use the bulleted list of questions to test your knowledge of this reading.

Bouchard, T. J. Jr., et al. (1990). Sources of human psychological differences: The Minnesota study of twins reared apart. *Science* **25):** 223–228.

- What kind of traits do the authors investigate, and which method do they use to estimate the heritability of these traits?
- What is the general range of their estimates for
    - physical traits, such as height and weight?
    - physiological traits such as brain wave characteristics?
    - mental traits, such as intelligence quotient (IQ)?
    - personality variables, such as social attitudes?
- Did the intra-pair correlation coefficients for the IQ of MZA twins depend on
    - the age of the twin pairs measured?
    - the length of postnatal contact time of MZA twins?
    - differences in socioeconomic status or education of MZA foster parents?
- What do the authors say about the striking similarity of MZA twins?

Feedback

On a scale from 1 to 10, the being the best, please rank the above reading for

1. Interest and Relevance to the topic (10 being most interesting/relevant) _____
2. Readability (10 being most clearly written, easy to understand) _____

In the remaining space, enter any comments thssat you may have on this reading.

CHAPTER FIFTEEN

HUMAN BEHAVIORAL GENETICS

Human traits are affected by genes as well as by environmental factors. Behavioral traits are of particular interest because of their legal, economic, and ethical implications. Behavioral geneticists try to estimate the heritability of human behaviors and, if possible, to identify individual genes that contribute to the variability of the behavior. With detailed genetic maps and sequence information from the human genome project, our ability to identify such genes has greatly improved.

A. Behavioral Genetics and the Media

To many people, the idea that human behavior is in part controlled by genes is intriguing and insulting at the same time. This ambivalence is sometimes reflected in media sensationalism. It is therefore important to keep the powers and limitations of behavioral genetics in proper perspective.

Media often report or imply that researchers have discovered "the gene" for intelligence, marital infidelity, criminality, alcoholism, homosexuality, and so forth. In response, one needs to reemphasize the basics that

- genes encode RNAs and proteins, not traits
- many genes are pleiotropic, many traits are polygenic (see Figure 3.9)

Heritability of some behavioral parameters, such as IQ, is politically controversial. IQ is positively correlated with socioeconomic status. Conservative politicians therefore welcome high estimates for IQ heritability (and emphasize methods that yield high estimates), because they support the conservative view that a stratified society is the natural order of things. Liberal politicians are likely to dispute the validity of the same data (and emphasize methods that yield low estimates for IQ heritability), because they support the liberal view that remedial education must be offered to level the playing field. The extent to which some behaviors, such as alcoholism or homosexuality, are genetically controlled affects the social stigma, insurance benefits, and legal status associated with these behaviors.

In the heat of political battle, it has happened in the past that public officeholders have misused scientific data and that scientists have resorted to fraud in attempts to promote their convictions. Both kinds of transgressions are counterproductive in the long run. For the common good, it seems best to *adhere to a division of labor*. The job of scientists is to explain the methods used to collect and process their data, as well as the validity and limitations of their conclusions. Other professionals can help put the discussion in perspective, for example, by pointing out that charisma and emotional presence are also positively correlated with socioeconomic status but are not covered by IQ tests. It is the job of elected officeholders, administrators, and voters to apply their ethics and judgment in translating scientific data into policies.

Some of the controversy surrounding behavioral genetics is rooted in the concern that studying a component of genetic control would infringe on the concepts of free will and personal responsibility. However, this concern can be resolved by keeping in mind that genetic predictions about human behavior are *statistical* in nature. They cannot, and do not aim to, predict the behavior of any given *individual*.

B. Methods Used in Behavioral Genetics

Behavioral geneticists use quantitative methods to estimate the heritability of certain behaviors and to identify individual genes involved in the control of these behaviors. Some methods for estimating heritability were already discussed.

- for discontinuously variable traits: measuring the *concordance ratio* (see Chapter 13)
- for continuously variable traits: determining *intra-pair correlation coefficients* for pairs of twins using the r_{MZA} method and the $(r_{MZT} - r_{DZT}) * 2$ method (see Chapter 14)

Additional methods for *estimating the heritability* of continuously variable traits use intra-pair correlation coefficients for adopted children and their birth parents (genetically related individuals who do not share a common environment) and for adopted children and their adoptive parents (genetically unrelated individuals who share a common environment).

Methods used to *map individual genes* that affect a specific behavior include the following.

- A **linkage analysis** is applied to *family pedigrees*. The analysis tests a model that links the behavior to the locus, allelism, and dominance of a gene. The quality of the model is measured by the *logarithm of odds (lod) score,* which indicates how much more likely it is to obtain the pedigree data under the linkage model than under the null hypothesis of no linkage between the behavior and the hypothetical gene.
- In **allele-sharing studies**, one tests whether *pairs of relatives* sharing a rare heritable trait also share the same alleles of any DNA markers more often

than expected by chance. If so, then one or more genes affecting the trait are likely to be located near these markers. What are the underlying assumptions of this method?

- In **association studies**, one simply scans DNA samples from a large number of *not necessarily related individuals* for genetic alleles, or DNA markers, that are most closely associated with the trait of interest. Associated genes are then tested for promoting the trait of interest. For markers, it is assumed that they must be close to a gene affecting the trait, because otherwise genetic recombination would have destroyed the linkage.

Association studies are less reliable than linkage and allele-sharing studies because humans tend to marry within their ethnic group. As a result, group-specific alleles may become associated with group-specific customs as both are propagated *independently* within the group. (Note the double meaning of the word "heritage" in this context!)

For example, a study of U.S. citizens is likely to reveal an association between the custom of eating with chopsticks and several genetic alleles, simply because eating with chopsticks is most prevalent among immigrants from East Asia, who will also have some group-specific genetic alleles and DNA markers because they have been reproductively isolated for many generations. The association between any such alleles/markers and eating with chopsticks would most likely be incidental, and the chances of finding a "chopstick gene" would be remote.

Another study of U.S. citizens may find an association between criminality (as measured by incarceration) and some genetic alleles or DNA markers that are common among African Americans. This can be expected because African American men are known to be overrepresented in prisons and their racial phenotype must have some underlying genetic alleles. However, since the overrepresentation of African Americans in prisons could be for cultural or socioeconomic reasons (racial profiling, a feeling of having nothing to lose, etc.), any observed associations with genetic markers could again be incidental rather than causal.

The use of genomic data (see Chapter 9) will facilitate studies of genetic alleles that seem to affect human behaviors. In many cases, such studies should lead to major improvements in the medical treatment of behavioral disorders. However, the strengths and limitations of the underlying methods must be kept in proper perspective. Examples discussed below will illustrate different methods for estimating the heritability of certain behaviors and to identify specific genes involved.

C. Cognitive Abilities

Behavioral geneticists have carried out numerous studies on special cognitive skills, especially verbal skills and spatial skills.

In tests of **verbal skills**, probands (the subjects of investigation) are asked to identify synonyms, to write down in a limited period of time as many words as possible starting with a particular letter, or to list all the things that have certain shape, and the like (Figure S15.a).

In tests of **spatial ability**, probands are challenged to do the equivalent of puzzles, to distinguish three-dimensional objects from their mirror images when shown in different perspectives (Figure 15.1).

FIGURE 15.1:
Tests of spatial ability
Redrawn from Plomin R. and DeFries J.C. (1998) The genetics of cognitive abilities and disabilities. *Scientific American* **May 1998:** 62–69

TESTS OF SPATIAL ABILITY

1. HIDDEN PATTERNS: Circle each pattern below in which the figure appears. The figure must always be in this position, not upside down or on its side.

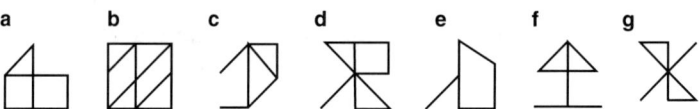

2. IMAGINARY CUTTING: Draw a line or lines showing where the figure on the left should be cut to form the pieces on the right. There may be more than one way to draw the lines correctly.

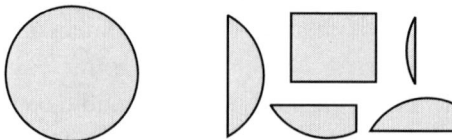

3. MENTAL ROTATIONS: Circle the two objects on the right that are the same as the object on the left.

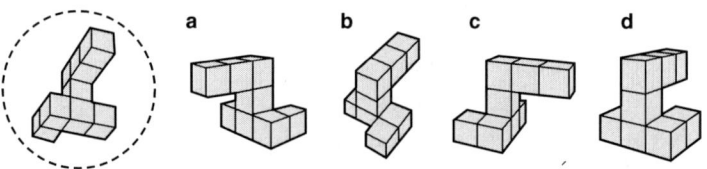

4. CARD ROTATIONS: Circle the figure on the right that can be rotated (without being lifted off the page) to exactly match the one on the left.

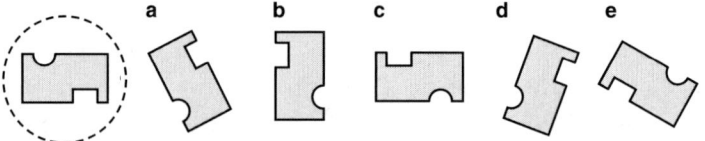

For both abilities, the correlation coefficients for pairs of MZT twins were significantly higher than those for DZT twins, with an average difference around 0.3 for verbal skills and 0.25 for spatial ability (Figure 15.2). Based on these data, the estimated *heritability* would be around 60 percent for verbal skills and around 50 percent for spatial skills, in agreement with the heritability estimate for the intelligence quotient based on the r_{MZT} minus r_{DZT} times two method (see Table 14.1).

Studies on verbal and spatial skills of adopted children show that they come to resemble their birth parents nearly as much as normal children do, while correlations between adopted children and their adoptive parents are small and do not increase with age (Figure 15.3).

D. Violence

Violence is affected by several genes and by environmental factors. Identifying at least some of these factors seems desirable in view of the large toll exacted by violence. Here, we will focus on the gene encoding **monoamine oxidase A**, the loss of which leads to extreme violence.

The discovery began with a *linkage study* on a large Dutch family, in which 14 out of 33 males showed borderline mental retardation and bouts of extreme violence including

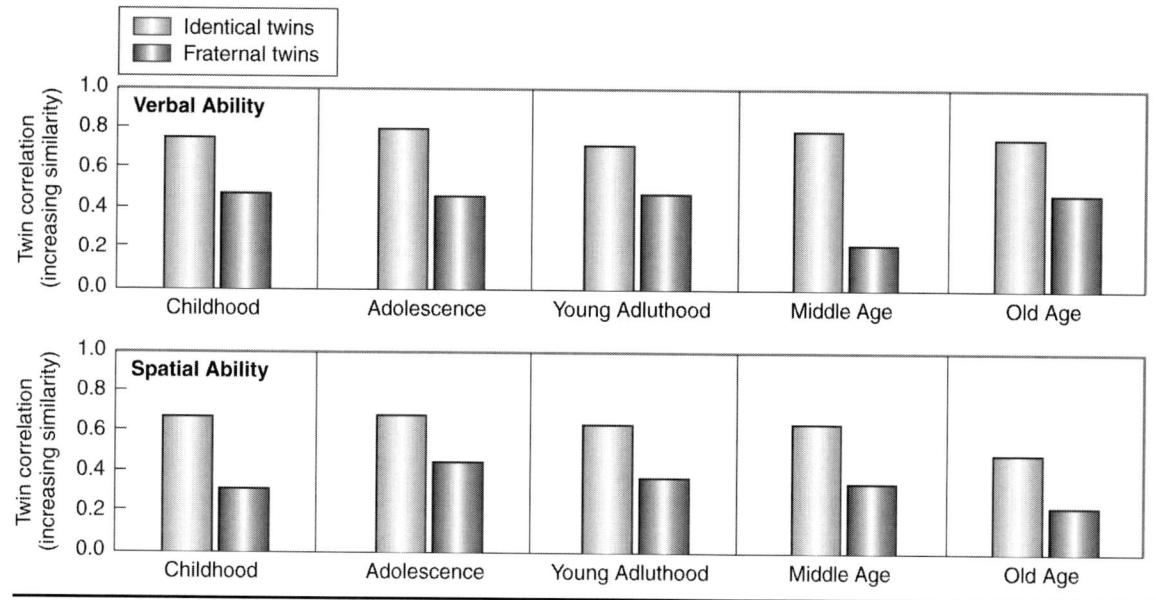

FIGURE 15.2:

Intra-pair correlation coefficients for cognitive skills of MZT and DZT twins. The greater similarity of MZT twins as compared to DZT twins is consistent throughout life.
Redrawn from Plomin R. and DeFries J.C. (1998) The genetics of cognitive abilities and disabilities. *Scientific American* **May 1998:** 62–69

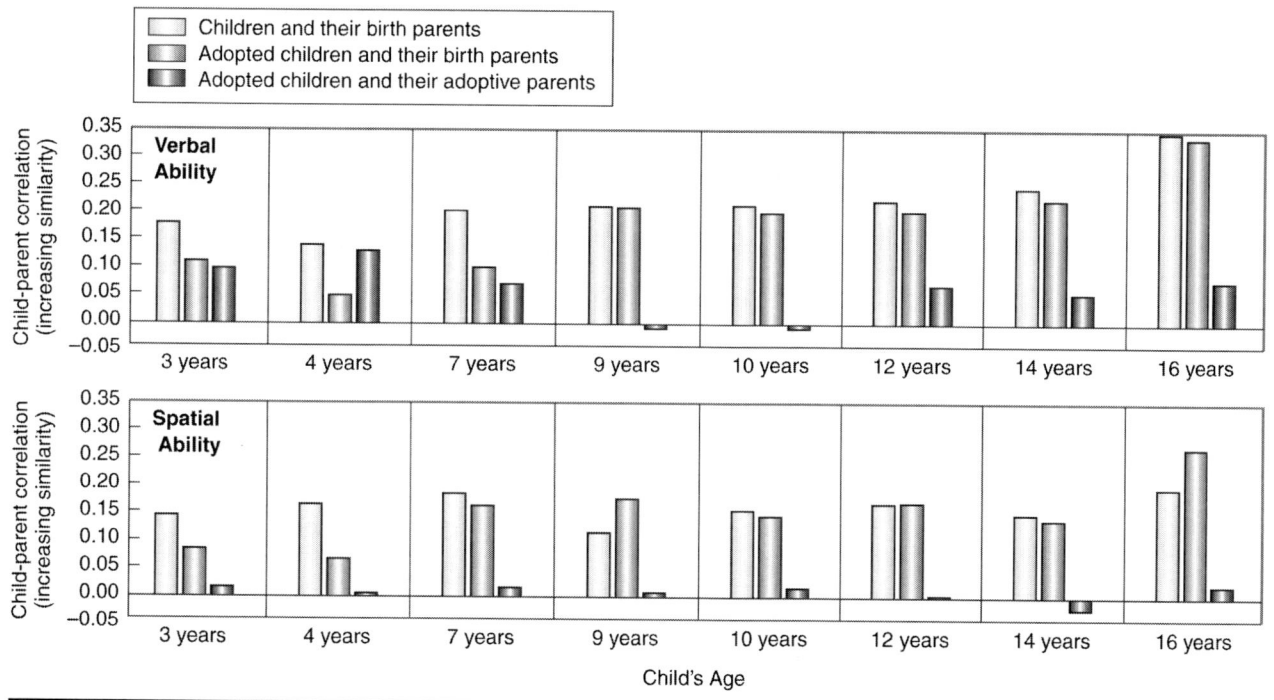

FIGURE 15.3:

Adoption studies on verbal and spatial skills. Adopted children come to resemble their birth parents nearly as much as normal children do, while correlations between the children and their adoptive parents turn out to be small.
Redrawn from Plomin R. and DeFries J.C. (1998) The genetics of cognitive abilities and disabilities. *Scientific American* **May 1998:** 62–69

arson, rape, and assault with a knife, a pitchfork, and a car. Females transmitting the syndrome showed normal intelligence and behavior. The simplest interpretation of this pedigree was that the abnormal behavior is caused by a loss-of-function allele of an X-linked gene. This would leave females of this family *heterozygous*, with one mutated and one wild-type copy of the gene. This genotype should result in at least the level of gene activity that is present in normal males. In contrast, males with a mutated allele would be *hemizygous* because no homolog is present on the Y chromosome. This genotype would leave males with a reduced or no gene activity depending on the nature of the mutation.

Scanning the X chromosome for DNA markers that co-segregate with the aggressive behavior turned up markers near two genes for monoamine oxidase, *MAOA* and *MAOB*. Cultured cells from each living affected male were deficient for the MAOA enzyme, whereas cells from unaffected males showed normal MAOA levels. DNA sequencing showed that each living affected male had the same mutation in the MAOA gene, which changes a glutamine codon to a termination codon.

Monoamine oxidase metabolizes serotonin, dopamine, and noradrenaline. Thus, the lack of MAO should raise the levels of these neurotransmitters. Transgenic mice deficient for MAOA are trembling and fearful as pups; adult males show enhanced aggression. Serotonin and noradrenaline show higher concentrations in the brains of these mice. The abnormal behaviors of the pups are reversed by serotonin synthesis inhibitor.

This particular study has benefited from three favorable circumstances. First, the existence of a single pedigree large enough for a conclusive linkage analysis. Second, the fact that the gene of interest was X-linked, which makes it easy to recognize in a linkage analysis. Third, the fact that the DNA markers co-segregating with violence were located near two genes known to affect neurotransmitters, so that a plausible hypothesis linking these genes to the abnormal trait presented itself.

This study has *not* discovered "the violence gene." The neurotransmitters metabolized by monoamine oxidase A do not act specifically in aggression, as indicated by the general retardation of the affected males. In other words, the MAOA gene is pleiotropic. Conversely, aggression is affected by other genes involved in the metabolism of serotonin, by testosterone, and by culture as well as personal history (see Chapter 21). The study shows, however, that lack of MAOA has particularly strong effects in those parts of the brain that are involved in violence. It is not unusual in organisms that the lack of a common ingredient causes a fairly specific biological effect. Do you remember another case?

A *combinatorial effect* of genetic and learned factors on violence was demonstrated in a longitudinal study of a cohort of young males. They were divided in three groups for having experienced maltreatment at ages 3 through 11: severe maltreatment, probable maltreatment, or no maltreatment. They were also divided independently into two groups for monoamine oxidase A (MAOA) level: high or low.

The young men were assayed for four measures of violence.

- conduct disorder (based on a psychological test)
- conviction for violent offenses (court records)
- disposition towards violence (another psychological test)
- antisocial personality disorder (testimony by people who knew them well)

By each of these measures, severe maltreatment at child age turned out to be the strongest risk factor for violent behavior as young adults (Figure 15.4). In the

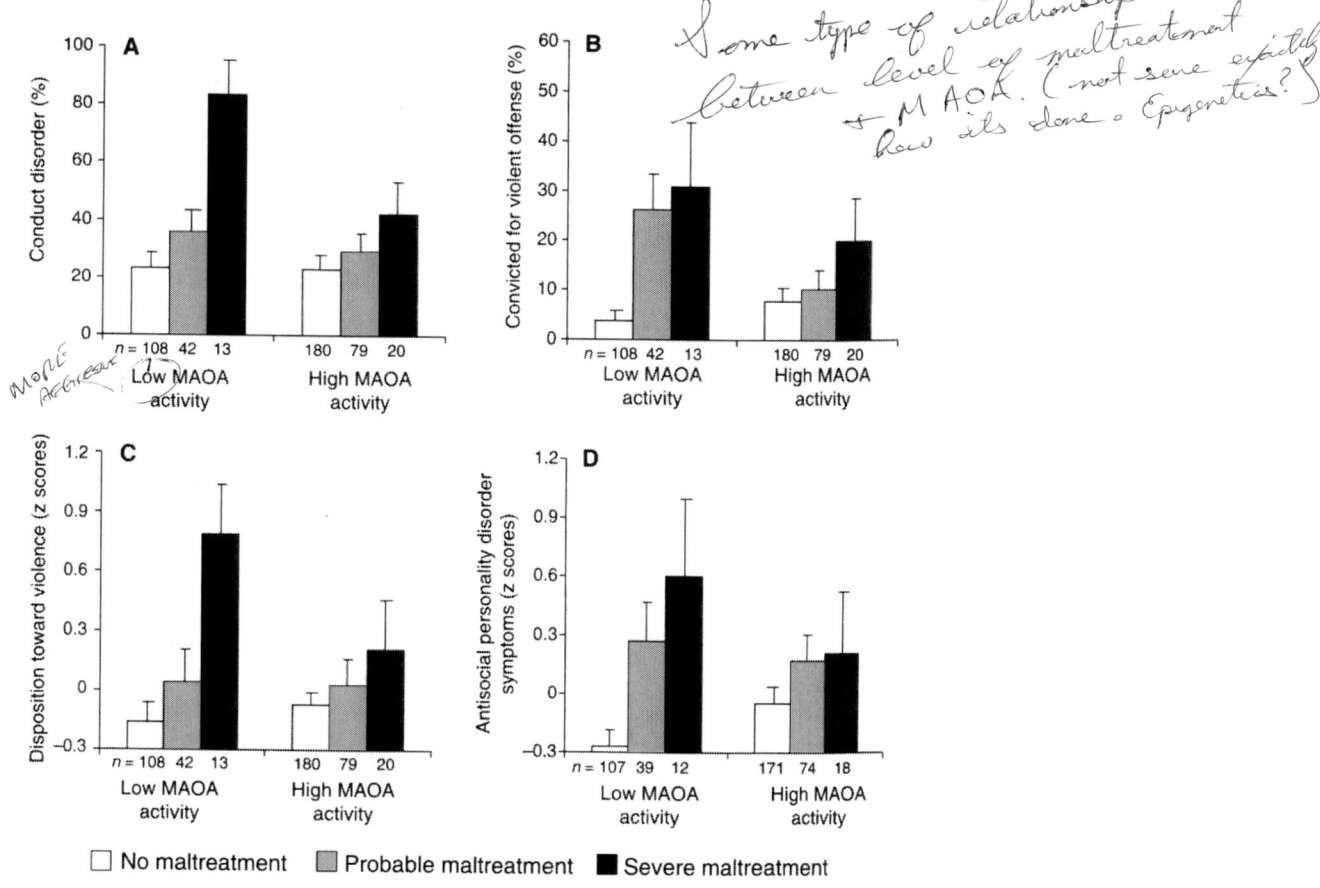

FIGURE 15.4:
Combined effects on violent behavior of young men of having suffered maltreatment during childhood and of having low levels of monanine oxidase A.
From Caspi A. et al. (2002) Role of genotype in the cycle of violence in maltreated children. *Science* **297**: 851–854

low and probable maltreatment groups, low MAOA activity had no significant effect. However, within the severely maltreated boys, those with low MAOA activity showed significantly higher levels of violence than those with high MAOA activity. The boys who were maltreated *and* had low MAOA activity constituted only 12 percent of the entire cohort but accounted for 44 percent of the violence convictions.

The results may be taken to suggest that low MAOA levels during childhood may predispose boys to respond to childhood abuse by developing aggressive behaviors later on. Alternatively, abuse suffered during boyhood may result in patterns of brain activity that increase the likelihood of aggressive behavior and are exacerbated by the lack of MAOA in juveniles.

E. Homosexuality

Human sexual orientation is usually rated by the **Kinsey rating scale** from 0 (exclusively heterosexual) to 6 (exclusively homosexual). The rating distribution varies with age and socioeconomic status of the probands but has generally the shape of a reverse J, with most humans being exclusively or predominantly heterosexual

(Kinsey ratings 0 or 1), 2 to 10 percent predominantly or exclusively homosexual (Kinsey ratings 5 or 6), and relatively few bisexual (Kinsey ratings 2 to 4). Based on this distribution, most authors treat sexual orientation as a *discontinuously variable* trait for the purpose of estimating heritability.

The heritability of homosexual behavior was assessed by twin studies. Gay men with MZ twins, DZ twin brothers, non-twin brothers, and adopted brothers were recruited through homophile publications. The sexual orientation of the probands was assessed by themselves and confirmed by Kinsey rating (mean 5.4). The sexual orientation of their male relatives was assessed by the probands and confirmed by self-rating of those relatives who responded. Among the relatives whose sexual orientation could be rated, the frequency of homosexuality or bisexuality was

- 52 percent (29/56) for MZ twins of gay probands
- 22 percent (12/54) for DZ twin brothers of gay probands
- 9 percent (13/142) for non-twin brothers of gay twin probands
- 11 percent (6/57) for adoptive brothers of gay twin probands

The low rate (9 percent) of homosexuality for non-twin brothers of *twin* probands contrasts with a rate of 22 percent for DZ twin brothers of gay probands, since in both cases the gay probands shared about half of their genetic alleles with their brothers. Similarly, in a separate study, 22 percent of non-twin brothers of *non-twin* gay probands were also gay. A reason for the discrepancy may be that twins are typically very close and thereby tend to exclude other siblings, who then might seek a different (non-gay) "niche" for themselves.

Corresponding studies on female relatives of lesbian probands gave similar results. Homosexuality or bisexuality was found in

- nearly 50 percent of MZ twins of lesbian probands
- about 25 percent of DZ twin sisters of lesbian probands
- about 16 percent of adoptive sisters of lesbian non-twin probands
- 12 to 35 percent of biological sisters of lesbian non-twin probands
- 2 to 14 percent of biological sisters of straight female probands

With concordance ratios (concordance rates for MZ twins divided by concordance rates for DZ twins) of two or greater for homosexuality and bisexuality, behavioral geneticists generally agree that sexual orientation is in part under genetic control. Beyond this conclusion, an attempt has been made to *map* one of the genes involved in male homosexuality. This study combined a *linkage analysis* with an *allele-sharing study*.

The first linkage analysis was done with 76 gay men recruited through an outpatient HIV clinic. Their pedigrees indicated the following frequencies of being gay among the probands' male relatives (Figure 15.5).

- 14 percent for brothers
- 7 percent for maternal uncles
- 8 percent for sons of maternal aunts
- around 2 percent for fathers, paternal uncles, and paternally related cousins

A second linkage analysis was done with a set of families, in which each gay proband had *two* gay brothers and no indication of paternally transmitted homosexuality (no father or son of a proband was gay). In these families, the frequencies of gay maternal

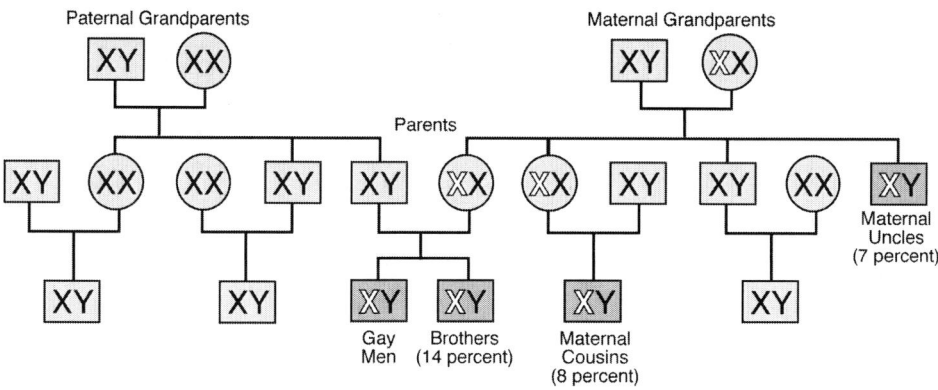

FIGURE 15.5:
Pedigrees of gay probands show high frequencies of being gay among their brothers, maternal uncles, and maternal cousins. The double-lined X chromosome may carry an allele promoting male homosexuality.
Redrawn after LeVay S. and Hamer D. Evidence for a genetic influence in male homosexuality. *Scientific American* **May 1994:** 44–49

uncles and cousins were even higher. Comparison of the two sets of data suggested that different types of genes contributed to male homosexuality but that an X-linked gene may be contributing strongly (see Figure 15.5).

In order to map the hypothetical X-linked gene, the investigators did an *allele-sharing study*, which tests whether relatives sharing a heritable trait have inherited identical alleles of genomic markers more often than expected by chance. If so, this would indicate that a gene affecting the trait is located in the same chromosomal region as the markers. A related study compared the alleles of X-linked *variable number tandem repeats* (VNTRs, see Chapter 9) in gay brothers. The comparisons were only informative if their mother was *heterozygous* for the VNTR (Figure 15.6). In these cases, the likelihood that both brothers would inherit the same allele *by chance alone* was 0.5. This was indeed so in a

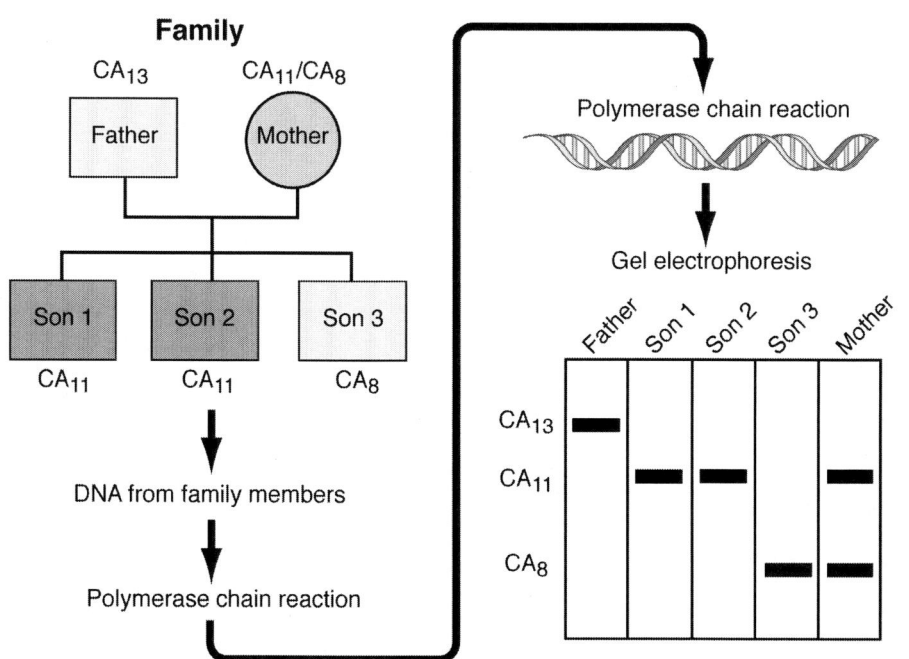

FIGURE 15.6:
Allele sharing study. Brothers were tested for sharing the same allele of an X-linked variable number tandem repeat (CA_n), for which their mother was heterozygous.
Redrawn after LeVay S. and Hamer D. Evidence for a genetic influence in male homosexuality. *Scientific American* **May 1994:** 44–49

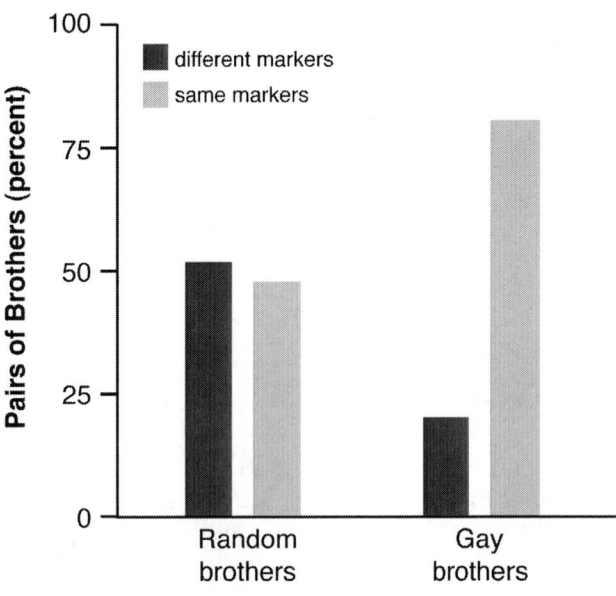

FIGURE 15.7:
Allele sharing statistics. In some families, gay brothers share molecular marker alleles in the Xq28 region more often than expected by chance.
Redrawn after LeVay S. and Hamer D. (1944) Evidence for a Biological Influence in Male Homosexuality. *Scientific American* **May 1994:** 44–49

control sample of brothers who were chosen randomly without regard to their sexual orientation.

However, in those families whose linkage analysis had suggested maternally inherited male homosexuality, there was an interesting exception for VNTRs in the Xq28 region of the X chromosome. Gay brothers shared the same marker alleles much more often, with frequencies greater than 75 percent (Figure 15.7). This result indicated that these marker alleles traveled together with an X-linked allele that promotes male homosexuality, and that by selecting brothers who were both gay, the investigators had also selected brothers who shared associated DNA marker alleles more often than expected by chance alone. This result was confirmed by two published follow-up studies from the same laboratory and by one independent but unpublished study. However, another independent and published study found *no* enhanced sharing of marker alleles in the Xq28 region of gay brothers.

In summary, the evidence from the twin studies for a significant genetic component to both male and female homosexuality is generally accepted in the scientific community. Discrimination against homosexual people is therefore inappropriate. However, the data linking male homosexuality to one or more genes in the Xq28 region have not been independently confirmed.

F. Novelty Seeking

Psychologists have distinguished four distinct domains of temperament—novelty seeking, harm avoidance, reward dependence, and persistence—and proposed that each domain may be based on different sets of neurobiological factors. Two independent *association studies* on large samples of unrelated people indicate that the normal trait of novelty seeking in humans co-segregates with a particular allele of the dopamine D4 receptor.

People scoring high on the **novelty-seeking** scale are characterized as impulsive, exploratory, fickle, excitable, quick-tempered, and extravagant, whereas those scoring low tend to be reflective, rigid, loyal, stoic, slow-tempered, and frugal. Two independent *association studies* published in a highly respected journal found an association between high scores in a psychological test for novelty seeking with the presence of a particular allele of the D4 dopamine receptor (Figure 15.8 and assigned reading).

G. Genetic Analysis of Other Human Behaviors

Many human behaviors seem to have genetic components, as indicated by twin studies. With new human chromosome maps and DNA sequence information, more genes that confer a predisposition to certain behaviors will be identified. This will

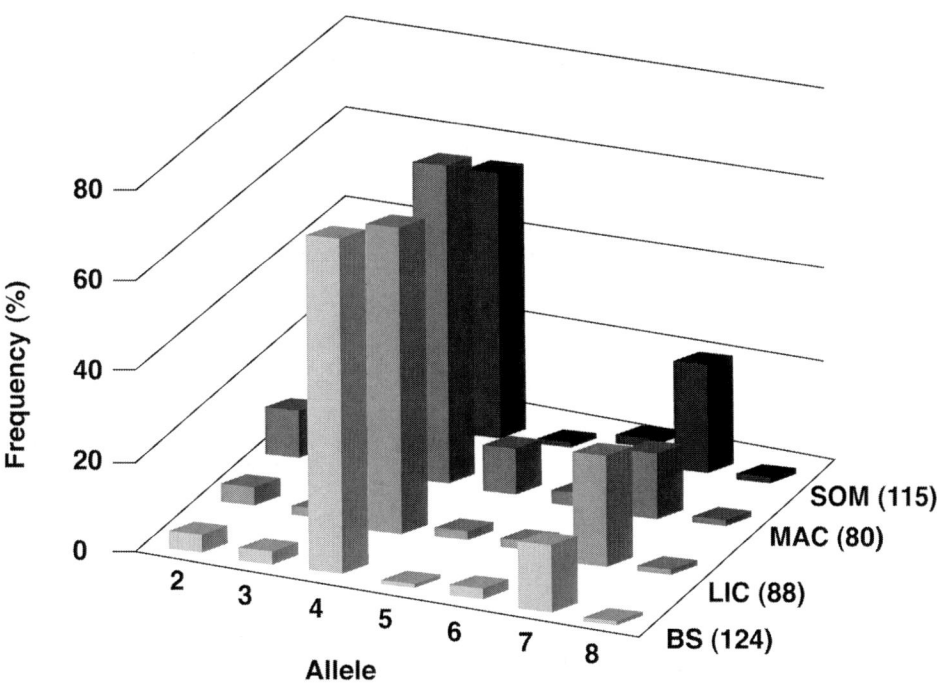

FIGURE 15.8:
Association study. Allele frequencies for the exon III 48-bp repeat region (2–8 repeats) of the dopamine D4 receptor gene in subjects from four different Israeli locations were analyzed as shown in Figure 15.6. Subjects with the 7 allele also scored high on questionnaires for "novelty seeking".
Redrawn after Ebstein et al. (1996) *Nature Genet.* **12:** 78–80

greatly enhance the development of better drugs for the treatment of many behavioral disorders, including very common ones.

Alzheimer's disease is a dreadful dementia that eventually befalls many seniors. As for many other disorders, it has been important to distinguish different forms of the disease. Early onset ("familial") Alzheimer's is strongly affected by allelism in the gene for apolipoprotein E. Late onset ("sporadic") Alzheimer's is affected by environmental factors, personal habits, the apolipoprotein E gene, and several other genes.

Schizophrenia is a mental disorder that cripples the lives of millions of people and their families. Roughly 1 percent of the world's population are stricken with symptoms including hallucinations, paranoia, aggressiveness, faulty logic, and impoverished speech. Many of them live in poverty or wind up in mental hospitals or jails. Medications exist but stop symptoms completely in only about 20 percent of all patients. Numerous twin, family, and adoption studies indicate that genetic factors can promote susceptibility for the disorder. Various reports have suggested that one or more susceptibility genes for schizophrenia may be located on chromosomes 1, 5, 6, 10, 13, 15, and 22, but no major contributing gene has been confirmed yet.

Attention deficit, hyperactivity disorder (ADHD) aggravates the lives of many students, parents, and teachers. Confirmed data again implicate allelism in the gene for dopamine receptor D4DR (see previous section). Dyslexia is a reading disability characterized by poor reading comprehension and trouble reading aloud. Twin studies indicate a significant genetic control, perhaps by some of the same genes that contribute to the variance of verbal skills.

Obesity is rapidly becoming a major health problem in developed countries. It is measured by the **body mass index (BMI)**, defined as body weight in kilograms divided by the square of height in meters. Nearly half of the U.S. citizens are obese (BMI > 30) or pre-obese (25 <BMI < 30). According to various estimates, 40 to 70 percent of the variation in human obesity is heritable. The remainder is ascribed to environmental factors that encourage overeating and discourage physical activity. Understanding the regulatory network that controls food intake and energy expenditure should enable us to develop safe and effective drugs that can help to control at least some forms of obesity.

The gene encoding *leptin,* a signal released from adipose tissue, has been cloned in mice and humans. The gene encoding the *leptin receptor* has also been cloned and its action on the brain is being studied in normal and diabetic mice. Mutations in human leptin and leptin receptor genes so far have not been found in obese people. It appears that several other genetic and environmental factors are involved in a complex regulatory network that we are only beginning to understand.

For many human mental disorders, twin studies indicate significant heritability. Yet attempts to identify specific genes that are critically involved have yielded spurious results that could not be independently confirmed. This is not surprising if one realizes that studies on human behavior have to deal with the complexities not only of genetic networks but also of the brain. Future genetic studies may be more effective if they use as endpoints of their measurements the *activities of certain brain regions* (as indicated by functional MRI), rather than human *behaviors.*

EXERCISE PAGE FOR CHAPTER 15

Student Last Name _____

Student First Name _____

Discussion Section _____

On the course web site, use the link on the **syllabus** to find a pdf file of your **assigned reading** for this chapter. The same reading is referenced below. Use the bulleted list of questions to test your knowledge of this reading.

Ebstein, R. P., et al. (1996). Dopamine D4 receptor (D4DR) exon III polymorphism associated with the human personality trait of novelty seeking. *Nature Genetics* **12**: 78–80.

- How did the authors define and measure the trait of novelty seeking?
- What kind of genetic polymorphism did the authors also measure? What was the most common allele, and what was the "7 allele"? Do the receptor proteins encoded by these alleles show any physiological difference?
- What associations did the authors find between D4DR receptor alleles and novelty seeking?
- How does an association study, as used here, differ from pedigree analysis and allele sharing studies discussed earlier?
- What potential problem with association studies do the authors discuss and how do they mitigate against it?
- In which situations is an association study promising?
- Which observations by others do the authors quote to buttress their hypothesis that there is a causal relationship between D4DR polymorphism and the trait of novelty seeking?
- Why did the authors emphasize the observation made in a parallel study that the association between D4DR receptor alleles and novelty seeking was found among siblings, of which one had the 7 allele while the other had not?

Feedback

On a scale from 1 to 10, the being the best, please rank the above reading for

1. Interest and Relevance to the topic (10 being most interesting/relevant) _____
2. Readability (10 being most clearly written, easy to understand) _____

In the remaining space, including the back of this page, enter any comments that you may have on this reading.

PART IV
BIOLOGICAL ASPECTS OF HUMAN BEHAVIOR

Part IV of this course is about genetic and hormonal effects on human behavior. To many people, the idea that human behavior is subject to biological factors is intriguing and offensive at the same time. This ambivalence is part of the philosophical question whether we want to see ourselves as being *above nature* or as being *part of it*. However, the biological underpinnings of human behavior also have practical ramifications. The extent to which behaviors, such as alcoholism or homosexuality, are genetic has ramifications on the social stigma, insurance protection, and legal consequences of these behaviors. Hormonal control of human behavior matters in a court of law, where crimes committed out of fear or jealousy may be treated differently from those committed in cold blood.

Scientists studying biological aspects of human behavior are often accused of "biological determinism." Critics are concerned that scientists make humans into creatures like ants, whose behavior seems completely determined by innate mechanisms. This view contradicts our personal experience of having a "free will," meaning an ability to choose between different courses of action, including options that will *not* promote our survival or procreation. Others charge that biologists underestimate the cultural aspects of human behavior.

Every biologist today understands that both animal and human behavior depend on genetically fixed as well as learned components, and that the two are often intertwined. This has been amply demonstrated at the level of individual neurons as well as whole organisms. Every anthropologist will also agree that cooperation and following social norms of behavior are critical to human survival. During hominid evolution, the connection between instincts and actual behavior has become a complex pattern of controlled release (at least as long as we are sober). Those of our ancestors who could not set aside eating and mating at least temporarily were likely to perish. As a result, we have acquired the ability to feel rewarded by following social norms. This ability allows humans to go on hunger strike, to risk their lives in battle, or to become suicide bombers. The neurological underpinnings of such choices are currently being investigated using brain imaging and other techniques.

The charge that biological study negates free will may also flow from a philosophical misunderstanding. Predictions of the behavior of humans and other complex systems are *statistical* in nature. They cannot, and do not aim to, predict the behavior of any given *individual*. This is perhaps most apparent for political elections. Pollsters become ever more adept at predicting statistically how voters of a certain age, sex, religion, socioeconomic status, and so on, will vote *as a group*. But even the most accurate statistical prediction will not foretell how a given *individual* will vote. Consequently, we are still holding elections. Likewise, it may become possible in the future to predict, on the basis of a genetic inventory, a person's *chance* of becoming a concert pianist, but it will still be up to that person to decide whether he or she will try and perhaps succeed.

The following chapters will emphasize human behaviors that are sexually dimorphic. Chapter 16 will lay the groundwork by reviewing the genetic and hormonal controls in the development of our reproductive organs. Chapter 17 will explore hormonal effects on human brain development and behavior. Chapter 18 will introduce the general concept that organisms behave so as to maximize their *inclusive fitness*, meaning the propagation of their genetic alleles. Chapter 19 will examine whether this principle applies—on average—to human courtship and parenting. Chapter 20 will continue with a discussion of human gender roles and some of the attempts that have been made to overcome the discrimination that is often associated with gender roles. The last chapter in Part IV, Chapter 21, will deal with aggression, cooperation, and the keen sense of "fairness" observed in humans as well as in non-human primates. This chapter will conclude with the *gene-culture-coevolution hypothesis*, which tries to explain the uniquely human ability to follow cultural norms of behavior.

CHAPTER SIXTEEN

DEVELOPMENT OF REPRODUCTIVE ORGANS

Sex determination is a central event in the development of an organism. In mammals including humans, sex determination is controlled by a gene on the Y chromosome known as SRY^+. After a sexually indifferent phase, *sex differentiation* begins with the development of testes in the presence of SRY^+ activity, while ovaries develop in the absence of SRY^+ activity. Subsequent sexual differentiation is controlled by sex hormones produced in the gonads and adrenal glands.

A. Sex Determination

Mammals exist is two sexes, male or female, which can be defined by different criteria. We will define **males** and **females** as individuals with testes or ovaries, respectively. **True hermaphrodites** are individuals that have both testicular and ovarian tissues. They are very rare in mammals, including humans, but common in other animals, especially in parasites such as roundworms or tape worms.

Males and females also differ in other **primary sex characteristics** such as penis versus clitoris and ductus deferens versus oviducts, which are formed during embryonic development. In addition, adult males and females from many species differ in **secondary sex characteristics**, such as body size, shape, coloration, weaponry, and behavior, which develop during the juvenile stage and after puberty. The presence of secondary sex characteristics is also known as *sexual dimorphism* (Figure S16.a).

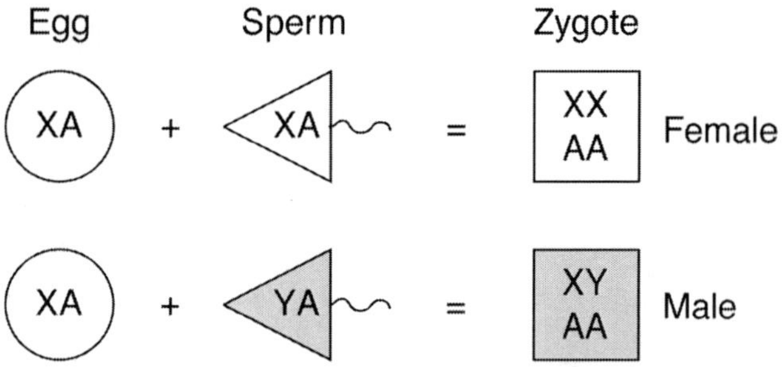

FIGURE 16.1:
Genotypic sex determination by mammals. Males typically have two sexually dimorphic chromosomes, designated X and Y, while females have two X chromosomes. "A" stands for one complete set of non-sex chromosomes, or autosomes. Thus, males produce tow types of sperm, with karyotypes XA and YA.
From Kalthoff, K. (2nd ed. 2001) *Analysis of Biological Dvelopment*, New York: McGraw-Hill, p. 694

The natural mechanism by which an organism becomes either male of female is called **sex determination**. Humans and other mammals have **genotypic sex determination**: Sex is determined *at fertilization* by genes that the fertilized egg inherits. This is in contrast to **environmental sex determination**, which occurs *after fertilization* by environmental factors. For example, the sex of many reptiles is determined by the environmental temperature during a critical phase of their development.

Mammals including humans have a pair of sexually dimorphic chromosomes called **sex chromosomes**, with **male heterozygosity** (XY males; XX females) and **male heterogamety** (sperm with X or Y chromosomes; eggs only with X chromosomes). Half of all fertilized eggs are therefore male (Figure 16.1).

The presence of a Y chromosome is necessary and sufficient for male development; in the absence of a Y chromosome, an embryo will develop as a female (Table 16.1). This is so for most mammals but not necessarily for other species with male heterogamety. For instance, in *Drosophila*, XO individuals develop as sterile *males*, while XXY individuals are fertile *females*. Sex in this species is determined by the *X:A ratio*, meaning the number of X chromosomes divided by number of sets of **autosomes** (non-sex chromosomes). Individuals with an X:A ratio of 0.6 or smaller develop as males, while individuals with an X:A ratio of 0.75 or greater become females.

The critical importance of the Y chromosome in human sex determination is indicated by the sexual development of individuals with unusual numbers of sex chromosomes, which arise from *chromosome nondisjunction*, that is, failure of homologous chromosomes to separate during meiosis. For example, women with only one X chromosome show **Turner's syndrome**, which includes short stature, lack of ovarian development, and hence, lack of sexual maturation at puberty. Men who inherit

Table 16.1: Sex Chromosome Number and Sexual Phenotype in Three Organisms with XX-XY Sex Determination			
Sex Chromosomes	**Human Phenotype**	**Mouse Phenotype**	**Drosophila Phenotype**
XO[1]	Sterile female[2]	Fertile female	Sterile male
XX	Normal female	Normal female	Normal female
XXX	Fertile female	Sterile female	Sterile female
XY	Normal male	Normal male	Normal male
XXY	Sterile male[3]	Sterile male	Fertile female
XYY	Fertile male	Semisterile male	Fertile male

[1] Only one X chromosome present
[2] Turner's syndrome
[3] Klinefelter's syndrome

two X chromosomes and one Y chromosome often show **Klinefelter's syndrome**, which includes a tall and lanky body build with narrow shoulders, little body hair, and reduced fertility. Men with one X chromosome and two Y chromosomes tend to be taller than average and may show minor developmental disturbances but are fertile.

The Y chromosome contains relatively few genes, which are required mostly for male sex determination and for sperm development. In contrast, the X chromosome contains more than a thousand genes, most of which are *not* involved in sexual development. Which X-linked genes have we discussed earlier?

B. The SRY1 Gene Is Necessary and Probably Sufficient for Testis Development

The sex-determining function of the Y chromosome probably relies on a single gene, designated SRY^+. The first visible difference between male and female embryos appears in the developing gonad. The genetic element on the Y chromosome that causes male development has therefore been called the **testis-determining factor (TDF)**.

Of critical importance for the analysis of TDF were human patients who sought medical attention because of infertility or irregularities in sexual development. The sexual phenotype of these patients did not seem to match their **karyotype**, that is, their set of chromosomes as visible under the microscope. These patients included

- **sex-reversed XX males**: sterile males who were phenotypically nearly normal but seemed to have two X chromosomes and no Y chromosome
- **sex-reversed XY females**: females presenting with Turner's syndrome who seemed to have one X chromosome and one Y chromosome

Most cases of sex-reversal can be explained by rare *crossover* events between X and Y chromosomes during male meiosis near the **pseudo-autosomal region**, which is nearly identical between the two sex chromosomes and located near the ends of their short arms (Figure 16.2). Y-specific DNA markers found in sex-reversed XX males were used to map TDF to a 35 kb segment near the pseudo-autosomal region of the Y chromosome (Figure S16.b). Sequencing of this segment revealed only one typical gene, which was termed the SRY^+ (for sex-determining region Y). Sequence analysis showed that the SRY^+ gene encodes a DNA-binding protein, indicating that SRY^+ controls the activity of other genes.

The SRY^+ gene *is necessary* for male development. A sex-reversed XY human female was found in which the *SRY* gene had a new point mutation; her father and brother both had a normal SRY^+ gene and were normal males. The paternity was confirmed by other DNA markers. Another sex-reversed XY human female was found to have a frame shift mutation in her SRY^+ gene. Corresponding observations were made on XY sex-reversed female mice. While the association between a mutated gene and a phenotype may be coincidental in any single case, the odds for this to happen in two or more unrelated individuals are extremely small.

Is the SRY^+ gene also *sufficient* for male development? This was tested by making XX mice transgenic for Sry^+, the mouse homologue of the human SRY^+ (see the assigned reading). Fertilized eggs were injected into one pronucleus with the cloned Sry^+ gene, as illustrated in Figure S16.c. Developing embryos were tested for being karyotypically XX and transgenic for Sry^+. From 11 such mice, three developed as males by the criteria of testis morphology, external genitalia, and copulatory behavior. Strictly, these results

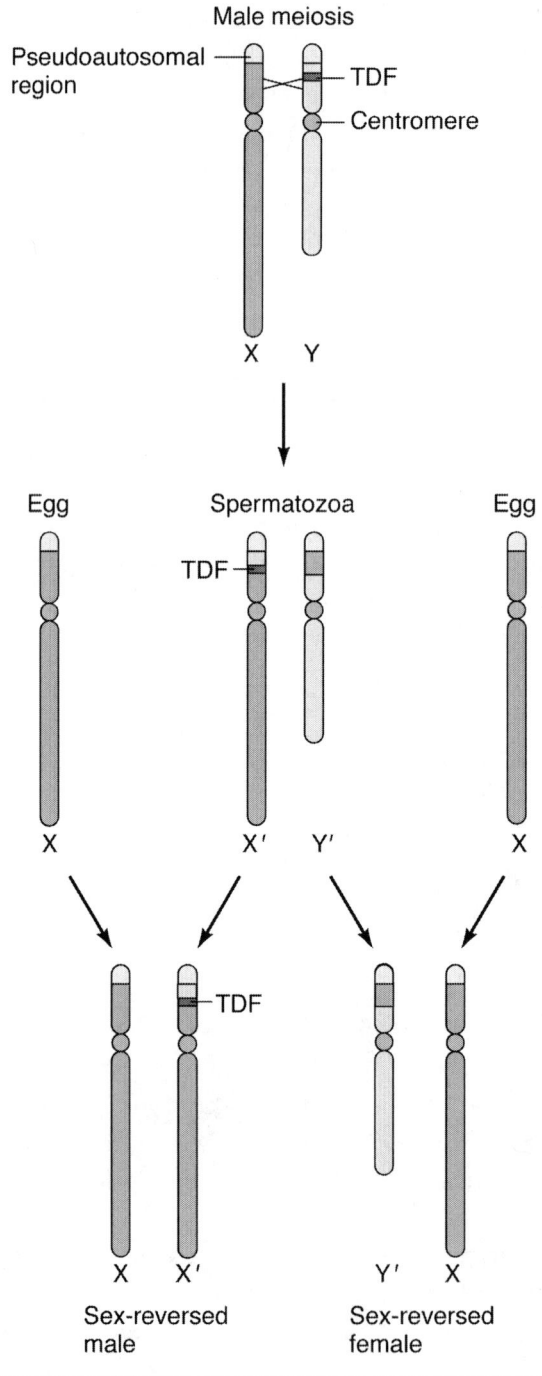

FIGURE 16.2:
Origin of sex-reversed humans by rare X-Y recombination during male meiosis.
The top diagrams show one chromatid each from the X chromosome and from the Y chromosome during meiotic prophase. If recombination events occur next to the pseudoautosomal region, and if the exchanged Y chromosome fragment contains the testis-determining factor (TDF), then the crossover generates an X-like chromosome (X') containing TDF and a Y-like chromosome (Y') lacking TDF. If sperm with such chromosomes fertilize then the resulting zygotes will develop into sex-reversed XX' males and XY' females, respectively.
From Kalthoff, K. (2nd ed. 2001) *Analysis of Biological Dvelopment*, New York: McGraw-Hill, p. 701

only show that the Sry^+ gene *can be* sufficient for male development. However, the investigators argue that, in the remaining eight cases, the transgene may have been lost from the developing gonad, or may not have been properly expressed. Such failures of transgenes injected into a pronucleus have indeed been observed in other experiments (see Chapter 11). Despite the ambiguity of these results, most researchers in the field assume that the SRY^+ gene alone *is* sufficient to act as TDF.

C. The Earliest Sexual Difference Appears in the Gonad

Early during development, the gonads of developing XX and XY embryos cannot be distinguished under the microscope. In humans, this **sexually indifferent stage** lasts through the sixth week of development. The first morphological difference between XX and XY mammalian embryos appears in so-called **supporting cells**, which are somatic cells surrounding the germ line cells in the developing gonad. In mouse embryos, the SRY^+ gene is transcribed in the supporting cells of male fetuses right before testis development becomes morphologically evident.

Beyond the indifferent stage, the SRY^+ gene acts as a switch that affects the further development of the supporting cells and the entire gonad (Figure 16.3). Supporting cells in which the SRY^+ gene is expressed differentiate as **pre-Sertoli cells,** and the gonad develops into a testis. If SRY^+ is absent or fails to be expressed, the supporting cells differentiate as **pre-follicle cells**, and the gonad develops into an ovary.

The first development of testicular versus ovarian structures after the indifferent stage is called **primary sex differentiation**. It is controlled by the presence versus absence of SRY^+ and occurs independently of sex hormones. Rather, primary sex

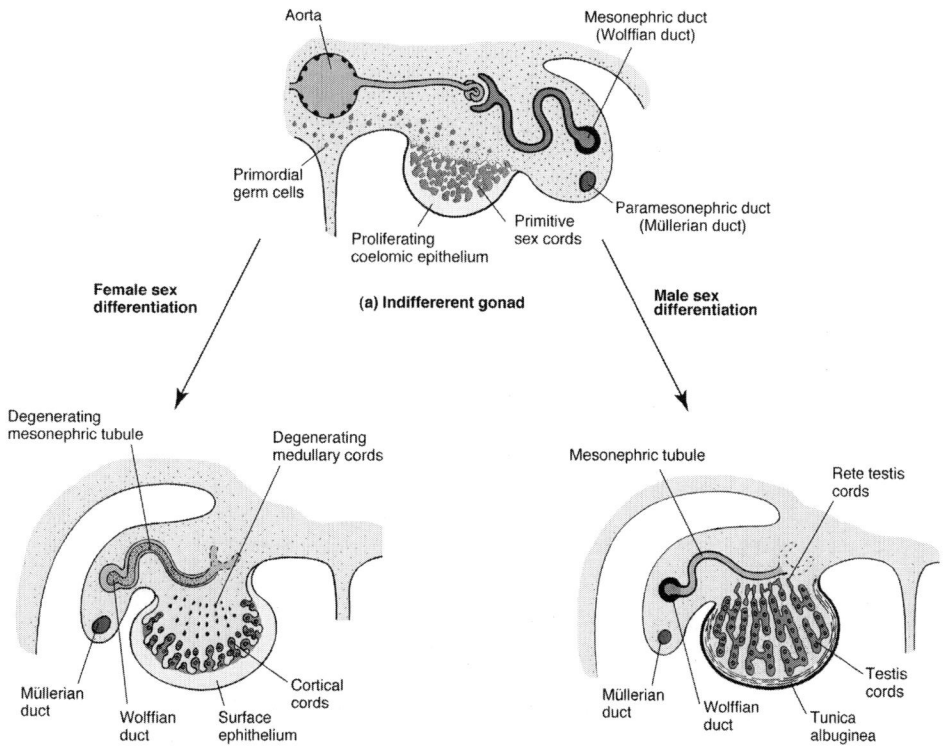

FIGURE 16.3:
Primary sex differentiation in the human embryo. All diagrams show transverse sections. **(a)** Indifferent gonad during 6th week of gestation. **(b)** Ovary during 7th week. Cortical portions of primitive sex cords form follicles. **(c)** Testis during 8th week. Medullary portions of primitive sex cords form testis cords. From Kalthoff, K. (2nd ed. 2001) *Analysis of Biological Dvelopment*, New York: McGraw-Hill, p. 698

differentiation includes the formation of those gonadal cells that will produce sex hormones during the following phase.

Under the influence of SRY^+, the **medulla** (central portion) of the gonad gives rise to horseshoe-shaped **testis cords**, which later form **seminiferous tubules**, where primordial germ cells develop into sperm (Figure 16.4). Two cell types in the testis are important for hormone synthesis.

- **Interstitial cells** (a.k.a. **Leydig cells**), located between the testis cords of the fetus and between the seminiferous tubules of the adult, produce *testosterone*.
- **Sertoli cells**, which develop from supporting cells in the fetal testis cords, produce *anti-Müllerian duct hormone (AMH)*. Sertoli cells are particularly sensitive to endocrine disrupting chemicals (see Chapter 25).

In the absence of SRY^+, the **cortex** (outer portions) of the gonad develop into **ovarian follicles**, each with an *oocyte* surrounded by **follicle cells** (Figure 16.5). Two cell types in the adult ovary are important for hormone synthesis.

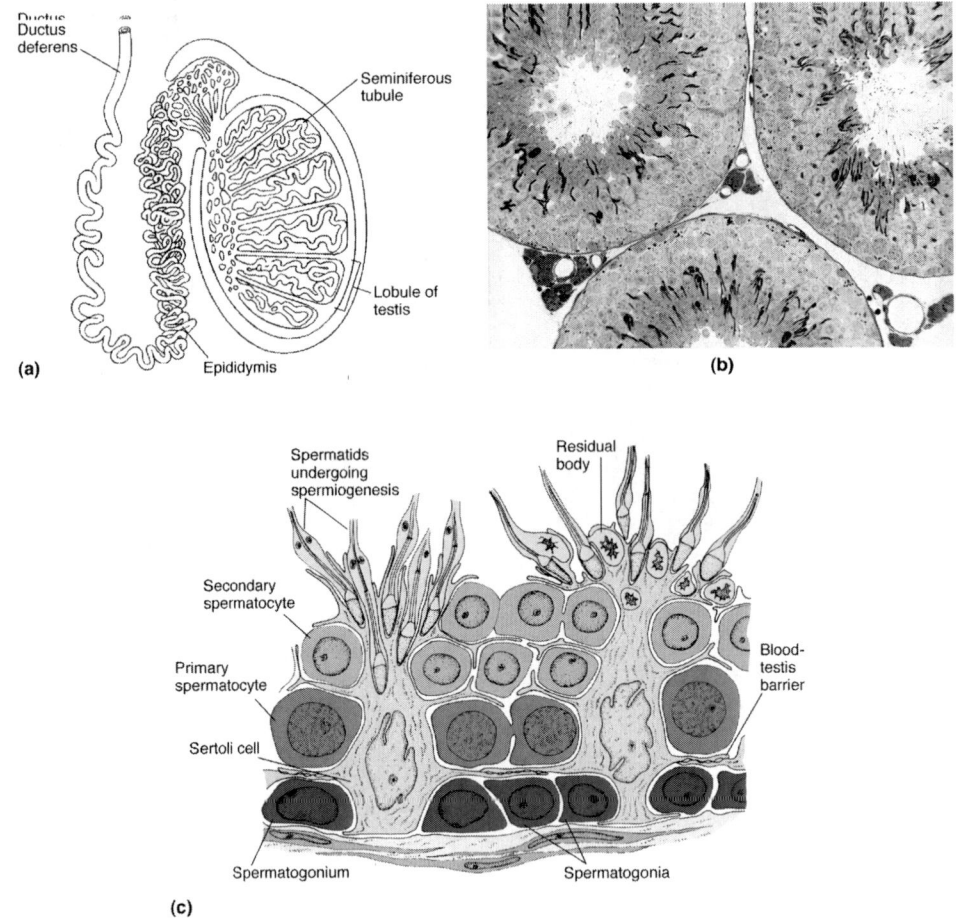

FIGURE 16.4:

Seminiferous tubule in mammalian testis. (a) schematic drawing of testis **(b)** photograph of seminiferous tubules in cross section. Cells between the tubules are the interstitial cells of Leydig. **(c)** drawing of a segment of seminiferous tubule. The lumen of the tubule is at the top of the drawing.
From Kalthoff, K. (2nd ed. 2001). *Analysis of Biological Development,* New York: McGraw-Hill, p.56.

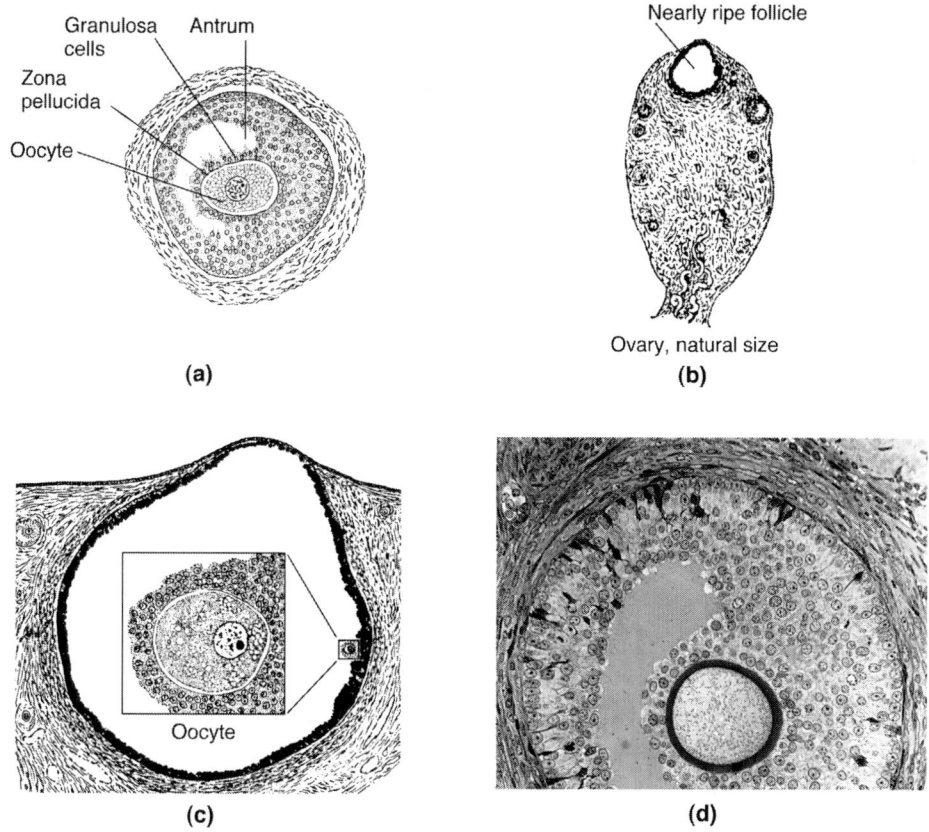

FIGURE 16.5:
Development of human oocytes in the ovary. (a) human follicle during antrum (fluid-filled cavity) formation, **(b)** total human ovary, **(c)** mature follicle, inset shows oocyte, **(d)** photomicrograph of a monkey follicle during antrum formation. The oocyte is surrounded by the zona pellucida (a clear layer appearing dark in this preparation) and by granulosa (somatic follicle) cells.
From Kalthoff, K. (2nd ed. 2001) *Analysis of Biological Development*, New York: McGraw-Hill, p. 70

- The follicle cells surrounding the oocyte, known as **granulosa cells**, synthesize *estrogen* from its precursor, testosterone.
- The **theca cells** surrounding the granulosa cells synthesize testosterone.

The indifferent gonad develops adjacent to the embryonic kidney (see Figure 16.3). Later, the ovaries descend somewhat from that position, whereas the testes descend further into the scrotum around the seventh month of gestation.

In summary, the functional parts of the testis and ovary develop from *complementary* portions of the indifferent gonad. In the presence of SRY^+ activity, the medulla of the gonad forms seminiferous tubules and interstitial cells. In the absence of SRY^+ activity, the cortex of the gonad forms ovarian follicles.

D. Development of the Genital Ducts and External Genitalia

In contrast to primary sex differentiation, which is controlled genetically, the following period of **secondary sex differentiation** is controlled by hormones, in particular by the male sex hormones testosterone and AMH. These hormones control first the development of the genital ducts and the external genitalia. Note that secondary sex differentiation begins much earlier than the development of *secondary sex characteristics*, such as breasts, facial hair, and so on, which appear at puberty.

As the gonads develop, a set of ducts that will transport either sperm or eggs also develops. These ducts originate from pairs of adjoining embryonic precursors, the *Wolffian ducts* and *Müllerian ducts*. Both ducts are present next to each other at the end of the sexually indifferent stage (see Figure 16.3). The **Wolffian duct** is also called the **mesonephric duct** because it originates as part of the *mesonephros*, an embryonic kidney. The mesonephros functions in both male and female embryos until it is superseded by the definitive kidney. Like other elongated anatomical structures, each reproductive duct has a **proximal** end (closer to the center of the body or to another point of reference) and a **distal** end (away from the center of the body or from another reference point).

In the developing female, the **Müllerian ducts** develop while the Wolffian ducts degenerate. The proximal portions of the Müllerian ducts form the paired **oviducts** (Figure 16.6). The distal portions of the Müllerian ducts fuse in primates and form the unpaired **uterus** and the upper portion of the **vagina**.

In the developing male, the entire mesonephros—including the Wolffian duct—becomes part of the reproductive system (Figure 16.7). Here, the proximal part of each duct forms an **epididymis**, a coiled tube adjacent to the testis that collects the developing sperm from the *seminiferous tubules* (see Figure 16.4) via the *efferent tubules*. The middle part of the Wolffian duct forms the **ductus deferens**, a long tube that

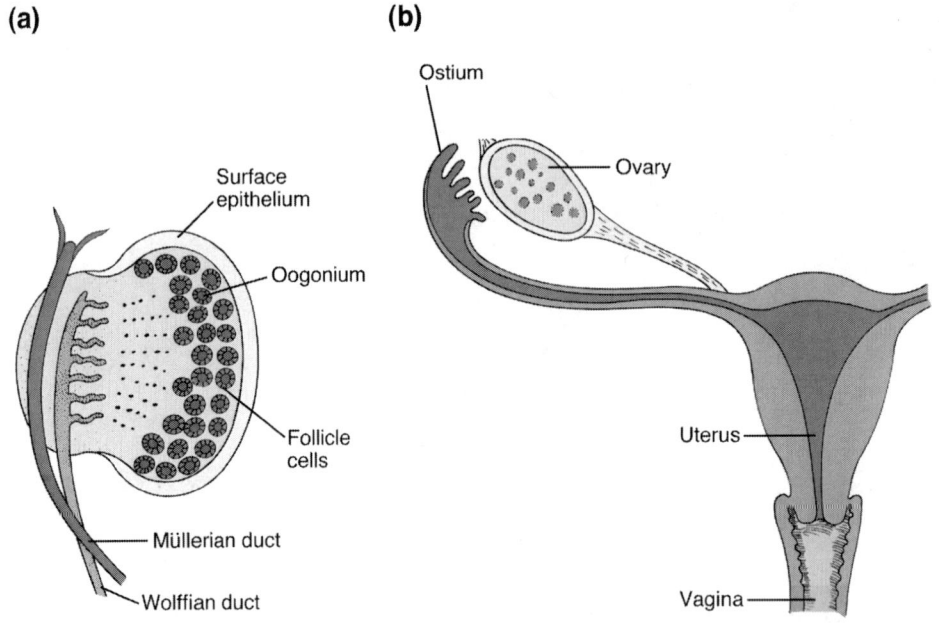

FIGURE 16.6:
Genital duct development in the human female. (a) Fourth month of gestation. Both Müllerian and Wolffian ducts are in place. **(b)** At birth, the proximal protion of each Müllerian duct has formed an oviduct. The distal portions of both Müllerian ducts have fused to form uterus and upper vagina. The Wolffian duct has degenerated.
From Kalthoff, K. (2nd ed. 2001), *Analysis of Biological Development,* New York: McGraw-Hill, p. 721

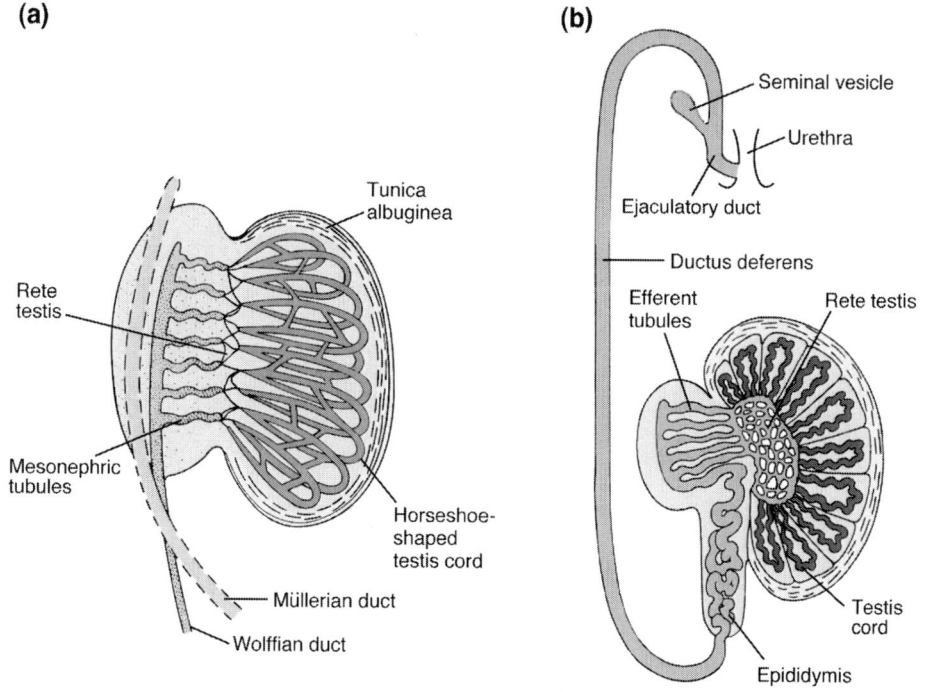

FIGURE 16.7:

Genital duch development in the human male. (a) Fourth month of gestation. The proximal portion of the Wolffian duct is embedded in the mesonephros. **(b)** The mesonephric tubules have become efferent tubules of the testis. The Wolffian duct has formed the epididymis, ductus deferens, seminal vesicle, and ejaculatory duct. The ducts from each side open into the urethra.
From Kalthoff, K. (2nd ed. 2001), *Analysis of Biological Development,* New York: McGraw-Hill, p. 721

transports the sperm to the prostatic gland at the base of the penis. The distal part of the Wolffian duct forms the **seminal vesicle**, which contributes much of the spermatic fluid, and the terminal portion of the ductus deferens, called the **ejaculatory duct**. The two ejaculatory ducts from each side open into the unpaired *urethra*, which also drains the urinary bladder. The Müllerian ducts of the male disintegrate.

Thus, as in the case of the gonad, the male and female genital ducts derive from *complementary* portions of the indifferent urogenital system.

In contrast to both gonads and genital ducts, the **external genitalia** of males and females grow in different ways from *the same* embryonic rudiments (Figure 16.8).

At the end of the indifferent stage, the external genitalia consist of

- the **urogenital groove**, in which the Müllerian ducts and the urethra with the Wolffian ducts end
- the **urethral folds** on both sides of the urogenital groove
- the **genital tubercle**, where the urethral folds unite anteriorly
- the **genital swellings** on both sides of the urogenital folds

This originally uniform set of embryonic rudiments turns into the very different external genitalia of males and females through sex-specific patterns of growth and morphogenesis, which are controlled by sex hormones.

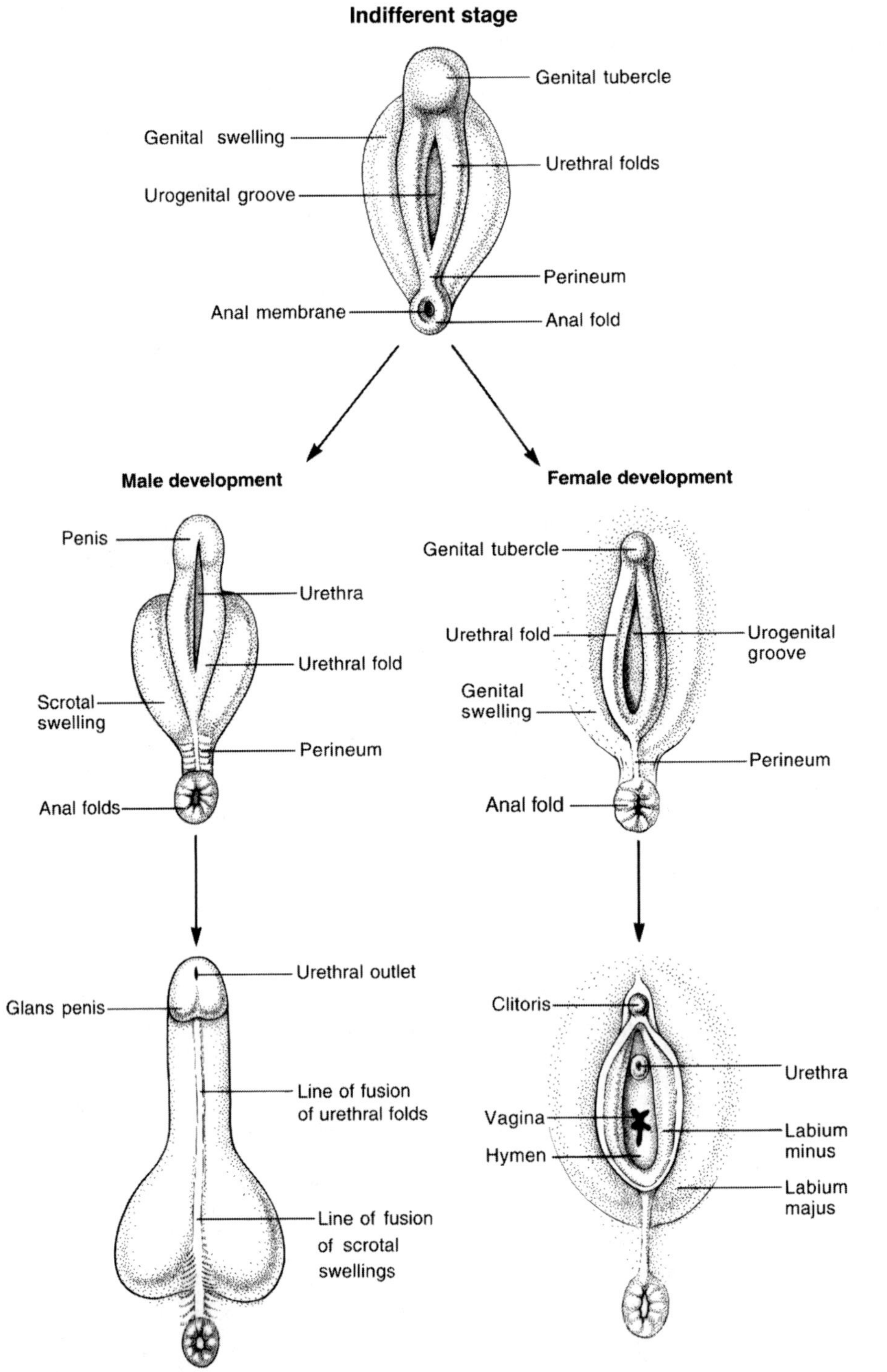

FIGURE 16.8:
Development of the external genitalia in the human. The genital tubercle forms either the corpora cavernosa of the penis or all of the clitoris. The urogenital groove becomes either the penile urethra or the vestibule of the vagina. The urethral folds form either the corpus spongiosum of the penis or the labia minora. The genital swellings give rise to either the scrotum or the labia majora.
From Kalthoff, K. (2nd ed. 2001) *Analysis of Biological Development,* New York: McGraw-Hill, p. 723

E. Synthesis and Action of Sex Hormones

A **hormone** is defined as a chemical signal that is produced in specific glands, is transported by the bloodstream, and acts on specific target tissues. The specificity of hormone action relies on the distribution of hormone receptors and other signaling components, which control the expression of stage-dependent and tissue-specific sets of target genes.

Most of the mammalian sex hormones are steroids (Figure 16.9).

- Four synthetic steps metabolize cholesterol to **progesterone.**
- Three additional steps transform progesterone into **testosterone.**
- Testosterone is converted by an enzyme, **5-α-reductase**, into **dihydrotestosterone (DHT)**. Testosterone and DHT are collectively referred to as **androgens** (male-generating hormones).
- Another enzyme, **aromatase**, converts testosterone into *17-β-estradiol*, commonly called **estrogen**.
- In addition to being a precursor of testosterone, progesterone is also metabolized into non-sex steroids, such as aldosterone and cortisol.

The principal organs of steroid hormone synthesis are the gonads and the adrenal glands, although some critical synthetic steps occur also in other tissues. The conversions of testosterone into either DHT or estrogen are controlled by the distributions of 5-α-reductase and aromatase. For example, 5-α-reductase in the indifferent external genitalia produces DHT, which promotes the development of male external genitalia. In females, the granulosa cells of ovarian follicle cells contain aromatase, which converts testosterone into estrogen.

Steroid hormone receptors act as *transcriptional regulators*, binding directly to regulatory regions of their target genes (Figure S16.d).

Anti-Müllerian duct hormone (AMH) is a glycoprotein (not a steroid), synthesized in the *Sertoli cells* of the testis (see Figure 16.4). AMH belongs to the transforming growth factor β (TGFβ) superfamily, acting on receptors in the plasma membrane of its target cells and controlling transcription indirectly.

By controlling different sets of target genes, sex hormones elicit different cell responses including

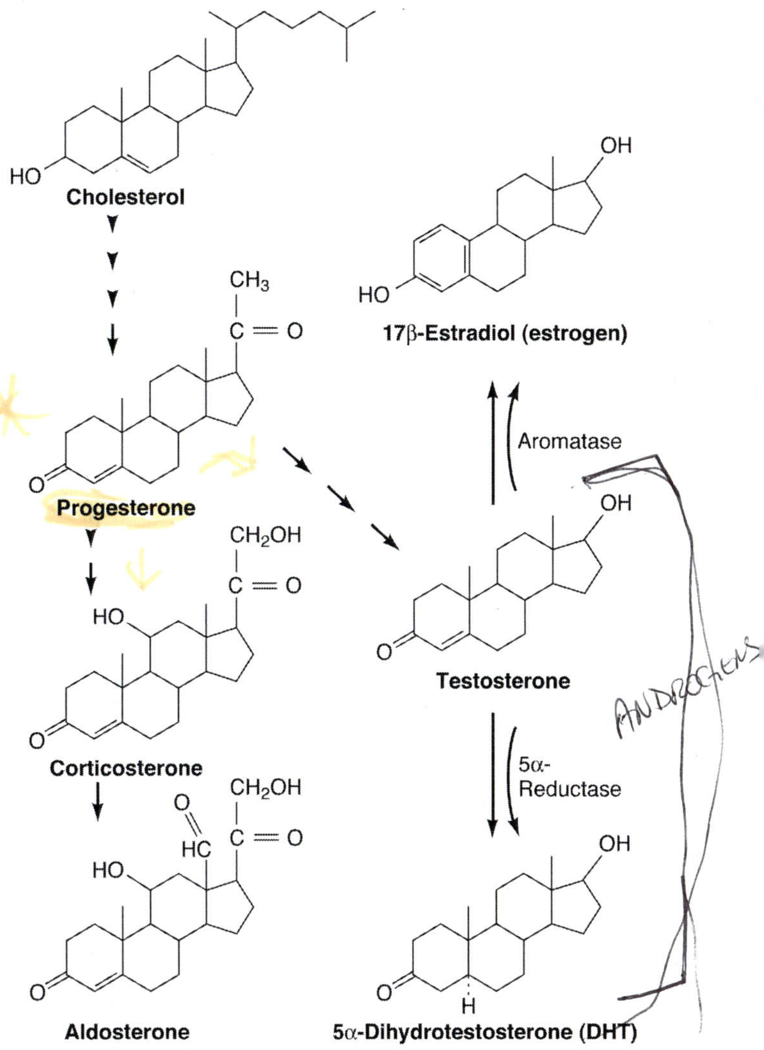

FIGURE 16.9:

Synthetic pathways for vertebrate sex steroid hormones. Testosterone is synthesized from progesterone, which in turn in metabolized from cholesterol. In the presence of the enzyme aromatase, testosterone is converted into estrogen. In the presence of 5α-reductase, testosterone is converted to 5α-dihydrotestosterone. Progesterone gives also rise to non-sex steroids incluing corticosterone and aldosterone.
From Kalthoff, K. (2nd ed. 2001) *Analysis of Biological Development*, New York: McGraw-Hill, p. 718

growth, division, adhesion, synthesis of specific proteins, tissue disintegration into cells, and cell death. These responses may lead to remarkable macroscopic changes in growth and development.

F. Hormonal Control of Fetal Sexual Development

The sexual differentiation of the genital ducts and external genitalia during fetal development is controlled by three sex hormones: testosterone, DHT, and AMH (Figure 16.10). As mentioned earlier, this process is called *secondary sex differentiation*, which should not be confused with the development of *secondary sex characteristics* after puberty.

The critical role in fetal development of blood-borne factors from the testis was first inferred from observations on farm animals. Female calves are masculinized ("freemartins") if they are connected through the uterine circulatory system with a male co-twin. The dominant role of male sex hormones was shown by experiments with fetuses of laboratory mammals. Gonad removal from XY fetuses leads to female development of genital ducts and external genitalia, whereas gonad removal from XX fetuses does not disturb prenatal female development. Thus, male secondary sex

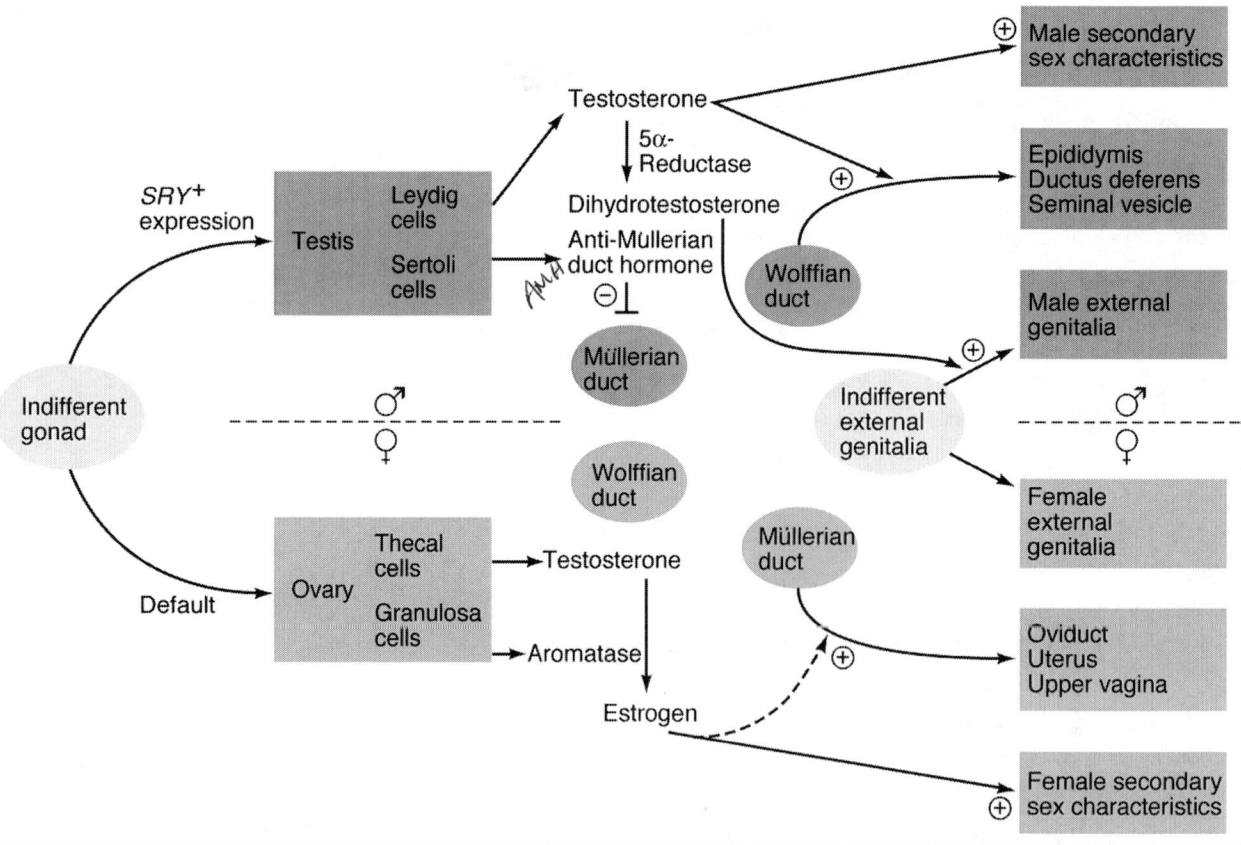

FIGURE 16.10:
Somatic sexual development in mammals. Primary (gonadal) sex differentiation depends on *Sry⁺* gene expression, whereas secondary sex differentiation is controlled by sex hormones.
From Kalthoff, K. (2nd ed. 2001), *Analysis of Biological Development,* New York: McGraw-Hill, p. 720

differentiation depends on testicular hormones, whereas female fetal development occurs mostly by default.

The function of more than one male hormone in controlling genital development was shown by comparing the results of two experiments. When a *testis* was transplanted into a female fetal rabbit, the Wolffian duct developed and the Müllerian duct regressed on the operated side. However, when *testosterone* was implanted instead of a testis, the Wolffian duct developed but the Müllerian duct did not regress. It follows that the normal disintegration of the Müllerian duct in the male depends on a testicular hormone other than testosterone. This additional hormone was termed *anti-Müllerian duct hormone (AMH)*.

The hormonal control of secondary sex differentiation was further elucidated by the study of human genetic disorders that interfere with the synthesis of sex hormones or with the function of their receptors.

The dominant effect of testosterone and its metabolites is demonstrated by **congenital adrenal hyperplasia (CAH)**, a disorder characterized by excessive growth of the adrenal glands. The growth is stimulated, through a feedback loop involving the pituitary gland, because the adrenal glands fail to metabolize progesterone into aldosterone and cortisol. The overly large glands produce elevated levels of progesterone, part of which is metabolized into testosterone. The most obvious effect in newborn girls with CAH is the ambiguous morphology of their external genitalia (Figure S16.e). For instance, the genital tubercle develops to an intermediate between a clitoris and a penis, and the urogenital groove may be closed. (This is called as hermaphroditism in the medical literature. However, biologists refer to it as *pseudohermaphroditism,* since true *hermaphroditism* is defined by the presence of both testicular and ovarian gonadal tissue.)

The effects of anti-Müllerian duct hormone (AMH) in the functional absence of testosterone and DHT is demonstrated by a human genetic disorder known as **androgen insensitivity (AIS)**. It is caused by a mutation in the X-linked gene encoding the *androgen receptor*, that is, the receptor for both testosterone and DHT. XY individuals with the mutant allele develop testes that produce androgens and AMH as in normal males. However, due to the lack of androgen receptor, the target organs cannot respond. The severity of symptoms in persons with androgen insensitivity syndrome (AIS) depends on how much residual androgen receptor activity is left.

In individuals with severe AIS, testes develop but do not descend. The Wolffian duct degenerates as the default program in the absence of androgen receptor signaling. The Müllerian duct disintegrates as a result of AMH action. Female external genitalia, including a shallow vagina, develop as the default program in the absence of androgen receptor signaling. At puberty, no menstrual cycle develops, but estrogen metabolized from testosterone is sufficient to promote full development of female secondary sex characteristics (Figure S16.f). The latter observation shows that estrogen receptors and any cofactors necessary for the development of female secondary sex characteristics are also present in males.

The hormonal control of external genitalia development by DHT is illustrated by a heritable form of male pseudohermaphroditism that is frequently observed in the Dominican Republic. The disorder is has been traced to a recessive mutation causing **5-alpha-reductase deficiency (ARD)**, which renders fetuses unable to convert testosterone to dihydrotestosterone (DHT). The lack of DHT inhibits the development of male external genitalia but has no known effects in girls. Boys with ARD are born with undescended testes and normal Wolffian duct derivatives, but their external genitalia range from ambiguous to apparently female. Consequently, boys with ARD have traditionally been raised as girls. However, at puberty, the high level of testosterone partially compensates for the lack of DHT in the external genitalia. As a result, a small penis develops, and testes may descend. Hence, boys with ARD are called **guevedoces**, which means "penis at 12 years."

G. Hormonal Control of Postnatal Sexual Development

Hormone production by fetal gonads is stimulated by *gonadotropins from the placenta*. Testosterone, DHT, and AMH levels are high in male fetuses, whereas estrogen synthesis is at low levels in both sexes and has only minor effects during fetal development of humans and many other mammals. Between birth and puberty, sex hormone levels are comparatively low.

With the onset of **puberty**, gonadal production of either testosterone or estrogen reaches high levels, stimulated by *gonadotropins from the pituitary gland*. Either sex hormone has major effects on physical as well as psychosocial development. The increased hormone levels affect the further development of primary sex characteristics including gonads, genital ducts, and external genitalia. Secondary sex characteristics, such as breasts, facial hair, width of pelvis and shoulders, muscle and fat distribution, tone of voice, also become more prominent. Along with the physical developments, there are physiological changes including menstruation in girls and erections in boys. Emotions, relationships, interests, and priorities change as well.

The power of sex hormones during puberty is particularly striking in *guevedoces*. They are traditionally raised as girls, helping their mothers around the house rather than being with their fathers farming or on jobs away from home. However, at puberty, guevedoces develop male secondary sex characteristics, and most of them assume a male sexual identity. They have erections, begin to masturbate, are attracted to women, and eventually have sexual intercourse. They are usually infertile, and some enter common-law marriages with women who have children from previous marriages. They aspire to male gender roles, and they often choose typically male occupations.

The guevedoces demonstrate that testosterone can have major effects on genital development even after puberty, and that late biological development as a male can promote a switch from female to male gender identity and gender roles. We will explore these topics further in Chapters 17 and 20.

EXERCISE PAGE FOR CHAPTER 16

Student Last Name _____

Student First Name _____

Discussion Section _____

On the course web site, use the link on the **syllabus** to find a pdf file of your **assigned reading** for this chapter. The same reading is referenced below. Use the bulleted list of questions to test your knowledge of this reading.

Koopman, P.; Gubbay, J.; Vivian, N.; Goddfellow, P.; and Lovell-Badge, R. (1991). Male development of chromosomally female mice transgenic for SRY^+. *Nature* **351**: 117–121.

- Which basic question did the investigators set out to answer?
- Outline the experimental strategy used in this study.
- How did the authors determine which of their experimental embryos were karyotypically XX and transgenic for Sry^+?
- What criteria did the authors use to identify developing mice as male?
- How do the authors explain the fact that only three out of 11 XX mice transgenic for Sry^+ developed as males? What additional test could have been done to test whether this explanation is valid?
- Can you think of an alternate explanation of why only some of the XX mice transgenic for Sry^+ developed into males?

Feedback

On a scale from 1 to 10, the being the best, please rank the above reading for

1. Interest and Relevance to the topic (10 being most interesting/relevant) _____
2. Readability (10 being most clearly written, easy to understand) _____

In the remaining space, enter any comments that you may have on this reading.

CHAPTER SEVENTEEN

HORMONAL CONTROL OF BRAIN DEVELOPMENT

The powerful effects of hormones on the development of primary and secondary sex characteristics have been discussed in the previous chapter. These effects are mediated by cell division, cell differentiation, and cell death, which in turn are explained by the molecular actions of hormones as regulators of gene expression. In particular, steroid hormones act by binding to receptors that act as transcription factors (Figure 17.1). It is therefore not surprising to see that hormonal control extends to brain development and behavior.

A. Sexual Differences in Behavior and Brain Activity

While many **differences in male and female behavior** are cultural, some of them are consistent across populations, including

- majority sexual orientation
- status-oriented aggression and risk-taking more prevalent in men
- depression and anxiety disorders more prevalent in women; stuttering, dyslexia, schizophrenia, and antisocial personality disorder more prevalent in men
- preferences for certain types of work and play, such as male preferences for dealing with tools and weapons and female preferences for clothes and child care

These are differences between *overlapping* distributions. In most populations, there will be *some* women who are more

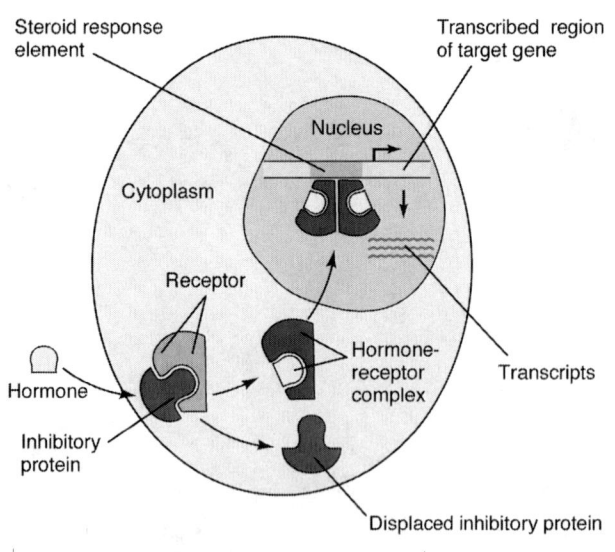

FIGURE 17.1:
Control of Gene Expression by Steroid Hormones. The target cells of steroid hormones have receptor proteins to which the hormones bind. The hormone-receptor complexes bind to response elements associated with target genes.
From *Analysis of Biological Development* by Klaus Kalthoff. Copyright © 1996 by McGraw-Hill. Reproduced with permission of The McGraw-Hill Companies.

aggressive than *average* men. Conversely, there will be some men who are more interested in clothes than average women. Nevertheless, the consistency across cultures of the differences in average behaviors suggests that there are some differences between the sexes with regard to brain activity, and possibly, brain anatomy.

With modern **imaging techniques**, including functional magnetic resonance imaging (fMRI), brain anatomy and brain activity patterns can be measured accurately. Generally, women use more brain regions than men to accomplish a given task. This may explain the long-standing observation that women recover better than men from localized brain damage caused by accidents or strokes: Women may have more brain areas that can compensate for the damage.

In an fMRI study of brain responses to viewing the faces of children, the investigators found sex-specific differences in the activation of the *fusiform gyrus* of the cerebral cortex. Generally, females showed greater activation than males. However, males showed greater activity than females in response to children's faces that resembled their own.

Another study, using positron emission tomography, compared the responses of heterosexual men, homosexual men, and heterosexual women to two human pheromones: 4,16-androstadien-3-one (AND) and estra-1.3.5(10),16-tetraen-3-ol (EST). The response patterns to AND and EST differed significantly between the three groups, whereas common odors were processed similarly.

A longitudinal MRI study has revealed **sex differences in the maturation patterns** of certain brain regions. For instance, frontal gray matter volume peaks around age 11.0 years in girls and 12.1 years in boys, whereas temporal gray matter volume peaks at about age at 16.7 years in girls and 16.2 years in boys. If these volumetric differences between the sexes were indicative of differences in functional maturation, this would be important for the timing of school curricula. A challenging new activity could be timely for the faster sex that has just completed the growth of related brain areas, while being frustrating for the slower sex.

The same longitudinal studies also provided a neurobiological justification for certain age limitations, such as for driving motor vehicles or drinking alcohol, which had so far been based on common sense: The dorsal lateral prefrontal cortex, important for controlling impulses, is among the latest brain regions to mature: It does not reach adult dimensions until the early 20s.

B. Distribution of Sex Hormone Receptors in the Brain

What could be the proximate causes for the observed sex-specific differences in brain growth, brain activity patterns, and behavior? Knowing that hormones are powerful regulators of both development and adult function, sex hormones are obvious candidates.

One possible explanation of the observed sexual dimorphisms in brain and behavior would be the assumption that males and females differ in the distribution of sex hormone receptors in the brain. For example, males may be more aggressive because they have more androgen receptors in brain regions that are involved in aggression. However, this hypothesis is inconsistent with related data. Figure 17.2 shows histological sections of lizard brain after *in situ hybridization* with labeled probes binding selectively to mRNAs encoding specific hormone receptors. As it turns out, the distributions of sex hormone receptors are *specific for each hormone*. For instance, the mRNA for androgen receptor is concentrated in brain regions involved in aggression and copulation, while the mRNA for progesterone receptor accumulates in areas involved in ovulation and sexual receptivity. Notably, *no obvious differences between the sexes* were found in these distributions.

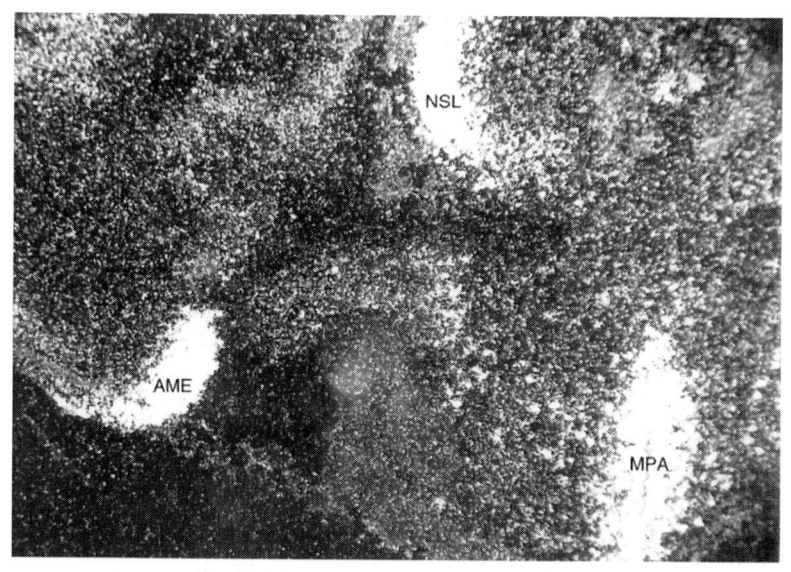

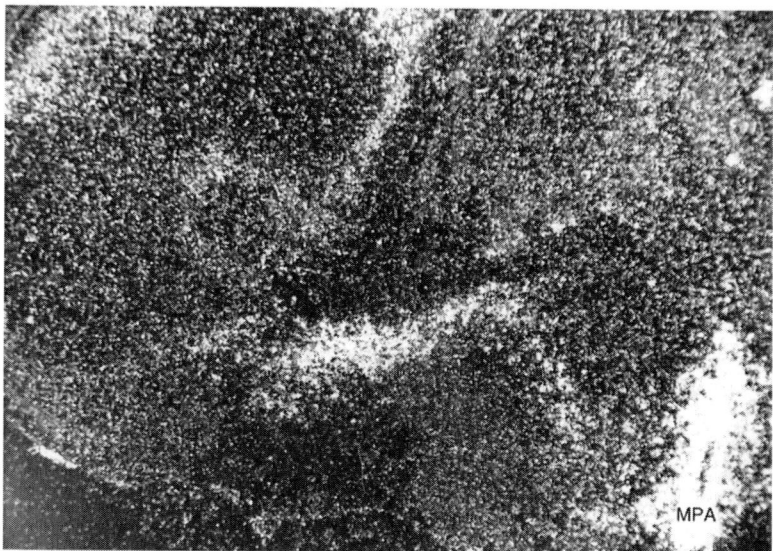

FIGURE 17.2:

Receptors for different sex hormones have distinct distributions in the brain. Photographs show frontal sections of lizard brains hybridized *in situ* with probes for mRNAs encoding specific hormone receptors. Androgen receptor mRNA (top) accumulates in brain areas involved in aggression and copulation, such as the external nucleus of the amygdala (AME) and the nucleus septalis lateralis (NSL). Progesterone receptor mRNA (bottom) accumulates in areas involved in ovulation and sexual receptivity. Other brain areas, including the medial periventricular area (MPA) show mRNAs for both receptors.
Image courtesy L. Young and D. Crews.

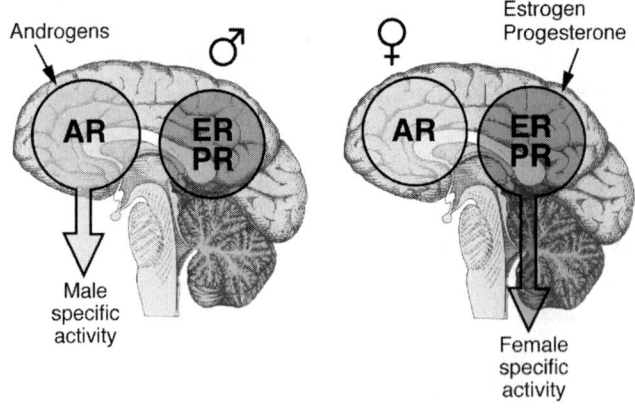

FIGURE 17.3:
Sex-specific patterns of brain activity caused by distinct distributions (simplified for conceptualization) of the receptors for androgens (AR), estrogen (ER,) and progesterone (PR). Even though the receptor distributions do not differ between men and women, different sex hormone levels cause sex-specific brain activity patterns.

However, the distinct distributions observed for each sex hormone receptor will translate into **sex-specific patterns of brain activity** due to the different concentrations of sex hormones in the bodies of males and females. Figure 17.3 illustrates this concept in a simplified way. Thus, *receptor-specific* distributions generate *sex-specific* patterns of brain activity dependent on which sex hormones are currently present.

If this hypothesis were to explain *all* sex-specific differences in animal behavior—meaning if all sexual dimorphisms in animal behavior were caused by different sex hormone levels *at the time of the behavior*—then

- an ovariectomized female injected with testosterone should behave like a normal male
- a castrated male injected with estrogen and progesterone should behave like a normal female

For the most part, these predictions have been confirmed by observations on appropriately manipulated animals (see Section C). However, there are exceptions. Ovariectomized and testosterone-injected females differ from normal males in some behaviors, and corresponding differences have been found between normal females and castrated males injected with estrogen and progesterone. These differences are ascribed to *prenatal* exposure to sex hormones that lead to *irreversible changes* in brain development (see Sections D through H).

C. Current Hormonal Levels Control Seasonal Singing and Brain Morphology in Songbirds

Male songbirds (canaries, zebra finches) at the beginning of their breeding season produce complex stereotypical sounds (songs), which establish territories and attract females.

Like other courtship behavior, bird song is strongly affected by hormone levels. There is a linear correlation between singing (number of songs per 15-minute interval) and testosterone concentration in blood serum; singing activity parallels the seasonal fluctuation of testosterone concentration.

Castration of males dramatically reduces their singing, and testosterone implants restore singing in castrates (Figure 17.4). Testosterone implants may cause males to sing out of season and may cause females to sing as well. These data indicate that testosterone or one of its metabolites (see Figure 16.9) is necessary, and can be sufficient, for singing in songbirds. Injections of testosterone metabolites into castrated zebra finches indicate that combinations of estrogen and dihydrotestosterone (DHT) restore singing to pre-castration levels. Estrogen and testosterone also promote different phases of song learning.

Certain nuclei in the brains of canaries and zebra finches are sexually dimorphic (Figure 17.5). Some nuclei are up to six times larger in males than in females during courtship. The same nuclei selectively bind radiolabeled testosterone and increase in size when females are injected with testosterone. The larger nuclei have more neurons, larger neurons, and more synapses. To put these results in perspective, it is important to realize that birds and mammals differ in the turnover of neurons in their brains: Neurons die while new ones are formed from precursor cells called neuroblasts much more rapidly in birds.

D. Prenatal Exposure to Sex Hormones Affects Adult Behavior and Brain Anatomy

Rodents have been favorite animals for laboratory studies on reproductive behavior because they show stereotypical, easy-to-score behaviors during copulation. Males mount (females and smaller males) from behind while receptive females show a behavior called **lordosis**: They arch their back to elevate rump and head while moving their tail to one side.

Lordosis depends on *both prenatal and adult* hormone levels, as indicated by several sets of experiments.

- Females cease to exhibit lordosis after removal of ovaries.
- Injection of estrogen and progesterone restores lordosis in ovariectomized females.

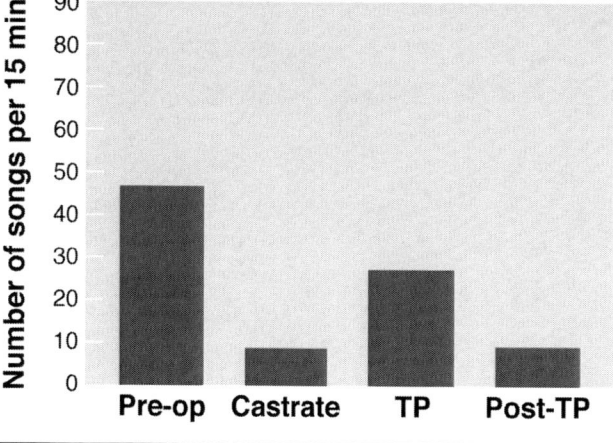

FIGURE 17.4:

Hormonal control of singing in male zebra finches. Frequency of singing is shown for normal males (Pre-op), castrated males (Castrate), castrates with implants releasing testosterone propionate (TP), and castrates after removal of TP implants (Post-TP).
From *Analysis of Biological Development* by Klaus Kalthoff. Copyright © 1996 by McGraw-Hill. Reproduced with permission of The McGraw-Hill Companies.

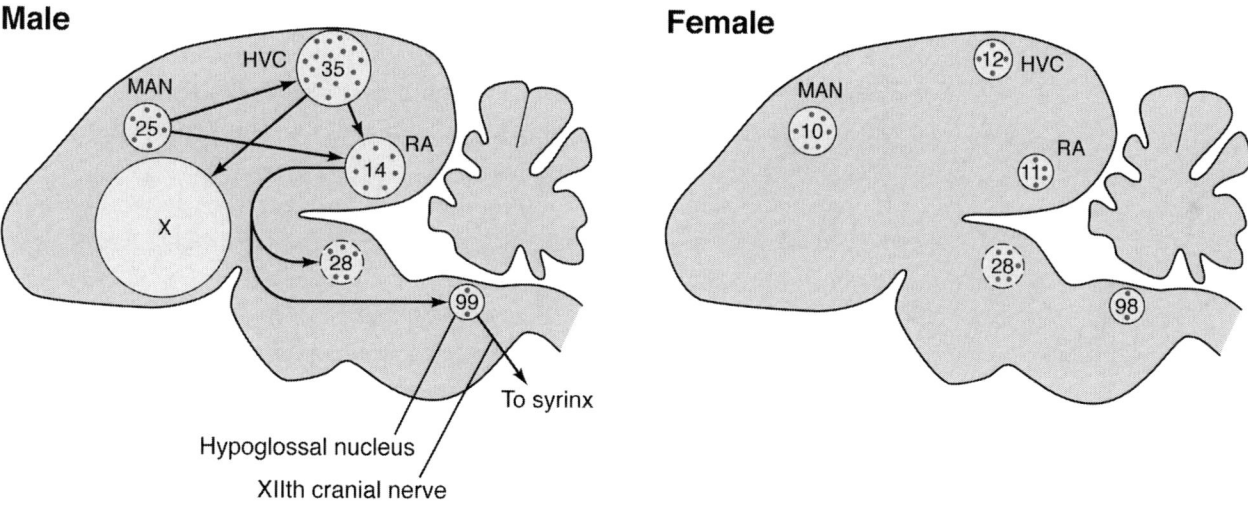

FIGURE 17.5:

Sexual dimorphism of control nuclei in the songbird brain. Drawings show paramedian sections, anterior to the left, dorsal surface up. The size of each circle is proportional to the volume of the corresponding brain region. Dots mark regions that become labeled after injection of radioactive testosterone. The number within each circle indicates the approximate percentage of labeled cells within each region.
From *Analysis of Biological Development* by Klaus Kalthoff. Copyright © 1996 by McGraw-Hill. Reproduced with permission of The McGraw-Hill Companies.

- The same hormone injections that restore lordosis in ovariectomized females *do not* promote lordosis in castrated males.
- Females exposed to testosterone prenatally *do not* show lordosis when injected with estrogen and progesterone and then mounted by a male.

Taken together, the results indicate that some sexual differences in behavior depend not only on current hormone levels but also on earlier events of sex differentiation.

In those mammalian species where the blood circulations of sibling fetuses are closely connected, the genital development and sexual behavior of each offspring are affected by the intrauterine vicinity of male versus female siblings. This was first observed in female cattle born with a male co-twin. Such "freemartins" have female (XX) karyotypes and external genitalia, but they are infertile and behave similarly to castrated males (steers). First observed early in the twentieth century, these characteristics were ascribed to permanently masculinizing factors that travel through vascular connections between the placentas of twin cattle.

Such observations on cattle and other farm animals were followed up in controlled experiments with laboratory rodents. In these animals, the Müllerian ducts form two uterine "horns" opening into a common vagina. During pregnancy, five to six fetuses are lined up serially in each horn. In experiments amounting to birth by Cesarean section, rodents can be removed from their mothers right before term and classified as

- 0M if they had no male neighbor *in utero*
- 1M if they had one male neighbor *in utero*
- 2M if they had two male neighbors *in utero*

When 0M females were compared in their development and adult behavior to 2M females, the latter showed masculinized external genitalia, took longer to reach puberty, had fewer reproductive cycles, were less attractive to males, and were more aggressive to other females.

Corresponding effects on males are observed in rodent species that produce fetal estrogen, such as mice. As compared to 0M males, 2M males had larger seminal vesicles, were more aggressive, and (somewhat counterintuitively) were sexually less active.

These results indicate that the sexual development of mammals depends not only on current but also on *prenatal* hormone levels.

In humans, certain spontaneous ultrasounds *produced* in the cochlea of the inner ear and propagated back to the tympanic membrane are masculinized in females with male twin brothers and in homosexual or bisexual females. However, the functions of these ultrasounds are unknown, and the development of human females with twin brothers seems to be normal in other respects. It seems that the vascular connection between the placentas of dizygotic human twins is not close enough for one twin to have significant hormonal effects on the other.

E. The Organizational Hypothesis

Having established a long-lasting control of prenatal hormone levels over reproductive *behavior* in rodents, researchers scanned the brains of rodents and humans for *anatomical brain* structures that were sexually dimorphic (different in males versus females) and susceptible to prenatal exposure to hormones. Such areas were found in the **hypothalamus**, a central region on the underside of the brain marked by the

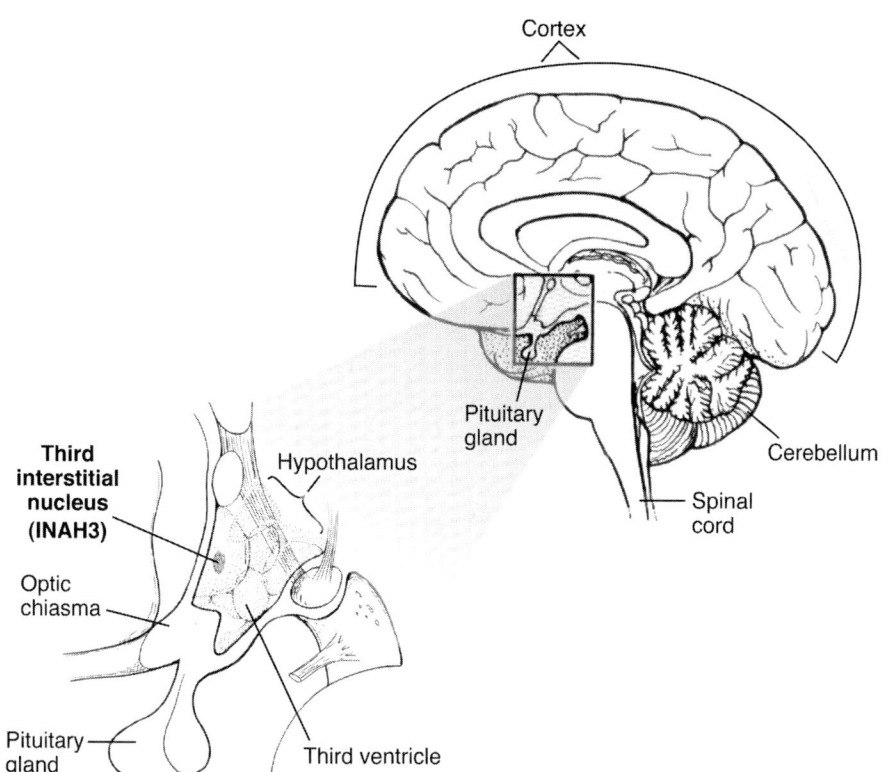

FIGURE 17.6:
The hypothalamus of the human brain surrounds a ventromedial extension of the third ventricle. The third interstitial nucleus of the anterior hypothalamus (INAH3) lies on both sides of the ventricle. Anterior to the left.

optic chiasma (partial crossing over of the optic nerves) and the attached pituitary gland (Figure 17.6). The hypothalamus controls many basic body functions including temperature, alertness, fluid uptake, hunger, and sex drive. It exerts its control functions through neural connections with other parts of the brain and the autonomic nervous system, and by stimulating the pituitary gland, which in turn controls other endocrine glands including the thyroid gland, the adrenal glands, and the gonads (see Figure 25.1).

The human hypothalamus contains a sexually dimorphic nucleus known as INAH3, for *i*nterstitial *n*ucleus of the *a*nterior *h*ypothalamus (see Figure 17.6). Histological examination of human brains *post mortem* showed that the average INAH3 size is about three times greater in men than in women, with considerable variation within each sex (Figure 17.7). However, the human INAH3 is small, and its sexual dimorphism would probably have gone undetected if there had not been an earlier study on a corresponding brain region in rats, known as the **sexually dimorphic nucleus in the preoptic area (SDN-POA)** of the hypothalamus (Figure S17.aa, ab).

The rat SDN-POA is larger in normal males than in females, and its development depends on **perinatal** (prenatal or shortly after birth) exposure to testosterone. The SDN-POA remains small in males castrated immediately after birth, and it grows to male size in females treated with testosterone shortly before and after birth (Figure S17.ac). The growth response of the SDN-POA to testosterone ends a few days after birth.

Testosterone seems to exert its effect on SDN-POA development through one of its metabolites, estrogen. This is indicated by the following observations. Male rats with genetic androgen insensitivity, which lack androgen receptors but do have estrogen and estrogen receptors, develop a male-sized SDN-POA that contrasts with their otherwise female appearance. Perinatal treatment of rats with large doses of estrogen also enlarges the SDN-POA (Figure S17.ad). Female rat fetuses seem

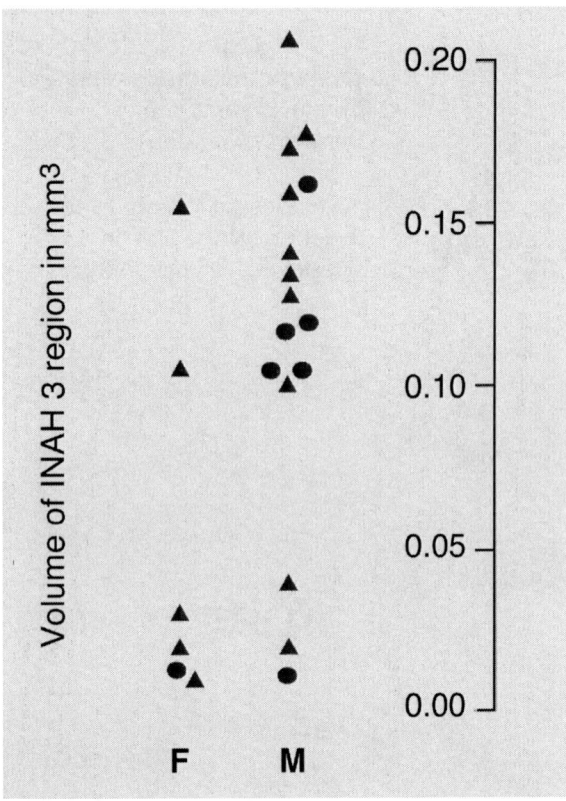

FIGURE 17.7:
Size of the human INAH3 nucleus in the brains of men and women. F: presumed heterosexual females M: presumed heterosexual males • infected with HIV ▲ not infected with HIV
Data from LeVay S. (1991) A difference in hypothalamic structure between heterosexual and homosexual men. *Science* 253: 1034–1037

to escape the masculinizing effects of estrogen by *alpha-fetoprotein (AFP)*, a blood serum protein that binds normal doses of estrogen, thus limiting its entry into the brain. Testosterone, which is not bound by AFP, circumvents this barrier.

The observed effects of sex hormones on vertebrate behavior and brain anatomy were summarized as the **organizational hypothesis:** Perinatal exposure to testosterone permanently alters the brain, causing it to function in male-specific ways later in life. The organizational hypothesis has received further support from the following observations on humans.

F. Congenital Adrenal Hyperplasia

Congenital adrenal hyperplasia (CAH) is a human genetic disorder, in which the adrenal glands cannot fully metabolize progesterone into cortisol and its downstream products including aldosterone. The lack of aldosterone triggers a feedback mechanism that causes the compensatory growth (hyperplasia) of the adrenal glands. The excess adrenal gland tissue produces more progesterone, part of which goes into the synthetic pathway for testosterone (see Figure 16.8). The most obvious effect in newborn girls with CAH is the ambiguous morphology of their external genitalia, which often show an enlarged clitoris and closed urethral folds (see Figure S16.e).

The physical symptoms of CAH in newborn girls can be corrected surgically, and the most of the resulting metabolic defects can be treated with cortisol. However, these treatments do not change a predisposition of girls born with CAH for "tomboyish" behavior. In contrast to their unaffected sisters, they

- prefer boys as playmates
- engage in rough-and-tumble play
- enjoy competing against boys in physical activities
- play as much with toy cars and toy guns as boys do and show little interest in dolls, clothes, jewelry, makeup, and their hair
- report fewer daydreams about marriage and having babies
- show greater frequencies of homosexuality and bisexuality as adults

The parents of CAH girls are either unconcerned about the girls' behavior or, if they treat them differently than their sisters, they often try to encourage more "feminine" behaviors. It seems therefore unlikely that the CAH girls' unusual behavior is caused by parental influence.

The most straightforward explanation of the CAH girls' behavior is along the lines of the organizational hypothesis, meaning that their perinatal exposure to elevated levels of testosterone has masculinized the development of their brains. Alternatively, one may assume that the tomboyish behaviors result from *pleiotropic* effects of the mutations causing CAH, or from *side effects* of the girls' permanent treatment with cortisol.

G. Progestin-Induced Pseudohermaphroditism

Observations similar to those made on CAH girls were made on girls with **progestin-induced pseudohermaphroditism (PIP)**, who were born in the 1960ies to women who were given injections with progestins to avoid miscarriages. Girls born to mothers so treated showed ambiguous external genitalia. The medical term for this condition is *progestin-induced hermaphroditism (PIH)*, based on the medical definition of *hermaphroditism* that nongonadal reproductive organs are ambiguous or incongruent with to the gonadal sex.

The ambiguous genitalia of girls with PIP were amenable to surgical correction as in the girls with CAH. The girls with PIP did not have to be treated with cortisol because, unlike CAH girls who themselves have a permanent genetic disorder, PIP girls have been affected by temporary hormone injections given to their mothers. However, the PIP girls grew up to show the same "tomboyish" behaviors as their CAH counterparts.

Progestins are steroids that have a low affinity for the androgen receptor, which is normally activated by testosterone or DHT. In this regard, the PIP cases are unintended human equivalents to the experimental injections of female rodents with testosterone, both leading to the same result of masculinizing genital morphology and behavior of female offspring. Therefore, the simplest explanation of PIP is again based on the *organizational hypothesis* that perinatal exposure to testosterone (and testosterone mimics) causes a permanent masculinization of brain development and behavior. Indeed, this seems to be the only plausible explanation of PIP because, other than the case of CAH, genetic pleiotropy and side effects of cortisol do not apply.

In summary, the behavior of girls with CAH or PIP is best explained as a result of perinatal activation of androgen receptors in the brain, in accord with the organizational hypothesis.

H. Sex Differences in Standardized Tests

There are sex-specific differences in the average performance of men and women in certain standardized tests. For instance, men tend to do better than women on spatial tasks, whereas women outperform men on perceptual speed (see the assigned reading).

Standardized tests of cognitive abilities have been given mostly to high school students since 1960. The results were analyzed for sex differences in terms of

- **effect size (d)**, expressed as the difference between mean scores for males and females, divided by the standard deviation of the entire population
- **variance ratio (VR)**, obtained as variance of male scores divided by variance of female scores
- **talent** or **non-talent**, expressed as percentages of males and females scoring in the extremely high and low ranges of the overall distribution (Figure 17.8)

On average, females did better on tests of reading comprehension, perceptual speed, and associative memory. Males did better on tests of mathematics, social studies, and certain vocational aptitudes (mechanical reasoning, electronics information, auto and shop information). The effect sizes (*d* values) of these differences were generally small (around 0.2) to moderate (around 0.5) except for the vocational aptitude tests, where they were large. Some of the sex differences have diminished since 1960. In particular, the earlier gap between boys and girls in average mathematics test scores have disappeared in countries with more gender-equal cultures.

There are also sex differences in variance, the variance being greater for males in almost all tests across the world. The variance ratio is around 1.15 for mathematics tests and near 2.0 for vocational tests. Generally, the variance ratio is higher in countries with higher average test scores.

Small differences in mean and greater variance in males compound to major differences in talent and non-talent (see Figure 17.8). In capabilities for which males have a greater mean value, their greater variance enhances male representation in the talent region. Conversely, in capabilities for which males have a smaller mean value, their greater variance exacerbates male representation in the non-talent region.

For mathematics, the ratio of male to female high school students was 1.6 in the top 5 percent, 2.1 in the top 3 percent, and 7.0 in the top 1 percent of the overall distribution. An even more dramatic ratio is seen in the Putnam Competition for U.S. college students in mathematics, which has been held since 1938. The entries are coded to conceal the gender and ethnicity of the contestants to the jurors. In 1997, Ioana Dimitriu, born and raised in Romania, was the first woman to win this prize. In 2004, 19 women finished among the top 200. (In this select group, mathematical ability was more than 4 standard deviations above the mean.)

For test scores in car mechanics, the ratio of males to females was greater than 8.0 in the top 10 percent and greater than 10 in the top 5 percent. Conversely, in reading comprehension, perceptual speed, and associative memory, the ratio of males to females was around 1.5 in the bottom 10 percent and 2.2 in the bottom 5 percent.

Comparisons of test scores between *normal* males and females say nothing about the origins of any sex differences found. Such differences could be either genetic or caused by gender stereotypes that lead boys and girls to emphasize or de-emphasize certain abilities during early stages of their mental development. Comparisons of test scores for special mental abilities between individuals with hormonal disorders (AIS, PIP, or CAH) and their unaffected relatives are more informative.

Relative to their unaffected sisters, the test scores for girls with CAH or PIP were generally shifted towards the male averages. Individuals with androgen-insensitivity

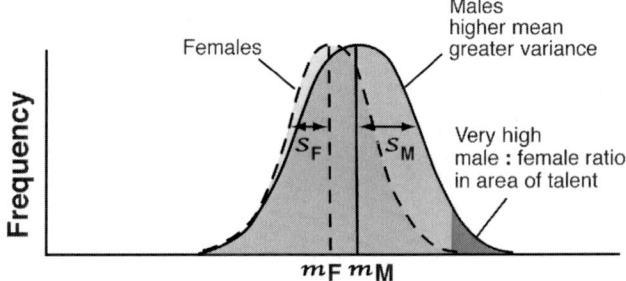

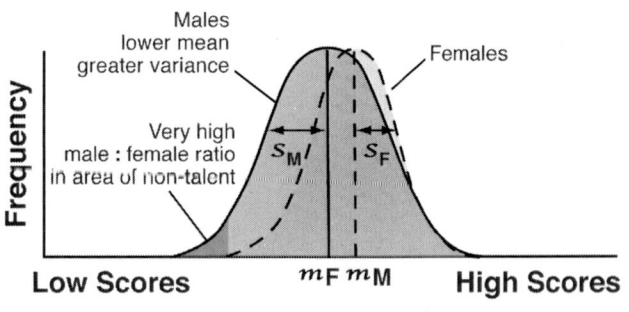

FIGURE 17.8:
Small differences in mean (*m*) and variance (shown as standard deviation, *s*) may compound to large differences in the areas of talent or non-talent.

syndrome (AIS) scored at the level of their female relatives in verbal skills but performed worse in spatial skills than both their male and female relatives. The simplest explanation for these results is that spatial skills are promoted by prenatal as well as current androgen levels. While normal females utilize the low androgen levels produced by the adrenal glands, AIS individuals cannot respond even to their higher androgen levels.

In summary, the test scores of individuals with AIS, PIP, or CAH indicate that prenatal exposure to sex hormones causes small differences in the spatial and language capabilities of men and women. These small differences are amplified in the extreme areas of talent and non-talent by the greater variance in the performance of males. The differences may also be affected by the timing of school curricula relative to sex-specific differences in brain development.

EXERCISE PAGE FOR CHAPTER 17

Student Last Name _____

Student First Name _____

Discussion Section _____

On the course web site, use the link on the **syllabus** to find a pdf file of your **assigned reading** for this chapter. The same reading is referenced below. Use the bulleted list of questions to test your knowledge of this reading.

Kimura, D. (1992). Sex differences in the brain. *Scientific American* **September 1992:** 118–125.

- In which types of tests do men and women, boys and girls perform differently on average?
- How large are these differences in terms of "effect size"?
- At which age do some of the differences appear?
- Does Kimura comment on the potential adaptive significance of these differences in voles and in humans?
- What evidence is there to indicate that the test scores are affected by current sex hormone levels?
- How do girls with CAH or PIP perform on these tests as compared to their unaffected sisters? How can these data be interpreted in terms of the organizational hypothesis and the hormonal imbalances caused by CAH and PIP?

Feedback

On a scale from 1 to 10, the being the best, please rank the above reading for

1. Interest and Relevance to the topic (10 being most interesting/relevant) _____
2. Readability (10 being most clearly written, easy to understand) _____

In the remaining space, enter any comments that you may have on this reading.

CHAPTER EIGHTEEN

INTRODUCTION TO SOCIOBIOLOGY

Edward O. Wilson, in a book entitled *Sociobiology* and published in 1975, proposed to merge the time-honored disciplines of ecology, evolution, behavior, and sociology into one new discipline, which he termed sociobiology. Some scientists prefer to call the new discipline *behavioral ecology*, because it deals with both social and nonsocial behavior. The subject of the new discipline is the evolution of animal and human behaviors under various ecological conditions.

A. The Basic Tenet of Sociobiology

The basic tenet of sociobiologists is that all organisms behave so as to maximize the propagation of their own genetic alleles. The sociobiologists' reasoning is simple: Genetic alleles that promote this kind of behavior have been positively selected for throughout evolution.

Basically, sociobiologists view life as a contest won by those individuals who propagate their own genes most effectively. In this view, animals and humans appear as strategists who are shrewdly calculating their risks and benefits and allocating their time and energy accordingly. *It is important to keep in mind that life strategies are not necessarily conscious.* The male bird singing "in order to" attract females may not know why he sings. The phrase "in order to" is simply a short way of proposing that bird song has evolved because it functions in attracting females and hence in passing on those genes that promote this behavior.

Sociobiologists, like other evolutionary biologists, distinguish two types of causes—*ultimate causes* and *proximate causes* (see Figure 5.7). This distinction, which we have made earlier in this course, actually goes back to Aristotle. Define the ultimate cause of a behavior in general terms. What is the ultimate cause of the birdsong mentioned earlier? Define the proximate cause of a behavior in general terms. Again, in the case of the singing bird, what is the proximate cause of the behavior?

B. Seemingly Altruistic Behavior in Animals

Altruism is defined as a costly or risky behavior carried out by one individual for the benefit of another. Sociobiologists have attracted attention by analyzing altruism, which seems paradoxical because it reduces Darwinian fitness. There are different types of altruism.

The most common type of altruism is based on *kinship*: The beneficiary is a relative. So, in anthropomorphic terms, this behavior could be called "nepotism." It is ascribed to **kin selection**, that is, the evolution of characteristics that promote the reproductive success of close relatives.

For example, meerkats live in family clans, in which females often "babysit" the young of the alpha female (Figure 18.1). In this and similar burrowing species, an individual may stand sentry while other clan members forage or play. If a predator approaches, the sentry gives a warning call, which sends everyone scrambling for a burrow. Male lions form alliances to defend their territory and their prides, although some alliance partners may never get to mate with any of the females. It turns out that the nonreproductive males usually promote the reproductive success of their

FIGURE 18.1:

Altruism based on kinship. The meerkat (*Suricata* suricatta) is a small mammal of the mongoose family, living in the desert of southwest Africa. Meerkats forage and play as a group with one "sentry", or a few of them, on guard watching for predators. Sentry duty is usually an hour long. If the guard spots danger, it barks loudly or whistles, and all group members disappear in burrows.
Image © Photodynamic, 2010, Shutterstock, Inc.

brothers. In some bird species, young adults stay with their parents and help them feed another brood. In all these cases, the altruists work for the benefit of close relatives.

Another type of altruism is known as **reciprocal altruism** because the two actors know each other and expect that favors will be returned: "I will scratch your back if you will scratch mine" (Figure S18.a). This behavior is common among social animals that can recognize and remember individual members of their group. Reciprocal altruism lasts only as long as favors are returned. Nonreciprocators will be met with pointed indifference, if not hostility.

There are cases of altruistic behavior between humans, and occasionally between animals, that do not seem to be based on kinship or reciprocity. The interpretation of such acts of kindness, which in humans tend to follow cultural norms of behavior, is still a subject of debate (see Chapter 21).

C. The Concept of Inclusive Fitness

In kin selection, the likelihood of altruistic behavior increases with the **relatedness coefficient (r)** between the individuals involved. So r is defined as the probability that a genetic allele of the **focal individual**, that is, the individual of primary interest (here: the altruist) will be present in the genome of another individual (here: the beneficiary). In this context, r is also simply called *relatedness*. For instance, in a diploid organism with sexual reproduction, the relatedness between parents and a given offspring is 0.5 because each parent passes on one chromosome of each pair of homologs (Figure 18.2). Under the same circumstances, the relatedness between siblings is also 0.5 because there is a 50 percent chance that they have inherited the same chromosome from any given pair of parental homologs. (We are ignoring here the possibility that genetic alleles may also be shared because parents were homozygous.)

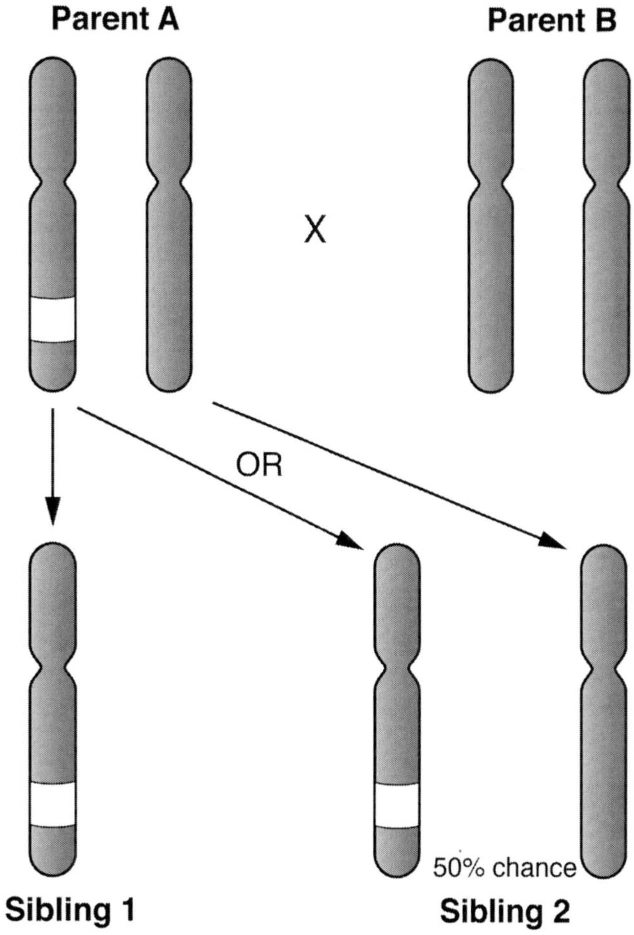

The incorporation of a quantitative measure of genetic relatedness allowed researchers to make many testable predictions about evolution and animal behavior. The British biologist William D. Hamilton pointed out that altruistic behavior should evolve if the cost (C) to the altruist is smaller than the benefit (B) to the beneficiary multiplied with the relatedness coefficient (r). Written as a simple equation

$$C < B \cdot r$$

it became known as **Hamilton's rule**. Another British biologist, J. B. S. Haldane, made the same point by saying that he would give up his life for the benefit of

FIGURE 18.2:
The genetic relatedness coefficient ("relatedness"), r is defined by the chance that a genetic allele of the focal individual (the one at the center of attention) is also present in another individual. For example, the relatedness between parent and child, and between siblings, is 0.5.

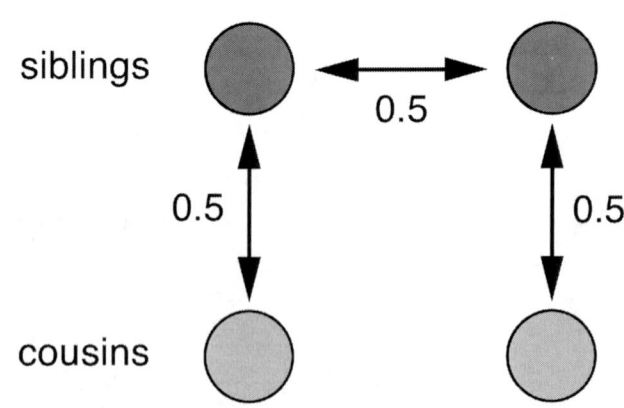

FIGURE 18.3:
J.B.S. Haldane: "I would give up my life for more than two brothers or more than eight cousins". This famous dictum implies the genetic relatedness coefficients (r) diagrammed here.

more than two siblings ($r = 0.5$) or more than eight cousins ($r = 0.125$). Implied in his statement is that genetic relatedness coefficients to a distant relatives are obtained by *multiplying* the relatedness coefficients to the intermediates, just as the probability of a complex event is computed by multiplying the probabilities for each step (Figure 18.3).

The **simple (Darwinian) fitness** is defined as the number of fertile offspring an individual produces.

The **inclusive fitness,** a concept introduced by W. D. Hamilton in 1964, encompasses the focal individual's own simple fitness, plus the incremental change (benefit) in simple fitness of any relatives aided by the focal individual, minus the decremental change (cost) incurred by the focal individual as a result of aiding each relative. Inclusive fitness can be expressed by the following formula:

$$FI = FS + \sum_{n=1}^{k} dFR_n \cdot r_n - \sum_{n=1}^{k} dFS_n$$

with

FI, inclusive fitness of the focal individual
FS, simple (or Darwinian) fitness of focal individual
k, number of relatives
dFR_n, increment in simple fitness of relative n
r_n, genetic relatedness of focal individual to relative n (weighting factor)
dFS_n, decrement in fitness incurred by focal individual by aiding relative n

Because of the genetic definition of relatedness, inclusive fitness measures how effectively an individual propagates his/her genetic alleles. Thus, the basic tenet of sociobiologists is that all organisms behave (consciously or unconsciously) in such a way as to maximize their inclusive fitness (FI).

The concept of inclusive fitness has resolved many puzzles, including many cases of apparent altruism, that could not be explained in terms of simple fitness. A striking example is the **honeybee**, which forms elaborate colonies with sterile female workers building and maintaining the nest, collecting food (nectar and pollen), and attending to the queen. The queen is the only reproductive female in the nest; being well-kept, she lives for several years and lays about a million eggs. Most of her offspring will again develop into sterile female workers, while a few will be reproductives—males called drones and future queens. The latter leave their native colony to mate with drones from other nests and then found new colonies. Drones do not contribute to nest maintenance and foraging, and only a few of them are allowed to survive.

The paradox that sterile worker bees have evolved although their Darwinian fitness is zero was not lost on Darwin himself. He argued that the sterile individuals labored on behalf of their own "stock," and that natural selection between stocks could resolve the apparent paradox of worker sterility. Without a valid concept of heredity, Darwin could not take the argument any further.

The key to understanding the extreme altruism of honeybees lies in their haplo-diploid sex determination (Figure 18.4). In this reproductive system, females are diploid, and oogenesis involves meiosis, whereas males are haploid and produce sperm mitotically. Fertilized eggs develop into females, while unfertilized eggs become males. Assuming that the queen bee keeps using sperm from the same drone, a female worker's genetic relatedness with a

- sister is 0.75
- brother is 0.25
- daughter or son would be 0.5

A worker bee—essentially fattening the queen to make more sisters, while killing most of her brothers—is then doing precisely what sociobiological theory predicts: propagating her own genetic alleles most effectively. In reality, a worker bee's reproductive "strategy" may be diminished by the fact that a queen often mates with several drones.

In addition to honeybees, there are other hymenopteran species (bees, wasps, ants) that are **eusocial** (forming large, sophisticated communities). It is estimated that eusociality has evolved 12 times independently, and 11 of these taxa are haplodiploid. An exceptional group are the isoptera (termites), which are eusocial but diploid.

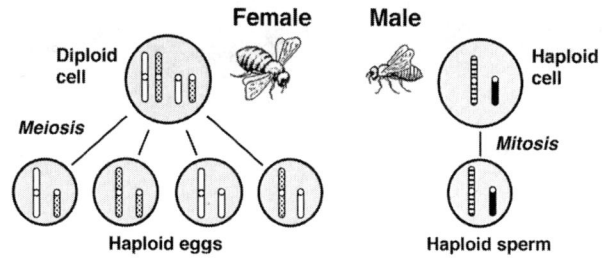

Focal Individual	Relative:					
	Sister	Mother	Daughter	Brother	Father	Son
Female	.75	.5	.5	.25	.5	.5
Male	.5	1.0	1.0	.5	--	--

FIGURE 18.4:
Genetic relatedness in a bee hive. Females develop from fertilized eggs, are diploid, and produce eggs through meiosis. Males develop from unfertilized eggs, are haploid, and produce sperm mitotically. A female worker and her sister, therefore, share *half* of their *maternally* inherited chromosomes (see Figure 18.2) and all of their *paternally* inherited chromosomes. Their genetic relatedness averages out to 0.75. (This is assuming that the queen bee keeps using sperm from one drone.) Other genetic relatedness coefficients compute in the same way and are tabulated.

D. Parental Effort and Its Distribution between the Sexes

The **reproductive effort** of an individual may be measured in *energy* spent on reproduction. Each expenditure causes a *loss of residual reproductive potential*.

The total reproductive effort can be subdivided into the **mating effort** (nest building, courtship, etc.) and **parental effort** (gametogenesis and, in some species, providing for the young).

An individual's choices for the allocation of reproductive effort to mating versus parenting define his or her **reproductive strategy**. Males and females of the same species may pursue different reproductive strategies.

Commonly, females spend a greater parental effort than males of the same species because eggs are large and expensive to make (Figure 18.5). They are designed to sustain a developing embryo until it is able to acquire nutrients. In contrast, sperm are essentially mobile genomes, numerous and inexpensive. Also, in most species that provide for their young, females do most of the work. This leaves males to invest primarily in their mating effort. In some species, these roles are reversed, and we will return to them later in this chapter.

Species with high requirements on parental effort (many birds, some fishes and mammals) tend to form **bonded heterosexual pairs** (marriages) to share and boost the parental effort. Such pair bonds are generally stable although extramarital matings do occur.

E. Parental Investment Affects Mating Behavior

In 1948, Angus J. Bateman quantified the reproductive success of fruit flies in the laboratory, taking advantage of dominant genetic markers that allowed him to distinguish offspring from multiple males and females kept together. He found that

- males were more aggressive in courtship than females.
- males showed greater variation in **reproductive success**, that is, in the total number of offspring per lifetime. Even though 21 percent of all males failed to leave any offspring, some males' reproductive success was far above average.
- copulation with multiple partners had only a modest effect on the reproductive success of females; their reproductive output less than doubled when they mated with more than one male. In contrast, males mating with multiple females had 3 to 4 times more offspring than their monogamous peers.

FIGURE 18.5:
Eggs are large and expensive to make. This particularly obvious in kiwis, flightless birds living only in New Zealand. In some kiwi species, eggs may reach one-quarter of their mother's weight.
Image © Steven Vidler/Eurasia Press/Corbis

Bateman interpreted his data in terms of the differing costs of making gametes, eggs being more expensive to make. On this basis, he argued that it was adaptive for a female to be choosey with her sex partners, because mating with an inferior partner would cost her a significant fraction of her reproductive potential. Males, on the other hand, would enhance their reproductive success by being aggressive and relatively indiscriminate in their mating behavior. Even if a male would copulate with a sick female or a female of a different species, he would not lose much.

F. Critique of the Bateman Principle

Bateman's observations on fruit fly behavior were scant, but his statements on coy females and competitive males were readily accepted in the (mostly male) scientific community, probably because they met with prevailing human gender stereotypes. Clearly, sexual behavior is an area where research is easily compromised by selective attention to data. Thus, Bateman's characterization of male sexual behavior as "undiscriminating eagerness" and female behavior as "discriminating passivity" became known as the **Bateman principle**. It has been criticized, especially by female scientists, as promulgating the human stereotype of ardent males and coy females.

One of Bateman's critics was Sarah Hrdy. In her field studies on langur monkeys (*Presbytes entellus*) in India, she noticed that when a new male rose in rank, even pregnant females would aggressively seek to copulate with him. The promiscuity of female langurs is at variance with the Bateman principle because langur females bear most of the burden of parental care. Instead, Hrdy proposed that langur females protect their young against infanticide from males by obfuscating the paternity issue.

Since DNA analysis has made it possible to determine paternity unequivocally, a degree of polyandry (one female mating with multiple males) has been found to be more common than expected on the basis of the Bateman principle. Because promiscuity increases the risk of exposure to parasites and injury, females who take this risk would be expected to gain some major benefit.

In certain insects and spiders, where males bring food items as nuptial gifts to females, the benefits of polyandry may be mostly nutritional. In prairie dogs (*Cynomys gunnisoni*), polyandrous females tend to have larger litters and healthier pups. Possibly, by setting up a competition among the sperm from multiple males, a female has her eggs fertilized by the most active or compatible sperm. In field crickets (*Gryllus bimaculatus*), polyandry may helps to avoid the costs of inbreeding. In an experiment to test this hypothesis, female crickets were allowed to mate with two brothers, or with one brother and one unrelated male, or with two unrelated males. The second and third groups had similar reproductive success, whereas eggs laid by females from the first group were less likely to hatch. The investigators concluded that, by some unknown mechanism, the females in the second group selectively allowed their eggs to be fertilized by sperm from the unrelated male.

Robert Trivers extended Bateman's concept, taking into account that in many animal species parental investment goes beyond gametogenesis to include gestation (or incubation), feeding, and protection. He proposed that the sex with the smaller overall parental investment should be more aggressive and risk-taking in their mating efforts.

As evidence, Trivers pointed to those relatively rare species, in which males make a *greater* parental investment than females. In these species, the common differences between male and female mating behavior are reversed as well. In certain sea horses, males have pouches on their abdomen or tail, into which females deposit their eggs for the males to fertilize them (Figure S18.b). The pouches also have morphological and physiological adaptations to osmoregulate, aerate, and even nourish the developing embryos. In these sea horses, females are bigger than males, are more brightly colored, and compete with one another for the privilege of getting their eggs raised in a male's pouch.

In the Panamanian poison arrow frog, *Dendrobates auratus,* males carry clutches of eggs and make sure the tadpoles reach a suitable body of water. Here again, females are more vigorous in their courtship, to the point of chasing males. In the moorhen *Gallinula chloropus,* males do about 72 percent of the incubation, losing almost 10 percent of their body weight per season. Females fight one another for access to males and court males more actively than vice versa. A similar sex reversal of both parenting and mating behavior was observed in the greater painted snipe, *Rostratula benghalensis.*

Taken together, observations on a wide range of animal species indicate that it is not necessarily the male, but the sex with the smaller parental effort, which is more aggressive toward individuals of the same sex, more active in courting the opposite sex, and less discriminate in mate choice. This trade-off between parental and mating effort, and the associated differences in mating behavior, are still called the *Bateman principle* or the *Bateman gradient*.

G. Males Gauge Parental Effort According to Likelihood of Paternity

For species with paternal care, sociobiological theory predicts that males gauge their parental efforts to the perceived likelihood of paternity. This prediction was confirmed by observations on reed buntings and bluegill sunfish (see the assigned reading).

In many bird species that form pair bonds to raise their young, extra-pair copulations nevertheless occur. In the reed bunting, pairs often raise two broods per season. This provides an opportunity to compare the degree of extra-pair paternity and the extent of paternal investment between two broods raised by the same pair in a similar environment. Paternity was established by *microsatellite (VNTR) analysis* of DNA, and paternal investment was measured by male feeds per chick per hour. There was a significant correlation between a male's actual degree of paternity for a clutch and his paternal investment in the same clutch.

The data indicate that reed bunting males are somehow able to assess their likelihood of paternity. Because males feed—to an extent—even broods consisting entirely of extra-pair offspring, they do not seem to be able to tell their own young from extra-pair offspring. Thus, they appear to have ways of gauging the marital fidelity of their mates during their fertile period.

In bluegill sunfish (*Lepomis macrochiros*), males provide the sole care for developing young in their nests (Figure 18.6 and the assigned reading). They gage the intensity of their parental care according to their perceived likelihood of paternity.

H. Male Reproductive Competition and Violence in Mammals

In most animal species, the *reproductive potential* of males is greater than that of females. Female reproductive potential then becomes a limited resource for males to compete over. The asymmetry in reproductive potential and the resulting competition are high in mammals, where the reproductive potential of females is limited by

- the duration and energetic cost of pregnancy
- the duration and energetic cost of nursing the young
- the demands of the young for protection, cleaning, and guidance, which fall mostly on the mother due to her proximity at nursing

The competition of males over females in mammals is reflected in various behaviors of males that enhance their reproductive success. Such behaviors include

- sperm competition
- guarding females, especially during estrus
- vaginal plugs
- fighting, raids
- infanticide

FIGURE 18.6:
Male parental investment depends on confidence in paternity. The bluegill sunfish (*Lepomis macrochirus*) male gauges his paternal care to the presence of "cuckolder" males at spawning, and to the smell of newly hatched eggs.
Photo courtesy of U.S. Fish and Wildlife.

Sperm competition occurs when a female copulates with multiple males and the males' sperm compete within the female's reproductive tract to fertilize her eggs. The testis size of males in different species is correlated with the prevalence of sperm competition. For example, chimpanzees and bonobos mate promiscuously, and their males have large testes. In gorilla families, silverback males monopolize the mating and have comparatively small testes. Males may increase the number of ejaculated sperm if they perceive a situation of sperm competition.

Guarding of females by dominant males is common in mammals, especially while the females are in estrus.

In mice and other rodents, males seal the vagina of females after copulation with a **mating plug**, a substance that prevents penetration by subsequent males for about 12 hours, which is enough time for the sperm from the first male to fertilize.

A direct form of male competition is **fighting** between individual males, between troops of males, or ambushes of a troop on single males. Such fights or raids are often deadly in lions, seals, wolves, hyenas, gorillas, and chimpanzees. Females of the raided group are taken over by the raiders. In other species, fighting between males is ritualized. Losers are denied access to females but may survive. The critical role of fighting is reflected in the costs that males incur for sustaining a big body mass and for growing heavy weaponry like deer antlers (Figure S18.c).

Males who take over females of another group often proceed to kill the females' young. Such **infanticide** promotes the reproductive fitness of the attackers:

- When female mammals are no longer nursing, they go back into estrus and will bear the conquerors' offspring without delay.
- Parental care is no longer spent on surviving offspring of the vanquished males.

This kind of infanticide has been reported for lions, langurs (Indian monkeys), baboons, gorillas, and other species. In lions, prides are taken over by coalitions of incoming males, which are then very intent, and often successful, in killing all present cubs (Figure 18.7). However, lionesses have been observed defending their cubs in several ways. They may take older cubs away from the pride to save them. They may also gang up on a new male, putting themselves at great risk of injury. Alternatively, a lioness may copulate with an incoming male even though she is already pregnant, in an apparent attempt to confuse the male into thinking her offspring is his own. However, offspring born immediately after a takeover will be killed.

A reverse form of infanticide is the selective cannibalism observed in spadefoot toads. Their tadpoles routinely eat smaller tadpoles but let go if the individual caught is a relative, which they seem to recognize by smell.

FIGURE 18.7:

Infanticide. When a lion *(Panthera leo)* pride is taken over by an incoming coalition of males, the raiders proceed to kill all present cubs.
Image © Acclaim Images.

EXERCISE PAGE FOR CHAPTER 18

Student Last Name _____

Student First Name _____

Discussion Section _____

On the course web site, use the link on the **syllabus** to find a pdf file of your **assigned reading** for this chapter. The same reading is referenced below. Use the bulleted list of questions to test your knowledge of this reading.

Neff, B. D. (2003). Decisions about parental care in response to perceived paternity. *Nature* **422**: 716–719.

- Which are the major parental efforts extended by males in this species?
- What are the two cues that males seem to use to gage the risk of reduced paternity?
- How did Neff manipulate the perceived paternity of parental males? What were the controls?
- How did he measure the "parental care" plotted in Figures 1 and 2? How did he avoid observer bias during these measurements?
- How did the behavior of the parental males change in response to Neff's manipulations?
- What did Neff conclude from his data?

Feedback

On a scale from 1 to 10, the being the best, please rank the above reading for

1. Interest and Relevance to the topic (10 being most interesting/relevant) _____
2. Readability (10 being most clearly written, easy to understand) _____

In the remaining space, enter any comments that you may have on this reading.

CHAPTER NINETEEN

SOCIOBIOLOGY OF HUMAN BEHAVIOR

Wilson's (1975) *Sociobiology* book had 25 chapters dealing with animals including nonhuman primates, and these chapters won nearly unanimous praise. In the 26th chapter, he extended to humans the concept that behavior has evolved to maximize inclusive fitness. This chapter was met with praise as well as condemnation. Wilson's scientific critics charged that he was overestimating the genetic aspect and underestimating the cultural influence on human behavior. His political adversaries were outraged (and some of them doused him with water at a scientific conference) because they felt he was dignifying repulsive human behavior, in particular "male chauvinist" behavior, by pointing out the adaptive value for which it may have evolved.

In the present chapter, we will derive some predictions about human behavior from sociobiological theory and test them against historical observations and recent data. As it will turn out, the predictions will be confirmed *statistically*. This does *not* mean that sociobiological theory can predict *individual* behavior. For example, those non-Jewish individuals who helped Jews to escape Nazi terror clearly did not enhance their inclusive fitness. Neither do people who adopt a child. Humans can behave according to cultural norms, and feelings of satisfaction from playing by the rules and punishing defectors seem to have evolved so as to optimize group survival (see Chapter 21). Now individuals *do have the option* to follow cultural norms irrespective of whether these behaviors enhance their individual fitness or not. However, the data indicate that maximizing individual inclusive fitness is *a significant component of average human behavior*.

A. Marriage Is a Universal Human Institution

Among all primates, humans have the most helpless babies and youngsters, who need the most extensive care and socialization. General sociobiology would therefore predict that humans form **marriages**, that is, long-lasting bonds between men and women who raise their common children.

A unique combination of human sexual characteristics (loss of estrus, permanent female breasts, modesty, etc.) can be interpreted as means of stabilizing a sexual bond between committed partners (see Chapter 2). This bond is critical because males gauge their parental investment according to their confidence in paternity.

Reinforcing the biological bond, virtually all human societies have the cultural institution of marriage, a covenant between husband and wife that involves

- a pledge to support each other and any common children
- persistence in time
- renouncement of extramarital sex
- societal initiation and permanent recognition
- some form of legitimization of any common children

Societies differ in the kind of mating pattern that they sanction. A survey of 849 human societies done in 1967 showed

- 708 **polygynous** societies (a man may have multiple wives, not vice versa)
- 137 **monogamous** societies (one man, one woman).
- 4 **polyandrous** societies (a woman may have multiple husbands, not vice versa)

Most primitive societies are polygynous, allowing high-ranking men to increase their reproductive success by having several wives, while most men live in monogamous marriages, and a significant percentage of men are left unmarried. This state of affairs leaves virtually no woman unmarried. In a sociobiologist's view, this is the most adaptive arrangement because it makes the best use of the limited female reproductive potential of a society.

Polygynous marriages can be tempestuous. Co-wives have conflicts, especially when their children compete for limited paternal resources. In many polygynous societies, this problem is mitigated by the custom that co-wives are often *sisters*. This solution is in accord with sociobiological theory because then each wife increases her inclusive fitness by promoting the welfare of her sisters' offspring.

The few societies with *polyandry* are societies of poor landowners. Polyandry is typically an arrangement between *brothers* who want to avoid subdividing the family estate to below the size that can support a family. If their economic fortunes improve, they acquire additional wives until every brother has one. Genetic relatedness between co-husbands again mitigates conflict from jealousy. A man who helps to raise his brother's children is at least propagating 25 percent of his own genetic alleles.

In summary, virtually all human societies sanction families, as sociobiologists would predict in view of the enormous parental costs of raising children. Details of marital arrangements, for example, the polygyny of most primitive societies and the fact that co-wives and co-husbands tend to be siblings, are also as would be predicted based on the principles of sociobiology.

B. Different Reproductive Strategies for Men and Women

Men have a greater *reproductive potential* than women. Pregnancy, nursing, and subsequent care of children take up major fractions of women's time and energy. A woman can therefore raise only a small number of children in her lifetime. Women are compelled by nature to make large parental investments, while men have the option of shifting their effort from parenting to mating. Thus, a man can have hundreds of children—if he can persuade enough women to have sex with him. According to the *Guinness Book of World Records,* the last Sharifan Emperor of Morocco, Moulay Ismail the Bloodthursty (1672–1727), had 888 children. Under these conditions, sociobiological theory predicts reproductive strategies for men and women that have some common elements but also different ones.

Both men and women should try to find mates with "good genes," that is, genes that will help the propagation of their own genes by raising many healthy and successful children, who in turn will have many grandchildren, and so forth. Men and women should look for mates who will be good parents, so that their children will do well and in turn will help their children to do well. However, because of the great difference in reproductive potential between men and women, their common goals may lead to different strategies.

A man, because of his large reproductive potential, and because nature does not *compel* him to provide parental care, is likely to be aggressive in his sexual pursuits. His greatest fear is to be cuckolded, in which case he would spend parental effort on a child that does not carry his genes. If propagating his genetic alleles were a man's top priority, his best strategy would seem to be ensuring the fidelity of his wife while impregnating as many other women as possible. However, such behavior would mean breaking the sexual bond that supports his marriage (see Chapters 2 and 13). He must also weigh the additional reproductive success from extramarital affairs against the risks of being killed or injured by a jealous husband, of being treated badly by his spurned wife, and of being cuckolded when he leaves her unguarded.

A woman can only have a small number of children, and nature exacts a substantial price from her for every child she bears. She therefore does well to choose a mate wisely, ensuring his willingness to join her in a long-term parental effort before risking pregnancy. If a woman had doubts about the quality of her husband's genes, then her best strategy would seem to be marriage plus occasional secretive affairs with men she presumes to carry better genes. Again, this would mean breaching the sexual bond that stabilizes her marriage. She must also weigh the potential improvement of her children's genes against the risk that her husband may discover her affairs and kill, hurt, or leave her.

The reproductive strategies predicted by sociobiology for men as well as for women entail the option of **extramarital sex** and children born out of wedlock. With the advent of molecular techniques, both male and female infidelity have become incontrovertible. Such studies indicate that, depending on time and location, between 5 and 30 percent of all newborn babies in England and the United States have been conceived adulterously.

Based on greatly uneven reproductive potential, sociobiological theory predicts that men should be less selective in the pursuit of sex partners than women. This prediction is confirmed by anecdotal evidence as well as by results of polls. College

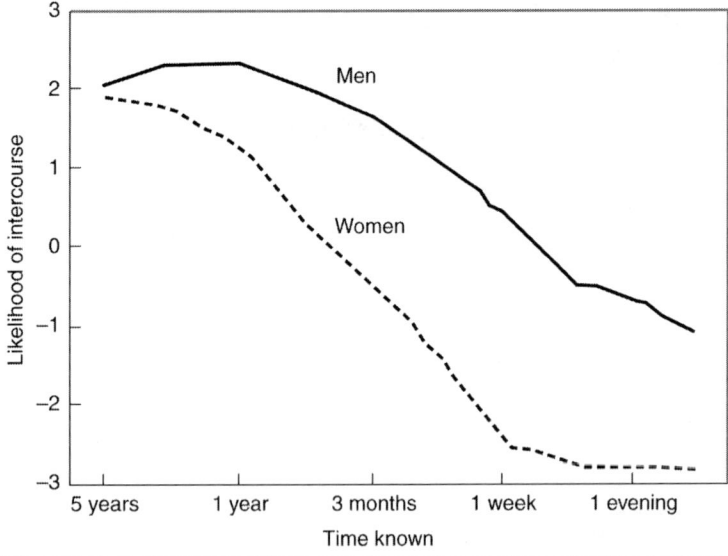

FIGURE 19.1:
Ready for sex? Subjects were asked to rate their willingness to have sexual intercourse with an attractive member of the opposite sex after being acquainted for different periods of time. Ratings were on a scale from 3 (definitely yes) to -3 (definitely no).
Redrawn after Buss D.M (1994) *The Evolution of Desire: Strategies of Human Mating.* New York: Basic Books. Figure 4

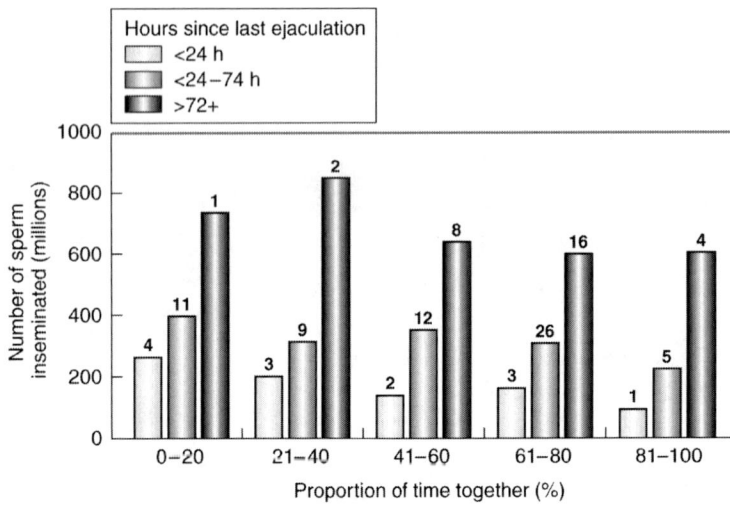

FIGURE 19.2:
Hedging against sperm competition. The number of sperm inseminated during intrapair copulations increases with time elapsed since the last ejaculation. A man also tends to inseminate *more* sperm if he has spent *less* time with his woman, which implies a higher risk of sperm competition. Numbers on columns indicate how many ejaculates were analyzed.
Redrawn after Baker R.R. and Bellis M.A. (1995) *Human Sperm Competition.* New York: Chapman & Hall, p. 206

students were asked how many sex partners they would like to have over certain time periods. While men, on average, desired about eight sex partners over the next two years, women wanted one. For the course of a lifetime, the average man reported the desire to have 18 sex partners, whereas the average woman desired no more than four or five.

College students were also asked to rate the probability that they would consent to sexual intercourse with an attractive member of the opposite sex after having known him/her for various time intervals. Generally, men declared themselves ready for sex much earlier than women did (Figure 19.1). Of course, answers people give on questionnaires may be confounded by gender stereotypes, unrealistic assessments of own feelings, and so on.

To overcome these limitations, two British biologists, Robin Baker and Mark Bellis, have counted sperm and trained couples to keep track of their orgasms and collect vaginal flowbacks. Their results indicate that both sexes have their ways of *unconsciously manipulating* the chances that a sexual intercourse will result in conception. The effects of these manipulations are as predicted by sociobiological theory.

The *number of sperm* ejaculated by a man in a copulation might be expected to *increase* with the amount time he spent with his woman before the act because her company has stimulated him for a longer time. Contrary to this expectation, men ejaculate *more* sperm as the time spent with their women *decreases*, and this trend is independent of the time elapsed since their last ejaculation (Figure 19.2). The sociobiological explanation is that men gage the number of ejaculated sperm according to the risk of *sperm competition*, which increases with the time his woman was away. The chance that she had sex with another man just before him may be small, but the stakes in terms of whose child he may help to raise are high, apparently high enough for this ability of men to gage their sperm count to evolve.

A woman has other (largely unconscious) ways of enhancing or dampening the chances that a sexual encounter will lead to fertilization. (These ways of course evolved before there was modern contraception.) One way is whether she has a *copulatory orgasm* and when she has it. If it occurs within a few minutes after male ejaculation, then it seems to help the uptake of sperm from the vagina into the uterus. *Non-copulatory orgasms* (from masturbation or spontaneous ones during sleep) also count, but in the opposite direction. They seem to suck up vaginal fluid into the uterus, lowering the uterine pH and creating a less hospitable environment for sperm from subsequent copulations.

Bellis and Baker found that women who practiced extramarital sex did so in ways that favored conception from the extraordinary partner (lover) rather than the ordinary partner (husband). These women gaged the occurrence and timing of copulatory orgasms so as to favor the of retention of their extraordinary partner's sperm as compared to their ordinary partner's sperm (Figure 19.3). Polyandrous women also had fewer non-copulatory orgasms before sex with their extraordinary partner as compared to their ordinary partner.

Others observed that mate preference of women changed with their *menstrual cycle*. During their fertile days, female probands preferred photographed faces of men who looked dominant and masculine, while the same women during their infertile

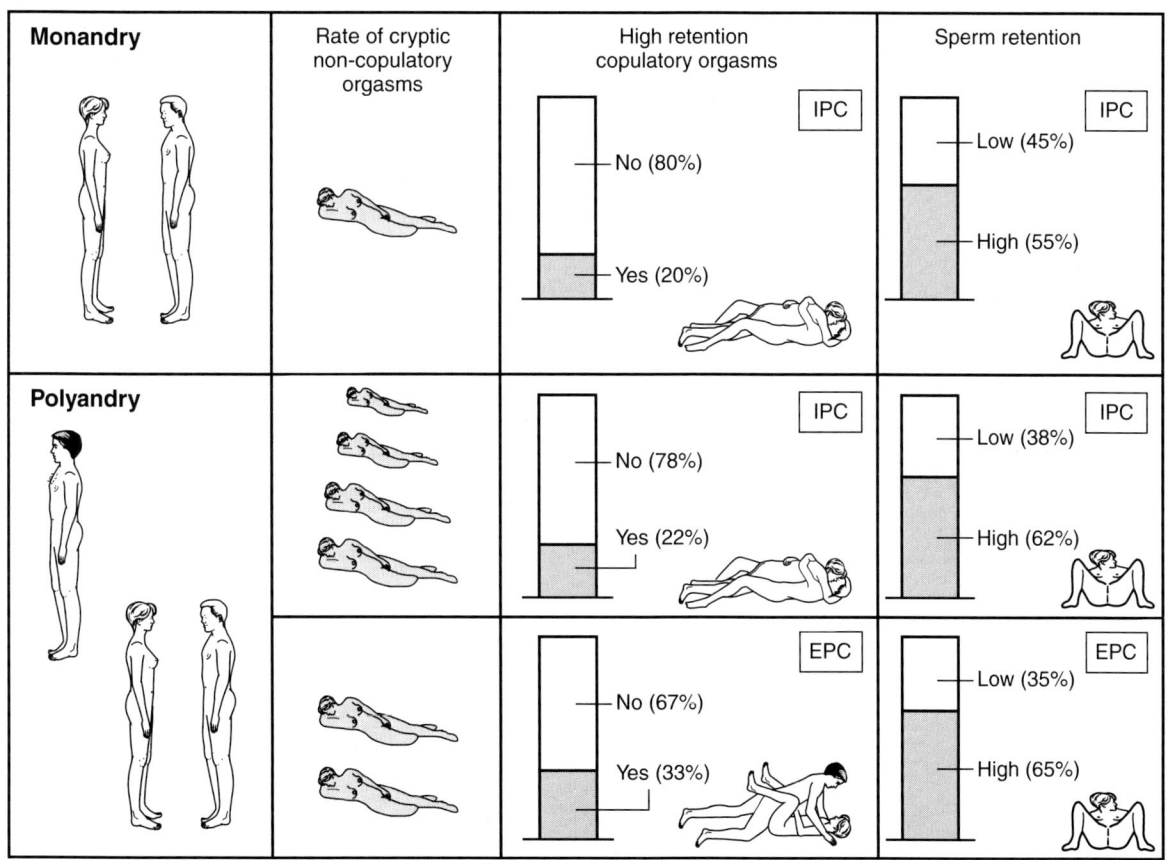

FIGURE 19.3:
Skewing the chances of conception. Polyandrous women favor sperm retention from extrapair copulations (EPC) as compared to intra-pair copulations (IPC) by timing their copulatory orgasms and by limiting non-copulatory orgasms. Redrawn after Baker R.R. and Bellis M.A. (1995) *Human Sperm Competition.* New York: Chapman & Hall p. 243

days favored faces of men who looked softer and more cooperative. This cyclic change would be predicted for a strategy of seeking the genetic contribution of a masculine lover while staying bonded to a softer husband who cooperates with child rearing.

The parallel existence of features stabilizing marriage through promoting a sexual bond and other mechanisms facilitating extramarital conception illustrate selective pressures acting in opposite directions. Both marital fidelity and infidelity have biological underpinnings.

C. Men Are More Prone to Violence and Risk-Taking

Sociobiologists would expect that men, because of their greater reproductive potential, compete over women. This should make them more prone to risk-taking and violence.

Human history is replete with accounts of men fighting over women. This competition takes place between communities as well as within communities. Between communities, male competition over females has led to raids and wars, which are still common among primitive peoples such as the Yanomamos. Within communities, a man's chances of winning contests over women depends on a range of attributes, the single most important one being **status**.

Men acquire status generally by leadership skills, and by demonstrations of physical fitness and risk-taking. In traditional communities, men acquired status by their prowess as hunters and warriors. Since agriculture allowed the accumulation of wealth, material possessions became indicators of a man's socioeconomic status. In modern societies, socioeconomic status still counts, especially for older men. Young men continue to acquire status by demonstrations of skill and risk-taking, which often revolve around *automobiles*.

Characteristically, young male drivers are more prone to reckless driving when they have male passengers than when they have female passengers or are alone, whereas female drivers are not evidently influenced by passengers. The driver death risk (driver deaths per billion miles driven) is about 3 times higher in males under age 25 than in females. A penchant of young men for risk-taking is also evident in *homicide* statistics. The leading cause is "trivial altercations" before an audience of acquaintances.

D. Men and Women Emphasize Different Indicators for a Mate's Reproductive Value

Sociobiological theory implies that there must be outward signals by which an individual can gage the reproductive value of a mate. For such signals to work, they need to vary from one individual to another, they need to be reliable indicators of reproductive value, and they need to be perceived in this way by the opposite sex. Typically, sexual selection is for traits that signal vigor or sexual receptivity of its carrier, such as peacock tails, deer antlers, the red vulvas of many female primates in estrus, and a wide range of behavioral displays. Are there similar outward signs of reproductive value in humans?

Both men and women value the **physical attractiveness** of a mate, but men more so than women (Figure 19.4). Men judge physical attractiveness more by optical cues, such as the "hourglass figure." This preference is compatible with sociobiological theory because visual attributes to which men respond are good predictors of a woman's health and fertility, and thus, a husband's reproductive success.

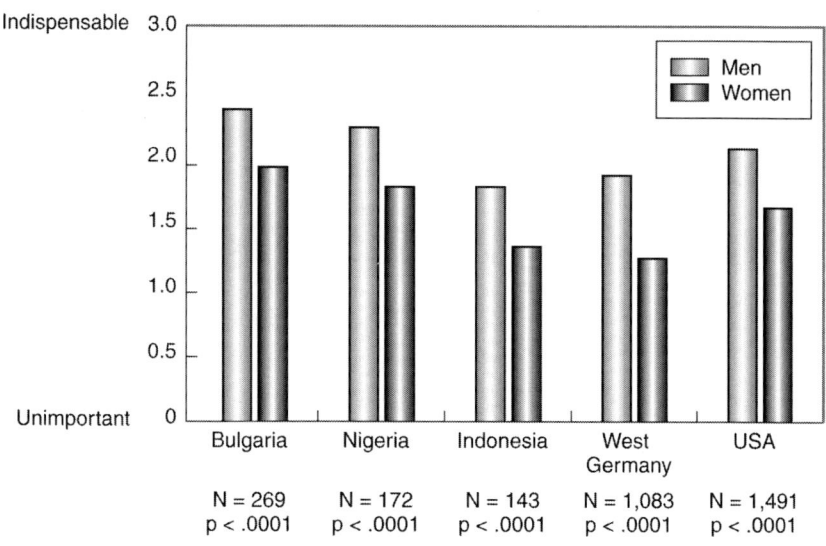

FIGURE 19.4:
Physical attractiveness of a mate or marriage partner is valued by both sexes. However, when asked to rate this variable in the context of 18 other variables men attached greater importance to this criterion than women.
Redrawn after Buss D.M. (1994) *The Evolution of Desire: Strategies of Human Mating.* New York: Basic Books, Figure 5.

Women rely more on the sense of smell for finding mates with "good genes." They are attracted to the scent of men whose *major histocompatibility* genes differ from their own. The adaptive value of this preference would be the greater immunological repertoire of their children. Women also prefer men with *symmetrical features*, such as ear length, elbow width, etc. Symmetry is a measure of developmental stability and correlates with longevity, health, and frequency of their partners' orgasms, which in turn enhance the chance of fertilization as discussed earlier.

Because symmetry is not always obvious, it is adaptive that women again use smell as a proxy. Double-blind tests showed that their preference for the scent of symmetrical men depends on their menstrual cycle (see the assigned reading). Normally cycling women preferred the scent of symmetrical men during their fertile days, whereas no such preference was seen during their infertile days. Likewise, no preference for the scent of symmetrical men was observed in women taking contraceptive pills. What do these results mean for women who meet their potential marriage partners while they are "on the pill"?

Women have also been known to look for a prospective husband's **financial prospects** (Figure 19.5). Again this is as predicted by sociobiological theory because a man's material resources enable him to support a family and indicate either that he is competitive himself or that he comes from an influential family.

My own polls of undergraduate students indicate that, in this segment of the population, the traditional preference of women for good financial prospects has all but disappeared. This departure may be limited to young women with college educations, who expect to earn a comfortable income themselves. Being based on responses to questionnaires, these data may also be affected by trends in self-perception. If confirmed by actual mate choice data, the ability of a substantial fraction of women to abandon the traditional preference for well-healed males would indicate that this mate choice criterion has been cultural rather than genetic.

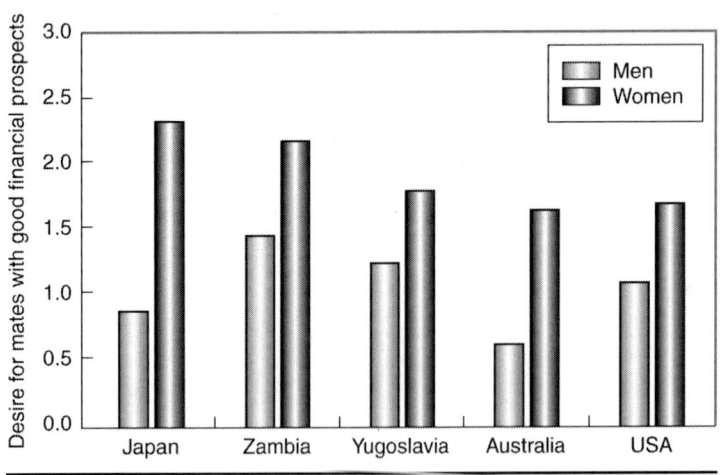

FIGURE 19.5:
Financial prospects of a long-term mate or marriage partner. When asked (in the 1990s) to rate this criterion on a scale from 0 to 3, women attached greater importance to it than men. Redrawn after Buss D.M. (1994) *The Evolution of Desire: Strategies of Human Mating.* New York: Basic Books, Figure 15.

The extent to which women's preference for wealthy men is culturally malleable is of great importance in the context of environmental degradation (see Chapter 24). The latter is caused by the growth of the human population, but even more by the "energy-burning lifestyle" of developed societies, which will be emulated by large developing societies. This disastrous trend would be very difficult to counteract if sexual selection would continue to favor men who flaunt big cars, powerboats, ostentatious houses, and so forth. For those women who take these items as signs of a desirable mate, it would be interesting to see whether their preference can be redirected to men with fuel-efficient cars, sailboats, ecologically designed houses, and so on.

E. The Waist-to-Hip Ratio Is Used by Both Sexes to Gage Reproductive Value

The best-investigated clue to reproductive value is the **waist-to-hip ratio (WHR)**, that is, the minimal girth around the waist divided by the maximal girth around the buttocks. Most authors consider the ideal WHRs to be 0.7 (or somewhat less) for women and 0.9 for men.

Male attraction to a female WHR of 0.7 seems to be strong and has been observed in different countries. This was shown clearly by Devendra Singh, who let men rate line drawings of women representing four WHRs (07. 0.8, 0.9, and 1.0) and three levels of body weight (normal, underweight, and overweight). By far the most favorable ratings for attractiveness and presumed health, youthfulness, and reproductive value went to the drawing representing normal weight and a WHR of 0.7 (Figure 19.6). The ratings by men of different age, ethnicity (U.S. European, African American, and Indonesian), and level of education were very similar.

In some populations, men prefer women who are overweight (for U.S. standards) and/or have WHRs around 0.9. In societies that value sons over daughters men prefer more "tubular" women. Their higher testosterone level reduces fertility while increasing the sex ratio (males per live births). Such women also have heavier and taller babies, traits that are traditionally valued for sons. Similarly, the Hadza men of Tanzania prefer plumper women, presumably because they are more capable of sustaining a pregnancy under conditions of food scarcity.

Two types of studies indicate a direct relationship between female WHR and fertility. Among women wanting to conceive naturally, those with normal WHR became pregnant more quickly, and delivered their first babies at an earlier age, than women with higher WHR. Among women in an artificial insemination program, the probability of conception decreased by 30 percent for every 0.1 increment in WHR.

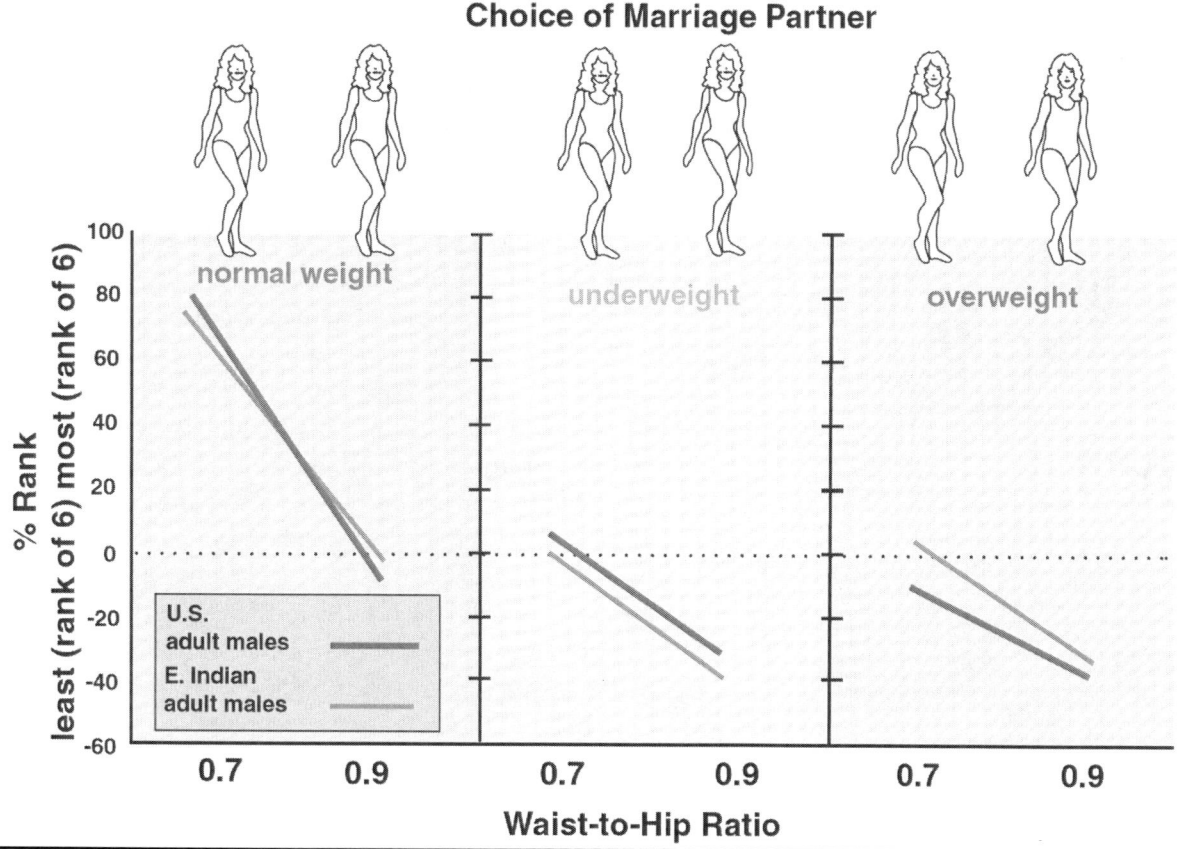

FIGURE 19.6:

Waist-to-Hip ratio. Male subjects were asked to rate their preference of marriage partners based on the following cartoons showing women of different body weight and waist-to-hip ratio.
From Singh D. (1993) Body shape and woman's attractiveness: The critical role of waist-to-hip ratio. *Human Nature* **4:** 297-321

A woman's WHR over 0.72 is also negatively correlated with *her children's* scores on cognitive tests (Figure 19.7). A link between the two variables may be certain fatty acids that promote female lower body fat as well as fetal brain development.

Women seem to be more variable in their response to male WHR than vice versa. In some African and Indian cultures, women prefer fat men because being fat is a sign of wealth. Indeed, young Dinka men of the Sudan have been reported to fatten themselves in order to become more attractive to women.

F. Men Go to Extremes to Avoid Being Cuckolded

Because men are less confident than women in their parenthood, sociobiologists predict that men are more intent on restricting their wives' sexual freedom than vice versa. This asymmetry is reflected in the "double standard" of sexual morals and adultery laws of most societies (see Chapter 20). Even today, many societies practice a truly barbaric way of controlling female sexuality: genital mutilation.

Genital mutilation is designed to destroy sexual interest of women or their physical ability to have vaginal sex. In a procedure euphemistically called female

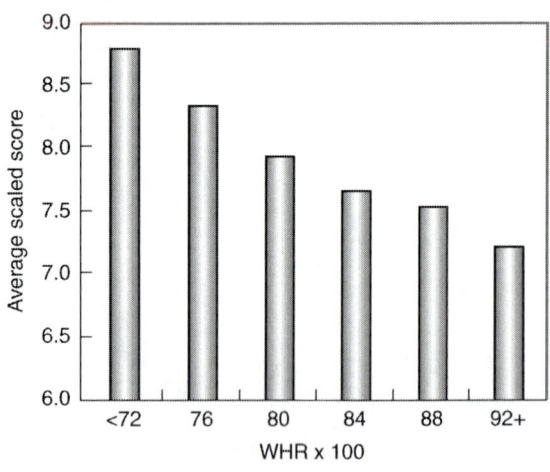

FIGURE 19.7:
Child cognition and maternal waist-to-hip ratio. Child cognition scores were found to correlate negatively with their mothers' WHR. From Lassek W.D. and Gaulin S.J.C (2008) Waist-hip ratio and cognitive ability: is gluteofemoral fat a privileged store of neurodevelopmental resources? *Evolution and Human Behavior* **29**: 26-34.

circumcision, the clitoris and labia minora are more or less completely removed. In another procedure called *infibulation*, the labia majora are sutured shut, except for a small opening to pass urine and menstrual blood. Infibulation makes sexual intercourse impossible. For marriage, the infibulated bride is cut open to a size that her husband finds pleasing, and then more cuts are made for delivery of a child. After birth, a wife may be re-infibulated and the whole process repeated, depending on her husband's whims.

The surgery is typically carried out by older women under primitive conditions, often causing infections resulting in infertility, life-long suffering, or death. These practices continue to exist in 23 countries in Africa, Indonesia, and Malaysia. All of these countries are male-dominated. The number of women subjected to genital mutilation has been estimated at more than 65 million in 1979, which would have translated into more than 5 percent of adult women at the time. The procedure is usually done to girls before they reach puberty. Many adolescent girls ask for it under peer pressure or familial pressure. A girl's refusal will prompt coercion, and if she persists, she risks being ostracized as "impure" or intent on promiscuity. The resulting disgrace would mean loss of marriageability and lack of material support.

The loss of fertility or life of the mutilated women should be powerful forces to work *against* genital mutilation. Also, one might expect that men would be deterred by the lack of sexual responsiveness on the part of their wives. Thus, the men who uphold genital mutilation in their societies pay a steep price in terms of both the reproductive success of their community and their individual pleasure. The price they pay reveals the strength of their desire to ensure their paternity.

G. Infanticide of Stepchildren

Loving babies and young children is one of the universals of human behavior. Their cute looks and behaviors compel adults to take good care of them—especially if they carry their caretakers' genes.

Children who live with a genetically unrelated stepparent are at risk of being abused or killed. In primitive societies, such as the Yanomamos of the Amazon, a man wedding a previously married wife may kill her infants or make her do it. In civilized societies, killing a child is rarely a deliberate act, but it happens. The risk of abuse or infanticide is much higher for a child growing up with a stepparent than for a child growing up with biological parents.

According to Canadian police records, stepchildren age 2 or younger are 70 times more likely to be killed by a stepparent than are children living with their natural parents (Figure 19.8). This discrepancy holds up even if only children living in two-parent homes and in equivalent socioeconomic groups are compared. Similar studies from England show that stepchildren age 4 and younger are 15 to 20 times more likely to be killed by a parent than are genetic children. In most cases of infanticide, children are killed by male stepparents.

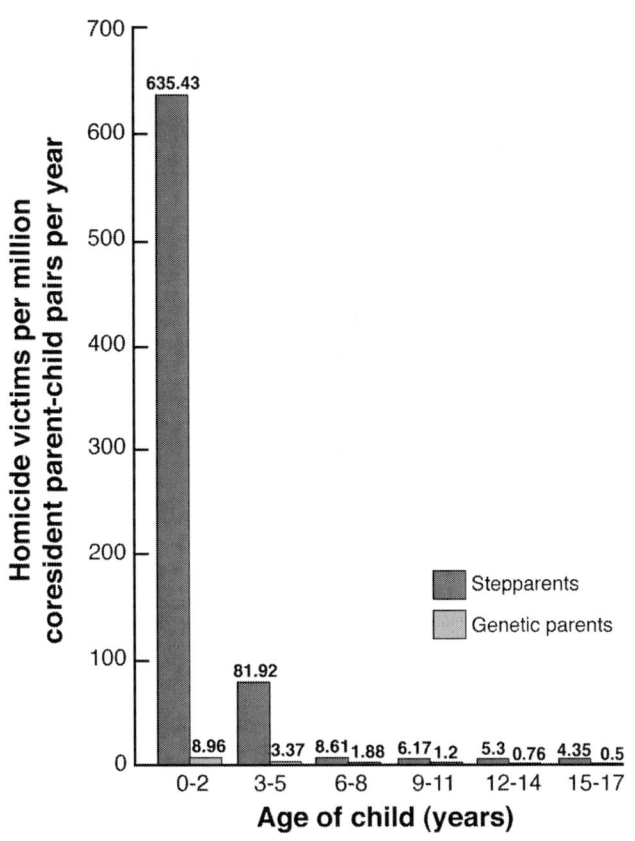

FIGURE 19.8:

Infanticide. According to police statistics from Canada, a young child is much more likely to be killed by a stepparent than by a genetic parent.
Redrawn from *Science* **261:** 987.

Infanticide statistics provide a grim explanation for the results of a study finding that children generally do not look sufficiently like their fathers for unbiased observers to identify them in photographic lineups with other men. There is one exception: One-year-olds look recognizably like their fathers. The dependence of a child's security on its genetic relatedness to its caretakers, in particular male caretakers, is exactly what sociobiological theory would predict.

H. The Ecological Problem of "Sociobiologically Correct" Human Behavior

The fact that humans on average behave as predicted by sociobological theory is deeply ironic. This kind of behavior has been adaptive for more than 99.5 percent of hominin evolution, while populations were small and survival was precarious. However, in combination with agriculture and modern medicine, "sociobiologically correct" human behavior has become unsustainable. Indeed, the human population has already grown to a size that may exceed the carrying capacity of Earth (see Chapters 22 through 24). This is a novel situation, which will require an enormous effort to manage.

EXERCISE PAGE FOR CHAPTER 19

Student Last Name _____

Student First Name _____

Discussion Section _____

On the course web site, use the link on the **syllabus** to find a pdf file of your **assigned reading** for this chapter. The same reading is referenced below. Use the bulleted list of questions to test your knowledge of this reading.

Gangestad, S. W., and Thornhill, A. (1998). Menstrual cycle variation in women's preferences for the scent of symmetrical men. *Proc. R. Soc. Lond.* **B, 265:** 927–933.

- Which earlier studies led the authors to expect that the preference of women for men with particular attributes, and not just sexual interest in general, changes during the menstrual cycle?
- Which earlier data do the authors quote as indicating that sexual responsiveness of women is strongly affected by the scent of men?
- What is fluctuating asymmetry, and how does it predict fitness components in animals and specifically in humans?
- Describe the "smelly T-shirt test" used by the authors. Was it designed to be "double blind"?
- Explain and interpret the data shown by the authors (Figures 1 and 2).
- What appears to be the adaptive value of the menstrual cycle variation in women's preferences for the scent of symmetrical men?

Feedback

On a scale from 1 to 10, the being the best, please rank the above reading for

1. Interest and Relevance to the topic (10 being most interesting/relevant) _____
2. Readability (10 being most clearly written, easy to understand) _____

In the remaining space, enter any comments that you may have on this reading.

CHAPTER TWENTY

SEX AND GENDER

The *sex* of a person is defined simply by the kind of gonads that are present: ovaries or testes. A person's **gender** is more complex: It encompasses his/her **gender identity** (whether an individual feels he/she is a man or a woman), **gender roles** (played in interactions with others), and **sexual orientation.** Each person is born with a sex and assigned a gender. For most people, but not for all, sex and gender are congruent.

Gender has both biological and cultural roots. Recognizing the biological underpinnings of gender should help to avoid futile attempts to change gender differences that have a long evolutionary history and are in part genetically controlled.

A. Gender, Stereotyping, and Discrimination

Recognition of a child's sex is typically the beginning of a gender assignment. At birth, each baby's external genitalia are inspected, and an announcement is made: "It's a boy!" or "It's a girl!"

Being a boy or girl means not only having different reproductive organs but also being raised to fulfill different gender roles, including

- wearing different haircuts and clothes
- playing with different toys
- showing different levels of noise, activity, and assertiveness
- placing different emphasis on subjects in school

For adolescents, gender expectations become even more typically divergent.

- Boys are expected to be unruly as well as physically and verbally assertive. Soft or gentle behaviors are discouraged.
- Girls are expected to be tidy, well-behaved, conciliatory, and socially sensitive. Coy, choosy, and refined behaviors are encouraged; dominance or aggressiveness are discouraged.
- Many gender roles revolve around the double standard in sexual mores, which ostracizes marital infidelity in women while condoning it in men.

The double standard is easily explained by sociobiological theory because men, due to their limited confidence in paternity, should be more intent on limiting the sexual freedom of their wives than vice versa. Indeed, adultery was traditionally defined as sexual intercourse between a *married woman* and any man other than her husband. The marital status of the man in an adulterous couple was considered less critical or even irrelevant. For example, the French revolutionaries, who rewrote many laws to abolish discrimination, stated explicitly that adultery laws are not to punish extramarital sex itself but to avoid the possible introduction of alien children into the family and even the uncertainty that adultery creates in this regard. Adultery by a husband with an unmarried woman has no such consequences and was therefore not punished. Thus, familial honor codes stressed the obligation of parents and husbands to guard the chastity of their daughters and wives. This obligation often entailed confinement of women to closely guarded quarters and extreme penalties for perceived lapses in chastity.

Male and female gender roles have led to discrimination, mostly against women. The fight against gender discrimination has prompted various attempts at overcoming gender differences surgically, through education, or by political means. Some of these attempts have been based on the assumption that men and women are alike except for the anatomy of their reproductive organs. Perhaps a more promising way of overcoming gender discrimination would be to recognize that men and women do not have to be alike to be equal.

B. Proximate and Ultimate Causes of Gender Differences

Gender differences have *proximate* and *ultimate* causes, which have been discussed in previous chapters and just need to be summarized here.

Proximate causes of gender differences include **sex hormones,** which have permanent effects on brain organization perinatally and generate different patterns of brain activity in adults (see Chapter 17). According to the organizational hypothesis, perinatal exposure to testosterone alters certain brain regions permanently, causing them to function in male-specific ways later in life. This hypothesis is supported by experimental data from rodents and by observations on humans with congenital adrenal hyperplasia and progestin-induced pseudohermaphroditism.

The behavior of men and women also depends on the current levels of their sex hormones. Courtship behavior and status-oriented aggression in men are promoted by androgens. Success in courtship or aggressive encounters in turn causes increased androgen production. Behavioral patterns of women, as well as their performance in standardized tests, change with their reproductive cycle.

Ultimate causes of human gender roles, according to sociobiological theory, include

- the enormous costs and duration of raising children, which make long-lasting families adaptive
- the large reproductive potential of men, which makes it adaptive for them to be competitive and take risks
- the large parental investment of women during pregnancy and nursing, which makes it adaptive for them to avoid conflict and to be prudent and selective in their mate choices
- nursing, which makes mothers the primary caregivers for young children

Both proximate and ultimate causes of gender have evolved over millions of years and are likely to be in part under genetic control. This degree of genetic control persists even in modern societies where commercial day care makes some of the older ultimate causes less compelling.

C. Surgical Reassignment of Sex and Gender

A biological component to human gender is indicated by two published cases in which sex and gender were reassigned surgically to two young boys whose penises had been lost accidentally in circumcisions. A role of brain organization in gender is also indicated by observations on adult individuals who desire different degrees of transgender and sometimes transsexual change.

One of the published accounts of surgical **sex and gender reassigment** is about Bruce, who was born in the early 1960s as one of a set of twin boys. After a mishap at circumcision had disfigured his penis, his parents consulted with pediatric surgeons and psychiatrists. At the behest of renowned pediatrician John Money, they decided to raise Bruce as a girl. This was done in accord with prevailing opinion at the time, which held that gender was molded entirely by education.

At 17 months, the child's name, hairstyle, and clothing were changed: Bruce was turned into "Brenda." At 21 months, the child was castrated, but the remainder of the penis was left in place to not interfere with the urinary tract. The parents were instructed not to divulge the treatment to anyone.

Brenda was encouraged by her parents to behave like a girl, while her twin brother was allowed to grow up as a normal boy. Brenda sometimes embraced her new identity and developed the feminine behaviors and interests encouraged by her parents, like keeping herself and her room neat and tidy. Psychiatrists took her compliance as confirmation of their belief that gender was all learned. Most of the time, though, Brenda rejected the dolls, sewing kits, and girls' activities proffered to her. She preferred her brother's toys, and when he refused to share them with her, she saved up money from her allowance to buy her own toy truck and toy machine gun. She liked to tinker with tools and gadgets. She played rough and tumble games with boys, even though her brother tried to exclude her. She got into schoolyard fistfights.

As a young teen, Brenda was teased by other girls for her masculine appearance, and for her interest in becoming a car mechanic. She fought those who teased her and was expelled from her school. Parents and physicians talked her into estrogen injections to curb her violence and help her to develop a more feminine appearance, but she refused to wear a bra and started overeating to conceal her breasts. Brenda began to urinate standing up, making messes and prompting other girls to lock her out of the girls' restroom. When she used the boys' restroom, she got into trouble with school administrators. Despite these problems, Money published a scholarly article pronouncing the experiment a great success. He pressured the parents to take the final step: constructing a vagina.

After Brenda rebelled and almost committed suicide, her father told her about her sexual history. She reported that everything clicked into place then, and that finally she understood why she had been feeling like a "captive animal" or "freak."

At age 14, Brenda demanded to have her breasts removed and to have plastic surgery to reconstruct a penis. She changed her name to "David." He received testosterone injections and worked out with weights to acquire a masculine body. He began to feel sexually attracted to girls. After four rounds of surgery, he became able to enjoy sex and ejaculate seminal fluid (without sperm). Although he felt insecure because of his imperfect penis, he converted a van into a windowless mobile home with a bedroom and bar for sexual encounters with girls. He became well accepted socially among boys as well as girls.

David embraced male gender roles and became the car mechanic he had always wanted to be. He married and—according to his wife—was "the one who wore the pants in the family." David became a media celebrity after author John Colapinto wrote a book about him entitled "As God Made Him." David said his biggest resentment was for the fact that because of his castration, he would never have children of his own. After a series of setbacks—death of his twin brother, losing his job, and separating from his wife—David committed suicide in March 2004.

A second case of sex and gender reassignment after a botched circumcision is described in your assigned reading at the end of this chapter. Here, the outcome for the patient was not without problems but not nearly as dramatic as in the case of Bruce/Brenda/David.

Many children are born with ambiguous genitalia, because of genetic disorders such as CAH (see Chapter 17), or because they have been exposed to endocrine-disrupting medications or environmental toxins during fetal development (see Chapter 26). A long-standing pediatric tradition for children born with a penis below a minimal size is surgery to construct female genitalia followed by hormonal treatment as needed for female development. The goal is to protect the children from ridicule and abuse. However, the surgical and psychological outcomes are not always satisfactory. Based on the organizational hypothesis, psychological problems are to be expected if the perinatal effects of testosterone on brain development are ignored.

Oddly, there does not seem to be a collection of patient histories available that could help pediatricians in weighing the pros and cons of surgical sex and gender assignment under different conditions. In the absence of such data, the standard of medical care for children born with ambiguous genitalia is still controversial.

A neurobiological basis for gender is also indicated by studies on **transgender individuals,** who have the biological characteristics of one sex but experience the gender identity of the opposite sex. Some men feel uncomfortable with their sex and gender and have a desire to make a **male-to-female (MTF) transition** by one or more of the following steps.

- dressing like women (cross-dressing)
- striving to assume more female (or in-between, androgynous) gender roles

E. Speculations about the Evolution of Gender Roles

While some roots of gender are biological, others are indeed cultural. In the course of hominin evolution, biological and cultural determinants of gender may have gone through a long periods of **mutual stabilization,** which should have been positively selected for because it helped to avoid ambiguous behaviors that might have reduced reproductive success. In more recent history, cultural changes such as agriculture and the **demographic transition** (see Chapter 22) have challenged traditional gender roles. The resulting changes have revealed that gender roles are in part cultural and may depend on economic development.

A long-standing foraging behavior of hominins has been epitomized as **man the hunter-woman the gatherer.** Existing primitive societies still have this division of labor, with women gathering plant food, eggs, and small animals near camp, while men periodically go out on hunting expeditions. This pattern may go back to *Homo habilis* times when males would scavenge rather than hunt. It may have been stabilized by the taming of fire, which could have made females the guardians of fire and the primary cooks. Cave paintings and carved figurines dating back more than 35 kya show the same gender roles, depicting men as lean hunters and shamans (Figures S20.a,b), while women are shown plump with much emphasis on sex, fertility, and motherhood (Figures 20.1 and S20.c).

The man the hunter—woman the gatherer pattern has been adaptive nutritionally because it provided meat, in a "feast or famine" pattern, on top of a stable supply of plant food and small animals. Leaving the gathering part to women allowed them to keep their children around and nurse them for extended periods of time (two years and more in primitive societies).

The man the hunter–woman the gatherer division of labor was also in tune with the reproductive needs of early hominin communities. It protected those who limited future group size—females and youngsters—from the perils of scavenging and hunting (Figure S20.d). At the same time, it channeled competition among males into adaptive forms as they gained status by honing their skills as hunters and warriors.

Given the long evolutionary life of the man the hunter–woman the gatherer foraging pattern, and its close fit with the biological underpinnings of gender, some of the related behaviors may have become genetically fixed. Indeed, in the assigned reading for Chapter 17, Doreen Kimura explains some of the differences in average test scores of men and women as adaptations to the man the scavenger/hunter-woman the gatherer/child caregiver division of labor.

The advent of agriculture has made the man the hunter–woman the gatherer foraging pattern obsolete and has allowed the evolution of larger and more complex societies. Some agricultural societies were **patriarchical,** meaning that men stayed in their native group and defended the territory and resources that they would pass on to their sons. In doing so, they monopolized economic clout and decision making. Other early agricultural societies were **matriarchical,** with female goddesses and women in positions of power (Figure S20.e). An advantage of this order was that matters of biological parenthood and inheritance are clearer for mothers than for fathers. Still other agricultural societies have left no signs of either male or female dominance.

The taming of horses, and advances in metallurgy (bronze and iron) have made warfare much more effective, while rewarding competitiveness and a taste for fighting,

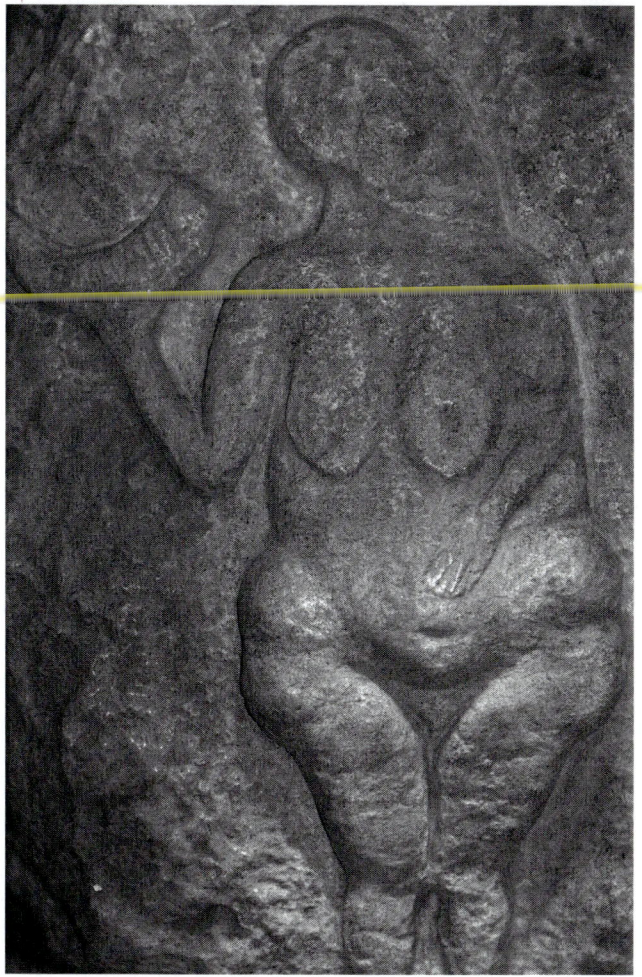

FIGURE 20.1:
Venus of Laussel, carved into sand stone by a Cro-Magnon artist about 22 kya. The animal horn in her right hand suggests some kind of ritual. The large breasts and buttocks show an emphasis on fertility - and hence, male sexual interest.
Image © Charles & Josette Lenars/CORBIS

to which men have long been preadapted due to their competition over women. A male preference for weapons and other tools, which may have coevolved with warfare, has apparently favored an almost universal ascent of male-dominated societies.

F. Gender Roles Are Accepted as Long as They Are Fair

Primates living in large communities have a keen sense of **fairness,** or aversion to disadvantageous inequity. This does not mean that each effort and benefit has to be split evenly, but they must add up to equitable packages. Studies on chimp colonies describe group life as a complex market in which all kinds of favors are constantly being traded: sex for food, grooming for loyalty, and so on. The two basic rules are "one good turn deserves another," and "an eye for an eye, a tooth for a tooth." So long as these rules are observed trades do not have to be in kind. Males in particular spend much time on forging and maintaining alliances for mutual advantage (see Chapter 21). Such alliances do not have to be symmetrical in every respect, so long as both partners feel they are getting a good deal overall. If one partner feels taken advantage of, he will leave the alliance.

The traditional gender roles of man the hunter–woman the gatherer would seem to be a fair arrangement, essentially balancing the hardships and dangers of hunting with the chores of housekeeping and raising children. With the advent of agriculture, the relative importance of hunting for food receded. By that time, the invention of the bow and arrow had also made hunting less dangerous. However, these developments did not make the traditional chores of women any easier. It would seem fair, then, that men would have shouldered much of the labor associated with agriculture. Actually, this seems to have happened in some societies but not in others.

Analysis of human skeletons from early agriculturalist societies in the Nile, Euphrates, and Tigris valleys shows signs of arthritis mostly in female skeletons, indicating that women bore the brunt of hard labor. In the same cultures, the signs of caries and other diseases are also more prevalent in the remains of women, indicating that their diet was inferior to that of men. In contrast, excavations of an early agricultural settlement at Çatalhöyük/Turkey yielded statuettes showing women in positions of power. Other female statuettes from the same site have grains embedded in their torso, apparently linking women to the growing of plants (Figure S20.f).

In her description of !Kung San people in Africa, Patricia Draper reports that gender roles became inequitable–and resented by women–as some communities made the transition from hunting and gathering to agriculture. In !Kung San bands who still lived as hunters and gatherers, men and women were equally relaxed and content about their division of labor. Boys and girls also seemed to assume their gender roles naturally and easily. However, in !Kung San communities who had become agriculturalists, women were doing the field work while men were just hanging out and laying plans to shape societal rules to their benefit. Not surprisingly, the women found these gender roles unfair and resented them.

It seems that different gender roles can be long lasting and acceptable to both men and women as long as they are *compatible with biological differences and also fair*. In other words, men and women do not have to be alike to be equal.

G. Reproductive Behavior May Be More Genetically Constrained in Men Than in Women

The extent to which gender roles are genetically fixed may be greater in men than in women.

Male competition over females, especially in a polygynous mating system, leads to uneven reproductive success among males. In a typical Yanomamo village, some men have four or more wives and more than 20 children, while about one-fourth of the mature men are childless. In contrast, the number of children per woman is more evenly distributed. Similarly uneven reproductive success for men was observed among the Kipsigi people of Kenya and other contemporary hunter–gatherer societies. The uneven reproductive success of men in extant hunter–gatherer societies probably represents a situation that has prevailed for more than 99 percent of hominin evolution. Assuming that some of the variation in reproductive behavior is genetic, it follows that selection for alleles that promote reproductive success has been stronger in men than in women.

With regard to several reproductive behaviors, women seem to adapt faster than men to particular societal conditions or historic changes. For example, women seem to be more variable in their response to male waist-to-hip ratio than vice versa (see Chapter 19). Among female college students in the United States, the traditional preference for wealthy mates seems to be diminishing. In developing countries, there is a negative correlation between the level of women's education and total fertility rate (see Chapter 22). No such correlation is observed in men.

A lesser degree of genetic fixation in women's reproductive behavior is critical to attaining two goals of worldwide importance.
- limiting the growth of the human population (see Chapter 22)
- overcoming sexual selection favoring males who flaunt material wealth in ways that promote the "energy-burning lifestyle" (Chapters 24 and 26)

EXERCISE PAGE FOR CHAPTER 20

Student Last Name _____

Student First Name _____

Discussion Section _____

On the course web site, use the link on the **syllabus** to find a pdf file of your **assigned reading** for this chapter. The same reading is referenced below. Use the bulleted list of questions to test your knowledge of this reading.

Bradley, S. J.; Oliver, G. D.; Chernick, A. B.; and Zucker, K. J. (1998). Experiment of nurture: Ablatio penis at 2 months, sex reassignment at 7 months, and a psychosexual follow-up in young adulthood. **Pediatrics** 102: p. e9. (Electronic article)

- What medical treatment did the patient receive and at which ages?
- At which ages was the patient interviewed, and what were the occasions for the interviews?
- During the interviews, what does the patient say about her preferences of games and toys while she grew up as a child?
- What does the patient say about her gender identity, gender roles, and sexual orientation as an adult?
- Which additional observation and test results do the authors report about the patient's vocational and recreational interests, and about her sexual orientation?
- Which possible reasons do the authors discuss to explain the different long-term outcomes of their case and the earlier (Bruce/Brenda/David) case as reported by Diamond and Sigmundson (1997)?
- Can you think of an additional reason for the different outcomes?
- Which two "schools of thought" on surgical gender reassignment among pediatricians become apparent in this article?

Feedback

On a scale from 1 to 10, the being the best, please rank the above reading for

1. Interest and Relevance to the topic (10 being most interesting/relevant) _____
2. Readability (10 being most clearly written, easy to understand) _____

In the remaining space, enter any comments that you may have on this reading.

CHAPTER TWENTY-ONE

AGGRESSION, COOPERATION, AND KINDNESS

Hunting is not aggression.

A. Defining Aggression

Aggression is defined as an action that causes injury, or implies a threat to cause injury, to a conspecific individual (Figure 21.1). This definition does not include inflicting death or injury on other species, as by hunting or grooming. Aggression also does not include actions carried out in defense of self or kin. Additionally, negligent or inadvertent actions causing injury are not considered aggressive.

Aggressive acts may occur

- between individuals of the *same community*. Such aggressions are common in social animals to *establish rank*.
- during *random* encounters of individuals from *different communities*.
- in an organized manner between different communities (*raids, war*). In the aftermath of war, the victorious group may proceed with *genocide*, killing the male survivors of the losing community and taking over their females.

B. Ultimate Causes of Aggression in Mammals

We find aggressive behavior interwoven with other behaviors in the contexts of territoriality, feeding, establishing rank, mating, and raising the young.

Territoriality is a common variant of aggressive behavior in many animal species. A territory is an area occupied exclusively by an individual or group. A territory invariably contains a scarce resource, such as a steady food supply, water hole, shelter, nesting

FIGURE 21.1:
Gorilla silverback threatening by baring his teeth
Image © Karl Ammann/CORBIS

FIGURE 21.2:
Stag fight. In the red deer (*Cervus elaphus*) and many other mammalian species, males fight over access to estrous females.
Image © Stefan Meyers/Corbis

site, and the like. The ultimate cause of territoriality is to secure the scarce resource(s) for the owner. A territory is staked out by various methods of advertising, such as singing or scenting (Figure S21.a). As needed, a territory is maintained by threatening or physically fighting off intruders. Crowding (small size of territories) increases the likelihood of aggression. Indirectly, then, territoriality may promote dispersal of the species over a larger range. Territories are maintained only as long as they are economically defensible: the increase in survival and reproduction must outweigh the energy expended and the risk of injury or death. For example, in many bird species, territories are defended during nesting but not at other times.

With or without territories, animals fight over **food**, especially over prized food items, such as meat for chimpanzees. The likelihood of aggression increases when food becomes scarce. The ultimate cause obviously is to avoid starvation.

Animals fight over **access to mates**. The sex with the greater reproductive potential, typically the males, compete more or less ferociously over access to females, which are the limiting factor for their reproductive success. Fighting of males over females, with the result of uneven reproductive success, is common in mammals (Figure 21.2; see also Chapter 18). Male chimps, gorillas, and orangutans fight fiercely over estrous females.

In social animals, individuals of a community often use aggression to establish a **rank order**. This may be a long linear sequence as in the proverbial pecking order of chicken (Figure 21.3), or essentially a contest over the top (alpha) position in the community. The principal benefits of high rank are preferred access to food and to mates. The ultimate cause (adaptive effect) of rank is to make resources available to the strongest individuals first and then to others depending on availability.

Finally, aggression is used in the **defense of offspring**. Especially species that raise only a few young, parents will defend their youngsters vigorously. The worst spot for a hiker or hunter to be in is between a female bear (mountain lion, etc.) and her cub.

Evidently, aggression is a common way of animals to promote their inclusive fitness.

C. Proximate Causes and Heritability of Aggression

Proximate causes of aggressive behavior in mammals include hormones and neurotransmitters.

It has been shown in many mammalian species that castration of males dramatically reduces their aggressiveness and that supplementation of testosterone restores it. Recent experiments indicate that testosterone intensifies neuronal activity in the brain along a pathway from the *amygdala* through the *stria terminalis* to the *hypothalamus*. In many women, irritability and other symptoms are heightened during the premenstrual phase of their reproductive cycle.

Serotonin generally dampens aggression in mammals. Pharmacological strategies of increasing serotonin levels, such as serotonin precursors or reuptake inhibitors, reduce aggressive behaviors in rodents. Another neuropeptide, oxytocin, stimulates maternal aggression (driving intruders away from pups).

Mice and other rodents can be bred for sex-specific aggressiveness. In mouse strains bred for testosterone-dependent aggressiveness (TA mice), males are more aggressive than males from other strains, but TA females show increased aggressiveness only when injected with testosterone. TA males are also more active than other males in revolving drums and score higher in maze tests.

These observations and experimental results show that aggression, like other behaviors, is subject to networks of genetic and hormonal control.

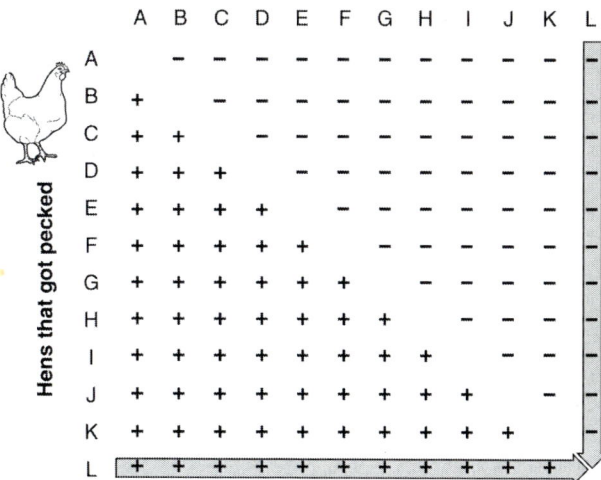

FIGURE 21.3:

Pecking order. Chicken establish a rank order of who pecks whom. Hen A pecks all others and is pecked by none. Hen L is pecked by all and pecks none. The other hens fall into a linear order in between.

D. Aggression in Humans

Humans show the same ultimate and proximate causes of aggression as other mammals.

Humans are territorial. Hunter–gatherers defend reliable food sources, water holes if water is scarce, hunting grounds, and fishing waters. Since agriculture, humans have protected their fields and herds. Trespassers have little or no protection under the law. Large-scale aggressions—riots, war, genocide—often break out over land use and under conditions of crowding. The human concept of property can be seen as an extension of territoriality.

Hungry or destitute people are more likely to be aggressive in begging, looting, stealing, or robbing. Under the law, crimes and misdemeanors committed out of hunger are treated more leniently than the same transgressions committed out of greed.

Humans are aggressive towards competitors for their mates. Under the law, even killing is treated more leniently if committed in a jealous rage. Aggression helps men to acquire status in their community, which in turn translates into mating opportunities.

Both men and women will be very aggressive in order to protect their children.

Hormones, neurotransmitters, and drugs have the same proximate effects on aggressiveness in humans as they do in other mammals.

In addition, **ecological circumstances** and **cultural rules** affect the level of aggression that humans display. An astounding example are the Semai people of Malaysia, who are extremely peaceful. Their language does not have a word for "to kill"; murder is virtually unknown, children are not spanked. However, when Semai men were recruited as soldiers by the British colonial government to fight a communist insurgency, they behaved very differently. They could be swept up in killing sprees which they described as "blood drunkenness." When the war was over and the Semai men went back to their settlements, they returned to their peaceful, nonviolent ways.

A similarly dramatic change in the level of aggressiveness was observed in the Maori tribes of New Zealand. During pre-European times, these tribes lived in a state of constant hostility and retribution. The preferred way for young men to acquire status was to participate in raids on neighboring villages, killing their men and dragging away their women. The introduction of muskets by British colonists exacerbated this state of affairs until about a quarter of the population had died in combat. Around 1840, the Maoris converted in large numbers to Christianity. Warfare among tribes ceased entirely. Similarly, !Kung San communities have changed from murderous to peaceful within 50 years as their population density decreased.

In addition to cultural effects on human aggression, **individual experiences** during childhood play a major role. This was demonstrated in a longitudinal study on a cohort of Danish boys, who were assayed for different measures of violence, ranging from psychological tests to convictions for violent offenses. By each of these measures, the strongest predictor of violent behavior as a juvenile was a history of maltreatment received during childhood (see Figure 15.4). In the group of severely maltreated boys, those with low levels monoamine oxidase A (MAOA) activity showed significantly higher levels of violence than those with high MAOA activity. MAOA inactivates serotonin and other neurotransmitters, but the relationship between serotonin levels in the brain and aggression needs further clarification.

In summary, aggression in humans is affected by many factors including

- hormones
- neurotransmitters
- ecological conditions and historical situations in which a community finds itself
- individual experiences of suffering from aggression by others and of learning how to use it for promoting one's own inclusive fitness
- cultural rules

How these factors translate into competing and reinforcing brain activities that lead to specific courses of aggressive or nonaggressive action is only partially understood and remains a wide and challenging field of investigation.

E. Coalitions and Alliances in Primate Communities

In many social primates, including chimpanzees, every individual has a **basic rank** based on his/her individual fighting ability. In addition, individuals have **dependent ranks,** which are conditioned on the presence of other individuals. Just by himself/herself, individual A may rank below individual B (basic rank). In the presence of individual C, however, A may rank above B (dependent rank). Dependent ranks reflect the existence of coalitions and alliances among individuals.

A **coalition** is defined by cooperation in an aggressive context; coalitions are for a limited duration and purpose. The simplest pattern is that of one individual interfering in an ongoing confrontation by supporting one of the two parties. The supported party may be the aggressor or the victim in the primary confrontation. Coalitions are often negotiated on the spot using certain utterances, facial expressions, and other gestures, including

- asking for support, often by stretching out a hand toward a third individual
- instigation, that is, stimulating a third party to act on one's behalf
- enlistment, in the extreme by dragging the third party to the site of confrontation
- third-party appeasement, for example, by kissing and embracing bystanders who might be inclined to support the opponent

A more permanent pattern of cooperation is called an **alliance**. It manifests itself in repeated coalition formation between two individuals. Alliances are often formed between *kin*. Alternatively, alliances are based on *reciprocity*, which may involve grooming, sex, food sharing, or other favors. For alliances to be stable, the benefits to the partners must be *equitable*. Individual A gages his support for individual B according to how much support B has provided for A.

> *Alliances: tended to be formed w/in families (not always).*

Success in coalition and alliance building requires mental abilities, including

- the ability to recognize individual members of the community
- a memory for past incidents in which assistance was received or denied
- an ability to assess another individual's alliances with and grudges against various members of the community

Frans de Waal (the author of your first assigned reading) has illustrated the principles of alliance building in his description of a dramatic episode in the chimp colony in Arnhem, The Netherlands, where he observed about 25 chimps living under semi-wild conditions on an island of about 1 ha size. The names of the three males in this episode are Luit, Nikki, and Yeroen.

> *Nikki couldn't take Luit. So he teamed up w/ Yeroen.*

Luit was the original alpha male. Nikki wanted to succeed him, but by himself was no match for Luit. So he formed an alliance with Yeroen, and with Yeroen's help, Nikki became the new alpha male. In return, Nikki was very tolerant of his ally's sexual activities, allowing Yeroen to increase his share of copulations from 9 percent to 39 percent. Yeroen also received the most "respect" (bows and pant grunts) from females.

Apparently feeling that Yeroen was overplaying his hand, Nikki formed a "non-intervention treaty" with Luit: If either male was in a contest with Yeroen over a female, the other would not intervene on Yeroen's behalf.

The tensions came to a head one day when Yeroen burst out in loud screaming against Luit, who was pursuing an estrous female. At the same time Yeroen held out his hand to Nikki in a request for support. But Nikki walked away from the scene. This led to a highly unusual surprise attack by Yeroen, who jumped Nikki from behind and bit him in the back. Two days later, both Nikki and Yeroen emerged with deep

> *Many interactions w/in chimp colony is about shifting alliances.*

wounds from an unobserved fight. Nikki behaved as the loser, while Luit regained his previous position as the alpha male.

F. Reconciliation after Conflict in Primates

In many primate communities, survival depends on mutual assistance. Under these conditions, aggression among members of the same community is constrained by the need to maintain cooperation. Relationships damaged by aggressive acts are often repaired by gestures of **reconciliation**. Depending on species, reconciliation behaviors may include touching, kissing, embracing, sexual intercourse, grooming, grunting, or holding hands (Figure S21.b). Bonobos in particular use sex for reconciliation. Aggression and reconciliation are often used in tandem to resolve conflicts.

The traditional notion that aggression promotes dispersal predicts *decreased contact* between individuals after fights. In contrast, reconciliation predicts *increased contact* between former combatants. Studies on social primates to test these conflicting predictions did reveal the widespread use of reconciliation. Reconciliation behaviors have a calming effect, as measured by the frequency of displacement activities such as self-scratching. The probability for revival of a conflict is lower for reconciled than for non-reconciled pairs. The probability of reconciliation increases if the combatants are genetically related or if they have high social or reproductive value for each other.

Reconciliation is often initiated by third parties. In chimp colonies, the alpha male breaks up fights and intervenes on behalf of females being bullied by lower-ranking males. In rhesus monkeys and macaques, similar peacekeeping roles are played by high-ranking females (Figure S21.c).

Reconciliation behavior is to a large extent learned. Long-tail macaques became three times more likely to reconcile after they had been trained to cooperate for obtaining food items. When youngsters from a reconciliatory species (stump-tail monkeys) were raised together with youngsters from a related but non-reconciliatory species (rhesus monkeys), the latter grow up to be more reconciliatory.

G. Cooperation, Fairness, and Social Pressure

Coalitions and alliances are examples of **cooperation**, that is, behaviors of two or more members of a species that enhance the inclusive fitness of each partner. Cooperation entails acts of *altruism* when individuals incur a momentary cost or risk to their reproductive fitness for the benefit of others (see Chapter 18). Thus, cooperation is most frequently observed among *kin*.

Cooperation also occurs between nonrelatives, for example as food sharing or mutual grooming in primates. Such cooperation is a form of *reciprocal altruism*, because a return of the favor is expected on the next occasion. Partners who do not reciprocate are met with pointed indifference or even hostility.

Finally, humans may cooperate in ways that cannot be explained by kinship or reciprocity because actors and beneficiaries are unrelated and may never meet again. Such cooperation may be out of civility (not jumping a queue), for the common good (donating blood, recycling, etc.), or in response to honor codes (valor in combat, offering oneself as a hostage). The costs of these behaviors range from trivial to heroic, and especially at the high end of this scale, such behavior seems to be uniquely human.

Researchers studying human altruism and cooperation often use simple games over money. In the **ultimatum game (UG)**, the experimenter gives a sum of money to one player, called the **proposer**, who needs to share the money with another player, called the **responder**. The proposer makes an offer, and if the responder accepts it, then the deal goes through. If the responder turns the offer down, then *neither* player gets *anything*. Both the proposal and the response are one shot; there is no opportunity for haggling.

In the basic versions of this game, the players are told that they will have only one interaction and will not see each other again. This rule, at least in theory, eliminates reciprocal altruism because there is no second encounter of the players. Nevertheless, almost all players are not strictly selfish. A selfish responder should take any offer because otherwise he will get nothing. Anticipating this, a selfish proposer should make a minimal offer, for example, 10 percent of the stake size. However, real players made offers of 25 percent or more. Lower offers were rejected with 40 to 60 percent probability by players from industrialized countries, while the rejection rates by players from primitive societies varied greatly in accord with the societies' customs.

The willingness of responders to refuse a low offer in the basic UG reveals a desire to **punish** selfish proposers. A desire to punish is also evident in advanced versions of the UG and similar games, in which players or third parties can impose penalties on tightwads and defectors *even if the punishers have to pay* for their actions.

Costly punishments are meted out even though the punishers know they will not benefit from the "lessons" they teach their opponents because they will never encounter them again. Nevertheless, punished tightwads and defectors behave less selfishly in subsequent rounds with other players.

The activities of face muscles and brain regions can be monitored in UG players. Unfair offers, such as 9:1 splits, trigger activity of the *levator labii* muscle (causing upper lip raise and nose wrinkle) just as tasting bitter or salty liquids. Unfair offers elicit brain activities also stimulated by pain, hunger, thirst, anger, and disgust. In contrast, mutual cooperation is associated with activation of brain areas linked with reward processing. The opportunity to punish activates the same circuits as tasting sugar, which gives new meaning to the old adage "revenge is sweet."

Group cooperation is upheld by effective sanctions against free-riders, so long as the sanctions are perceived as fair and appropriate. If the punishment is perceived as selfish, then it spoils the willingness to cooperate. Sanctions must be upheld by a sufficient number of individuals who punish deviants and reward cooperators, even at a monetary cost to themselves.

The willingness to administer costly punishment was found to varying extents in different human societies. The inclination to punish was positively correlated with altruistic behavior as measured in a modified UG in which the proposer simply dictates the share of the stake to the responder.

H. The Culture-Gene Coevolutionary Model

Many results from UG and related games are compatible with the following **culture-gene coevolutionary model**. Societies allow variation in the **genetic predisposition** of its members for feelings of reward, disgust, fear, guilt, and so forth, which drive the administering, anticipation, and avoidance of punishment. These predispositions must be strong enough for *a few* society members to be willing to punish at their own cost. "Punisher" societies, because they enforce cooperation for the common good, will then outcompete "slacker" societies that allow their members to free ride or disconnect. A corresponding argument could be made for genetic variation in feelings of **reward** experienced from "being a good citizen."

Implied in the culture-gene coevolutionary model is the idea of **group selection**, meaning that genetic alleles are positively selected in evolution because of the benefits they bestow on *groups* rather than on *individuals*. (You win by being a team player.) This model is unlikely to work for cooperation in animals because selfish mutants or migrants will outbreed cooperators faster than group selection can take hold. In humans, however, the dominant roles of language and culture make it possible to exert **social pressure** by punishment and rewards, which may reduce the success of migrants and deviants to the point where group selection can prevail (Figure S21.d). Group selection for effective cooperation could conceivably shift the distribution of genetic alleles that control an individual's feelings of reward, guilt, and so on, about following (or not) societal norms. Societies in which the distribution of related alleles is just right to produce the optimum level of cooperation at minimal cost should be most successful.

A strong component in the culture-gene coevolutionary model is a **sense of fairness**, which makes people feel angry if they think they are taken advantage of. A similar *aversion to disadvantageous inequity* has been demonstrated in nonhuman primates. In experiments with female capuchin monkeys, the animals turned down a perfectly good reward, with gestures of indignation, just because a partner got something better (see the assigned reading at the end of this chapter). Corresponding experiments with chimpanzees gave similar results, with the additional observation that tolerance for inequity increased with social closeness. (Likewise, humans "keep score" of favors among acquaintances but tend to be more generous with family and friends.) Variations of a sense of fairness could also have evolved in other mammals because they motivate individuals to leave unsatisfactory relationships if better ones are available.

I. Kindness and Compassion

As a rule among nonhuman primates, the strong bully the weak, and the sick and elderly receive little care. However, instances of kindness and compassion have been observed, although these behaviors are generally reserved for kin and close associates.

Long-term observers of chimpanzees, including Wolfgang Köhler and Jane Goodall, have noted instances of **kindness** and **compassion**, such as comforting the injured, protecting the weak, celebrating births, and grieving at deaths. In the Arnhem colony, Frans de Waal repeatedly observed acts of mischief and kindness: A male chimp was tricked into a ditch by his buddies and rescued by a female who let down a rope for him to climb out.

De Waal also describes a capuchin monkey born in the wild with paralyzed legs. He could climb but not jump, so he needed to be carried from tree to tree. Having too little exercise, he became obese so that he must have been quite a burden. Nevertheless, his buddies carried him until he was 17 months old, much older than capuchin youngsters are normally carried by their mothers. He then disappeared from observation.

On 16 August 1996 a dramatic incident happened in a zoo at Chicago. A three-year-old boy fell 25 feet down into the gorilla enclosure. There he lay motionless, bleeding from his forehead. He was saved by a gorilla female, Binti, who handed the boy over to paramedics, even though she had to antagonize larger gorilla males in the process. Part of an explanation for Binti's behavior may lie in her biography. Binti was raised like a girl by a human foster mother. Thus, Binti may have come to regard humans as kin and to treat them accordingly.

In a laboratory experiment, Jules Masserman and coworkers trained rhesus monkeys to obtain food pellets by pulling one of two chains. One chain simultaneously caused an electric shock to another monkey. Most monkeys, when they saw the connection, refrained from pulling the shock chain, or either chain, even if this meant they were starving for several days. This protective behavior was more likely when the operator and the victim had been cage mates, or when the operator had previously been subjected to electroshocks.

Evidently, one does not have to be human to be humane. But how do we explain kindness and empathy in animals? Anecdotes about humane behavior of animals are always about social animals, where group members are often genetically related, and reciprocity is a realistic expectation. Under these conditions, kindness may have originated as an "emergent" property, one that arises spontaneously as animals play and bring forth new variants of behavior. In human societies, kindness may have evolved according to the *culture-gene coevolutionary model* reviewed earlier (section H).

EXERCISE PAGE FOR CHAPTER 21

Student Last Name _____

Student First Name _____

Discussion Section _____

On the course web site, use the link on the **syllabus** to find a pdf file of your **assigned reading** for this chapter. The same reading is referenced below. Use the bulleted list of questions to test your knowledge of this reading.

Brosnan, S. F., and de Waal, F. B. M. (2003). Monkeys reject unequal pay. *Nature* **425:** 297–299.

- Which experimental setup and routine did the authors use to measure how important it was for the monkeys that they were treated fairly?
- Explain the data shown in their Figure 1. What do "non-exchanges, NT, RR, ET, IT, EC, and FC" mean?
- What data are shown in Figure 2, and how do the authors explain them?
- How do the authors comment on the fact that only capuchin females, but not males, reject unequal pay?
- Do the behaviors reported here for capuchin females remind you of similar behaviors you have observed in humans?

Feedback

On a scale from 1 to 10, the being the best, please rank the above reading for

1. Interest and Relevance to the topic (10 being most interesting/relevant) _____
2. Readability (10 being most clearly written, easy to understand) _____

In the remaining space, enter any comments that you may have on this reading.

PART V

HUMAN POPULATION GROWTH

AND ASSOCIATED PROBLEMS

Humans, like other species, have been striving to maximize their inclusive fitness. This innate drive, boosted by cultural achievements such as agriculture, industrialization, and medical care, has led to a frightening growth of the human population. The human biomass on Earth is more than 100 times greater than that of any other large mammal. This situation is creating ecological problems for which we are genetically and culturally unprepared.

The problem was pointed out more than 200 years ago by Thomas Robert Malthus, an English clergyman who became an economist and demographer. In 1798, he wrote *An Essay on the Principle of Population as it Affects the Future Improvement of Society, With Remarks on the Speculations of Mr. Godwin, M. Condorcet, and Other Writers*. In this essay, Malthus predicted that the human population would outrun its food supply, and that living conditions for humans could not be improved without limiting reproduction by late marriage and sexual abstinence. The swipe against Godwin and Condorcet was to express Malthus' deep skepticism against their overly optimistic views that technological progress and educational reforms would bring major improvements to human life.

The deleterious effects of human overpopulation are exacerbated by over-consumption in the most developed countries, which is wasteful in itself and sets an unattainable goal for developing countries. Over-consumption is stoked by two powerful forces: sexual selection and relentless advertising. As we have discussed in Part III of this course, men want beautiful women, and women want wealthy men. So a flood of commercials keeps telling Jane that she will turn the head of every John if she wears apparel from designer X. All the while, it is drummed into John that every Jane will be his if he drives up in luxury car Y.

To satisfy their needs and wants, a growing human population is turning Earth into a conglomerate of farms, factories, and shopping malls. Their run-off is poisoning us. They all consume energy, and currently we generate most of it by burning fossil fuels: coal, oil, and natural gas. The carbon dioxide released in the process is causing global warming. This and other ways of environmental degradation have set us on a collision course with the survival of ecosystems in which plants and animals support one another and eventually us humans. An enormous number of currently living species is threatened with extinction while hundreds of millions of humans are starving.

Managing these problems will be the greatest challenge of our century. Short-term solutions include improved agriculture, reduced carbon emissions, and bioconservation. Long-term solutions will have to come from reversing human population growth and curbing over-consumption.

CHAPTER TWENTY-TWO

HUMAN POPULATION GROWTH

Since Malthus published his essay on human population growth in 1798, the problem has acquired an unprecedented urgency: The world population, which was 0.9 billion in his time, will probably surpass 9 billion by 2050. Raising the standard of living in developing countries to that of developed countries, according to leading ecologists, could destroy many of the ecosystems that keep us alive. Are we headed for a crash?

A. Mathematical Models of Population Growth

In the context of this chapter, **growth** (*G*) is the change of the number of individuals in a population over an interval of time.

$$G = \Delta N/\Delta t$$

Δ: designates a small interval of the following variable
N: number of individuals present
t: time

The term can be used for any population of organisms, although in this chapter we will apply it mostly to all humans living in a specific country, a group of countries, or the entire world. Growth can be positive (>0) or negative (<0).

Under optimum conditions, a population's growth is limited only by species-specific factors collectively called the **biotic potential**. Growth is then proportional to the number of individuals already present (Figure 22.1).

$$G = \Delta N/\Delta t = r \bullet N$$

r: growth rate

The **growth rate** is the fractional change in the number of individuals over a time period, relative to the number of individuals present at the beginning of the period. Again under optimum conditions, the growth rate is independent of environmental factors and can be written as

$$r = \Delta N/N \bullet \Delta t$$

This type of growth is known as **exponential growth** because the above equation is mathematically equivalent to the following one, which makes it clear that the number of individuals (N) increases as an exponential function of time (t).

$$N(t) = N_0 \bullet e^{rt}$$

N_0: initial number of individuals (at t 0)
e: Euler's number ($e = 2.718...$)

In a linear coordinate system, a plot of N over t is a J-shaped curve (see Figure 22.1). Exponential growth starts slowly but picks up speed with time. For instance, a one-time deposit into a savings account with compounding interest grows slowly at first but then appreciably. (Disregard the offsetting effect of inflation.)

Sooner or later, the growth of a natural population will be limited not by its *biotic potential* but by *environmental factors,* such as available food and water supplies, predators, parasites, etc. The maximum population size that a habitat can sustain indefinitely defines the habitat's **carrying capacity**. A growth pattern that takes this limitation into account is the **logistic growth function**.

$$G = \Delta N/\Delta t = r \bullet N \bullet (K-N)/K$$

K: carrying capacity

Note that
for $N \ll K$: $(K-N)/K \sim 1$, and $G \sim rN$ (exponential growth)
for $N = K$: $(K-N)/K = 0$, and $G = 0$ (no growth)

A plot of population size (N) over time (t) according to the logistic growth function is an S-shaped curve, with the initial section looking like an exponential growth curve and the last section being a horizontal line indicating the environment's carrying capacity (see Figure 22.1). Real biological populations may oscillate around (periodically overshoot and undershoot) the carrying capacity of their environment. For instance, predators and their prey will often cycle, with the predator population lagging behind the prey population.

However, no population can exceed its habitat's carrying capacity for long.

B. Past and Current Human Population Growth

In the past, the human world population has grown steadily, first very slowly and then at a dramatically accelerated pace (Figure 22.2). A few milestones of this growth are

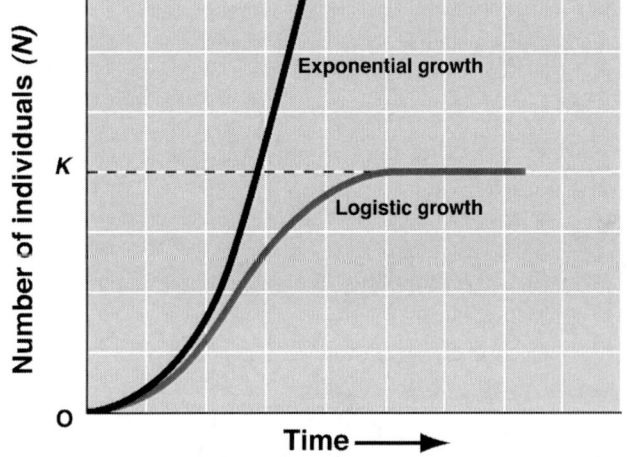

FIGURE 22.1:
Exponential and logistic growth. In logistic growth, the population size levels off as N approaches the habitat's carrying capacity (K).

- less than 10 million before agriculture
- 1 billion in 1820
- 2 billion in 1927
- 4 billion in 1974
- 5 billion in 1986
- 6 billion in 1999
- 6.9 billion in 2010
- 8.0 billion expected in 2025 according to UN "medium variant"
- 9.2 billion expected in 2050 according to UN "medium variant"

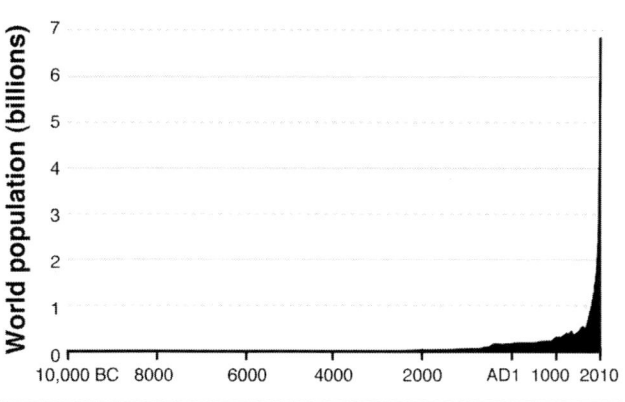

FIGURE 22.2:
Explosive growth of the human world population.

At the current relative growth rate, the world population grows

- each day by 208,000 (another Laredo)
- each month by 6.3 million (another Boston)
- each year by 75 million (another Germany)
- every 4 years by 300 million (another United States)

The relentless growth of the human world population is frightening (Figure S22.a). Why aren't our numbers leveling off, as one would expect based on the logistic growth function? Humans have used use technology to constantly push up the Earth's carrying capacity (K). Unfortunately, as we will see in the following chapters, some of the measures we have taken to increase K may not be sustainable.

In order to analyze and predict human population growth, demographers use different terms, including the following.

Population Growth Rate (PGR):	growth rate expressed in percent per year
Doubling Time:	time during which a population doubles in size
Total fertility Rate (TFR):	**average number of children born to a woman during her lifetime**
Birth Rate:	average number of births per 1,000 people per year
Death Rate:	average number of deaths per 1,000 people per year

For many purposes, the most useful measure of population growth is the TFR. It is critical that TFR values worldwide reach the **replacement level**, at which a population with an equilibrated age structure renews itself but does not grow. Under conditions of adequate nutrition and medical care, the replacement level is pegged at **2.06**, rather than 2.00, because not every girl born will reach reproductive age. In some developing countries, TFR replacement levels are higher than 2.06, but as living conditions in these countries improve, their replacement levels will decrease.

TFR values need to be watched very closely because a TFR even slightly above replacement level means that the population is still growing exponentially, and anything that grows exponentially is headed for a crash unless it is corrected in a timely manner. Current TFR values are distributed very unevenly, with the highest TFRs occurring in the less developed and least developed regions, which already account for most of the world population (Table 22.1, Figure S22.b).

Table 22.1: Distribution of World Population and Fertility Rates		
Region	Population (billions)	Total Fertility Rate
World	6.9	2.6
More developed	1.2	1.6
Less developed	4.8	2.5
Least developed	0.9	4.4

Data projected for year 2010 by UN Population Division

C. Estimates of Future Population Growth

Future trends of the TFR are the key to predicting the future size of the world population, which is a difficult task. The Population Division of the United Nations is publishing such projections and revising them from time to time. The twentieth round of projections was published under the title *World Population Prospects—The 2008 Revision*, online at http://esa.un.org/unpd/wpp2008/peps_documents.htm.

By 2050, the world population is expected to be between 8.0 billion ("low variant") and 10.5 billion ("high variant"), with a "medium variant" of 9.2 billion considered most likely (Figure 22.3). Even the high variant assumes that total fertility rates will decline worldwide. If TFRs were to remain constant, the world population would reach 11.0 billion in 2050 and keep growing.

The population of the more developed regions is anticipated to change little over the next 50 years. The TFR for this group of countries is expected to rebound from its current low of 1.6. In many developed countries, decreased fertility of the long-term residents will be compensated by immigration and higher fertility of new immigrants. For example, the U.S. population is expected to increase from 303 million in 2005 to 404 million in 2050. However, in some developed countries, including Japan, Germany, and Russia, the populations are expected to be smaller in 2050 than today.

The population of the less developed and least developed regions is projected to rise from currently 5.7 to 7.9 billion in 2050 according to the medium variant. This projection assumes that TFRs will continue to decline. In the absence of further decline, the population of these countries alone would reach 9.8 billion. People in the least developed are likely to live in a vicious circle of high TFRs, abject poverty (living on less than a dollar per day), environmental degradation, infectious diseases, and tribal warfare.

Any projections about the future growth of Earth's population are extremely sensitive to small changes in the future TFR. An upward deflection of the TFR from 2.06 to 2.17, which means only a 5 percent increase, would have the world population zoom to 21 billion by 2150 and would keep it growing exponentially. Conversely, if the worldwide TFR were to approach a level of 1.96 instead of 2.06, again a correction by only 5 percent, then the world population would return to 5.6 billion by 2150, the same level we had in 1994.

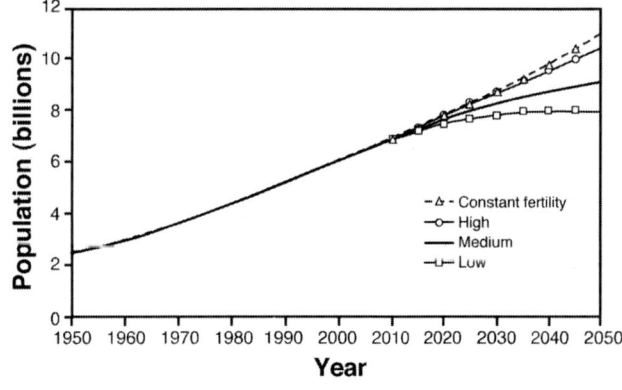

FIGURE 22.3:

Historical and projected world population based on different assumptions about future total fertility rates (constant, high, medium, or low). Based on *World Population Prospects-The 2008 Revision*. United Nations Population Division.

Thus, just a tick up or down from replacement level means the difference between self-destruction and a high but possibly sustainable level.

Because the worldwide TFR is currently above replacement level, the fraction of women at childbearing age will be overproportionately high for many years to come. The world population will therefore continue to increase significantly for several decades *even after a TFR has come down to replacement level* (Figure 22.4). Even if someone could wave a magic wand and bring the worldwide TFR to replacement level instantaneously, the world population would still not level off until it is near 10 billion.

The lag time of population growth, and the sensitivity population growth to very small changes in TFR values, leave us with two clear messages.

- Current TFR values (worldwide and in developing countries) are absolutely not sustainable.
- Every little decrease in TFRs will count, especially in those countries that currently have large populations.

D. The Explosive Growth in World Population Is Part of a Major Demographic Transition

In the past two centuries, all countries that are now considered "developed" have undergone a change known as the **demographic transition** (Figure 22.5). It can be divided into four stages.

- Stage one (**traditional stage**): birth rate and death rate are both high, while population growth is near zero.
- Stage two (**mortality transition**): the death rate falls (mostly due to better hygiene and improved medical care) while the birth rate remains high, so that the population will grow rapidly.
- Stage three (**fertility transition**): while the death rate remains low, the birth rate falls so that the population growth will slow down.
- Stage four (**developed stage**): birth rate and death rate are equally low so that the growth rate is near zero or slightly negative.

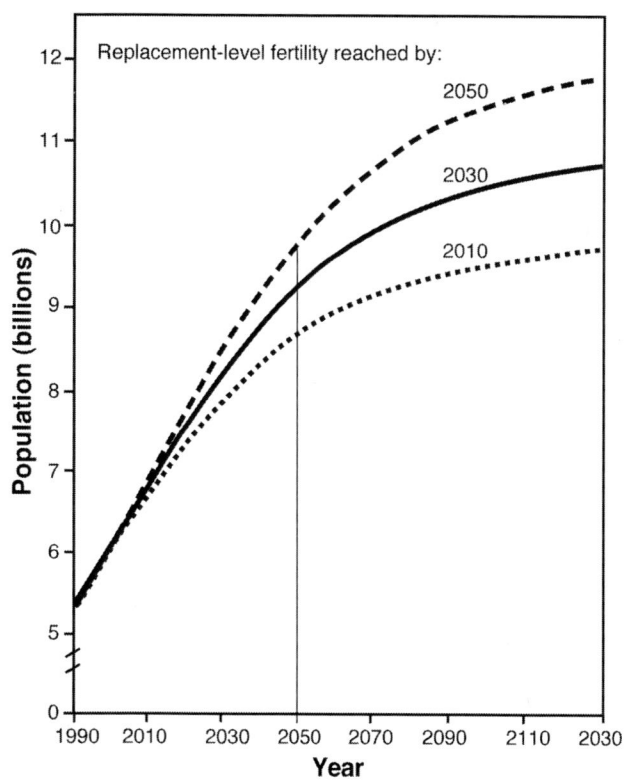

FIGURE 22.4:
Ultimate human population size, depending on when worldwide TFR reaches replacement level. Note time lag between reaching replacement TFR and leveling off of population curve. The year 2050 (vertical line) is the endpoint of the UN projections shown in Figure 22.3. Data source: Potts M. (2000) The unmet need for family planning. *Scient. Amer.* **Jan. 2000**: 88–93

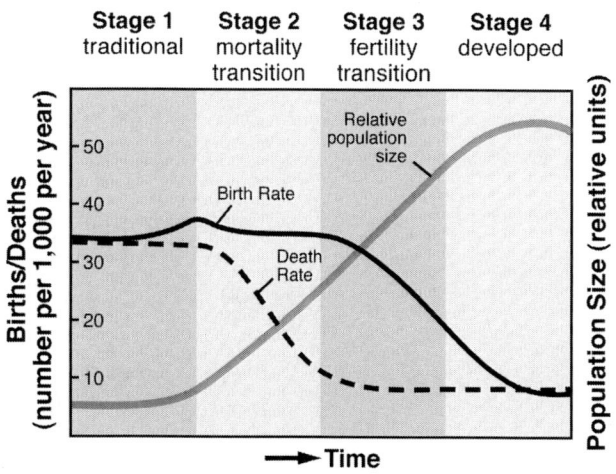

FIGURE 22.5:
Demographic transition

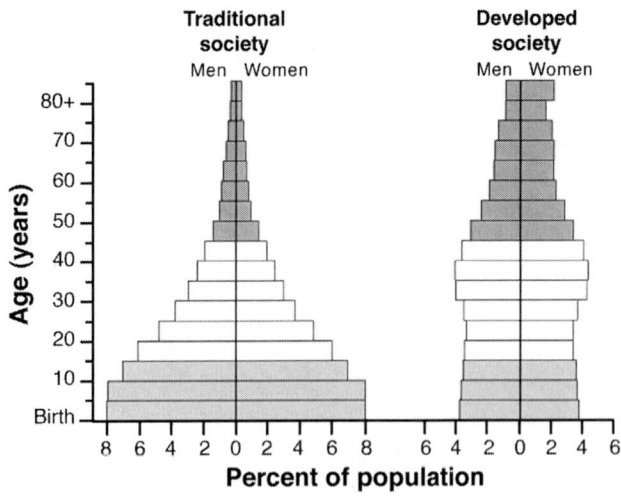

FIGURE 22.6:
Age distribution in traditional and developed societies. Ordinate indicates age groups in 5 year increments. Abscissa indicates size of each age group as percentage of total population, separately for men and women.

Both stages one and four are stable, but the population size is much higher at stage four. Other differences are that a population at stage one has a short average life span (less than 35 years) and an overly large fraction of young people, whereas at stage four a population has a long average life span (more than 70 years) and a more even distribution of young and old people (Figure 22.6).

With regard to human population growth, the critical aspect of the demographic transition is the length of time that elapses between stages two and three, because this is when the most explosive growth occurs. This means the only ethical way of curbing population growth is to accelerate stage three, the fertility transition. What encourages couples to make the fertility transition?

Industrialization and urbanization. Traditional farmers tend to have many children because they are useful helpers, there is always enough homegrown food to feed another mouth, and raising children is very compatible with farm life. In contrast, in a city household, there are fewer chores that children can do, food for each additional mouth drives up grocery bills proportionately, and raising children competes with job opportunities outside the home.

Social security. Traditionally, parents look at their children as caretakers for their old age. With increasing trust in governmental and corporate pension plans, there is less need for parents to hedge their bets by raising another potential caretaker.

Health care. Better hygiene and medical care instill confidence in parents that their first children will survive, thus encouraging them to have fewer children.

General modernization. Modern life offers people a wider range of interests and tends to break up traditional societies, in which raising many children was the only way for women to gain status and satisfaction in life.

Exactly which of these changes is most compelling for people to have fewer children is a matter of much debate. Industrialization is generally regarded as the most important factor, so much so that it is often used as a synonym for the demographic transition, and many sociologists have recommended industrial development as the most effective contraceptive. However, this view may be overly influenced by the history of England and North America, where the demographic transition occurred in step with industrialization.

In other countries, such as France and much of Eastern Europe, birth rates started to decline before industrialization. More recently, China, Sri Lanka, and the Kerala State of India have experienced sharp drops in birth rates without industrialization. Conversely, countries in South America have undergone much industrial development without decline in birth rates. These examples illustrate that industry is neither necessary nor sufficient for birth rates to decline. It appears that better hygiene and medical care, governmental birth control programs, education of women, and social security are equally strong factors that affect people's reproductive decisions, especially if they are propagated and backed up by credible governments.

E. Industrial Development as a Contraceptive Is Problematic

The question to which extent industrial development is or is not necessary for the fertility transition is of central importance, because industry tends to place such a burden on the environment that ultimately it may be counterproductive.

The pollution of air, water, and soil caused by the Industrial Revolution in Europe was atrocious (Figure 22.7). Emissions of dust, soot, chlorine, ammonia, carbon monoxide into the air led to an increase in pneumonia. Industrial wastes released or spilled into rivers killed the wildlife living therein or made it unhealthful for human consumption. Hundreds of thousands of acres of land were poisoned with industrial chemicals and littered with mine tailings and with abandoned industrial facilities. Some of these problems can be mitigated with today's technologies (and the willingness to pay for them). However, many problems of industrialization seem to be repeating themselves today in developing countries such as China (Figure S22.c).

Worldwide industrial development would certainly mean a huge increase in energy production (see Chapter 26). Most of our current energy comes from burning fossil fuel; this activity most likely causes global warming (see Chapter 24). Increasing this activity would exacerbate global warming to an extent that most people would find unacceptable. In view of the serious problems associated with worldwide industrialization, it is imperative to find less harmful ways of hastening the fertility transition in developing countries. We will focus here on birth control and women's education.

FIGURE 22.7:

Industrialization and Pollution. Historically, industrialization has been associated with pollution of air, water, and soil. Some of the damage has been reversed by environmental legislation.
Image © Robert Crum, 2010, Shutterstock, Inc.

F. Contraceptive Use Slows Down Population Growth

Women worldwide have begun to want fewer children, and modern contraceptives are providing the means to limit the number of children who are conceived. Both trends may be combining just in time to save the world from the worst in overpopulation.

Effective methods of **contraception** include sexual abstinence, vasectomy, tubal ligation, intrauterine devices, injections or pills preventing ovulation or implantation, diaphragms, and condoms from latex or polyurethane. The introduction of modern birth control in developing countries has been highly effective in reducing the overall fertility rate (Figure 22.8). In addition, there is every indication that more couples, or at least more women, would be using contraceptives if they were available. The World Health Organization estimates that one quarter of all pregnancies worldwide are unwanted. The number of married women worldwide who do not wish to have more children but do not use any modern contraceptive is estimated to be 120 million.

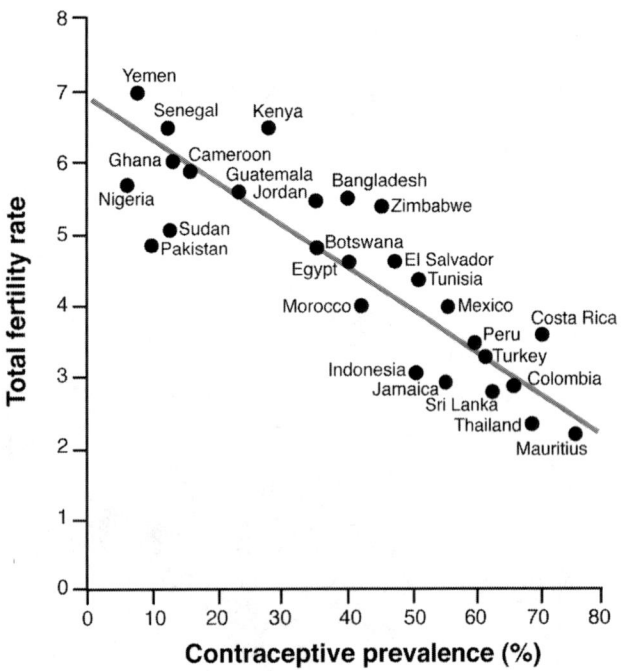

FIGURE 22.8:

Contraceptive prevalence and total fertility rate in developing countries, 1984–1992. Contraceptive prevalence is the percentage of women 15–44 years of age who are married and use contraception. Based on data published by Robey B. et al. (1993) in *Scient. Amer.* **Dec. 1993:** 60–67

An estimated 200,000 women perish each year from unsafe abortions, and many more are seriously injured. Thus, there is a great personal and societal need for modern contraceptives.

Unfortunately, the use of contraceptives is limited by various circumstances. Many poor people cannot afford to buy modern contraceptives or face major hurdles (medical prescription, spousal approval) if they want to acquire them. Most men dislike condoms because they reduce sexual pleasure. The use of hormonal contraceptives should be accompanied by regular physical checkups, which are often unavailable.

Pharmaceutical companies have been reluctant to launch new contraceptives since they are expensive to develop, liability suits loom large, and even a successful product sold to poor people would not generate much revenue. Many governments are reluctant to get involved because they do not want to alienate constituents who oppose some or all forms of family planning on religious grounds.

In the context of religious opposition to family planning, it is incumbent on biologists to reject the **incorrect use of biological arguments** in the political arena. The Catholic church teaches that the only god-approved purpose of sexual intercourse is procreation. This is religious dogma, and it is the church's prerogative to base it on whatever *religious* sources it deems appropriate. However, the Catholic church also tries to co-opt the authority of *science* by pointing out that most animals have sex only in the context of procreation and by implying that humans should follow the same *natural* order.

In humans, bonobos, and some other mammalian species, sexual intercourse has recreational and bonding functions (see Chapters 2 and 13). In humans, these functions stabilize marriage, an institution that the Catholic church professes to hold in high regard. Unfortunately, based on their teachings about the nature of sex, the same church discourages the use of modern contraceptives. This teaching is not only scientifically incorrect but also historically misplaced. It originated during the traditional stage of demographic development when high birth rates were balanced by equally high death rates (see Figure 22.5).

Continuing to force a link between sex and procreation today is frivolous and irresponsible, given the abject poverty and poor health of most humans living in countries with high TFRs. The last thing the world needs today is unwanted children!

G. Women's Education Promotes Birth Control

Given the problems of developing new contraceptives, it is important to fine-tune the existing ones and to educate the users and their governments. Health and humanitarian aid workers have long noticed that women in developing countries are more eager than men to embrace the use of modern contraception. The ability of women to prevail in this potential conflict seems to depend on their education.

Many successful government campaigns for birth control have included **educational programs for girls and women**. The spectacular success of Thailand in lowering its TFR is credited, at least in part, to the high educational standards of Thai women. The Kerala state in Southern India stands out from the rest of the country by its low birth rates. Kerala's TFR was already decreasing in the 1960s and 1970s, and it even fell all the way to 1.8 in 1991, in contrast to the rest of India, where the average TFR was much higher. What made the difference in Kerala was not economic development, because the state was and still is poor. Some analysts point out that traditionally the literacy of a woman in Kerala was considered to be a plus in marriage negotiations.

In developing countries, there is a negative correlation between TFR and women's education, as measured by literacy or years of schooling (Figure 22.9). There is no such correlation for men. While the negative correlation between TFR and women's education does not prove a causal relationship, there are several apparent reasons why educated women would have fewer children.

First, education promotes goals in life other than raising children. At the same time, education opens up job opportunities and endows women with greater income-generating potential, which is not fully realized while she bears children. A woman's unrealized earning potential raises the implicit costs of having children. Education and job opportunities tend to delay marriage and thus shorten the period of childbearing. Once married, educated women are in a better position to resist pressures from their husbands for more children, because they no longer have to barter sex and childbearing for support. Finally, educated women are more adept in using modern contraceptive methods.

Education of girls and women also leads to marked improvements of health and well-being for mothers and children. Each pregnancy is a drain on a woman's

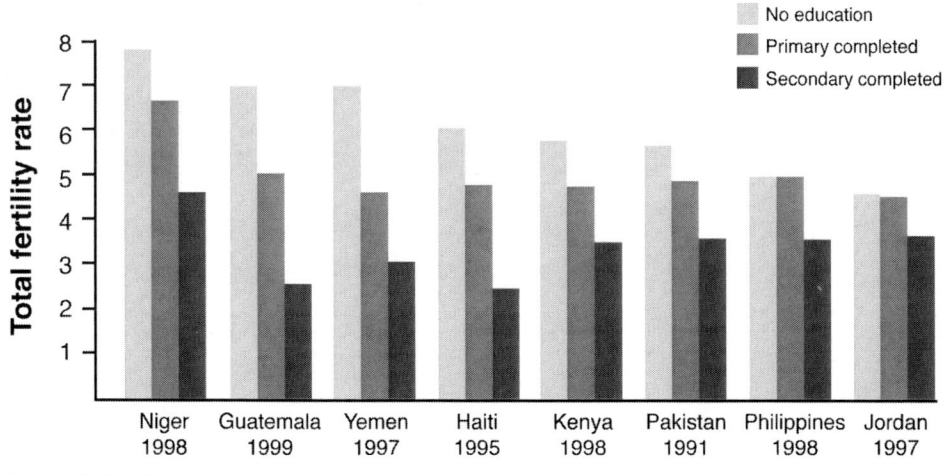

FIGURE 22.9:

Women's education and total fertility rate in developing countries in the 1990s. In each case, there is a negative correlation between women's level of schooling and TFR. Data source: *Demographic and Health Surveys, 1991–1999*, Population Reference Bureau. Found at http://www.aag.org/education/center/cgge-aag%20site/population/lesson3_page2.html

body: each birth is a risk to her life and to the welfare of her existing children. Educated women not only are better prepared to manage the stress and risk of childbearing, but they also need to go through the process less frequently if nearly all of their children survive to adulthood.

Thus, for many interrelated reasons, the combination of modern birth control and women's education seems to be the most promising way to meet one of the greatest challenges facing our species today, to sharply reduce population growth.

H. Limiting Population Growth Means Departing from Inherited Patterns of Behavior

Overpopulation was not a problem while humans lived as hunters and gatherers. Indeed, given the enormous challenges of early hominin life, giving birth to many children was a matter of survival as a species.

The advent of agriculture has allowed human communities to grow into empires with organized religions, armies, and tax collectors. Such empires were intent on settling new territories and waging war as needed. Many political and religious leaders have historically propagated population growth as a way to achieve military and political dominance, and some still do.

Current resource calculations indicate that the human population is outgrowing the carrying capacity of Earth (see Chapter 23). Virtually all areas of Earth that can be settled have been settled. Wars have become enormously expensive and destructive. They no longer have the adaptive function of promoting dispersal.

Ironically, then, we have to rein in some of our most successful and genetically most hardwired patterns of behavior—the drive to procreate—in order to approach TFR values at replacement level or less worldwide, and as quickly as we can. If we fail to reach this goal in a planned way, hunger, disease, and violence will do it for us.

EXERCISE PAGE FOR CHAPTER 22

Student Last Name _____

Student First Name _____

Discussion Section _____

On the course web site, use the link on the **syllabus** to find a pdf file of your **assigned reading** for this chapter. The same reading is referenced below. Use the bulleted list of questions to test your knowledge of this reading.

Potts, M. (2000). The unmet need for family planning. *Scient. Amer.* **January 2000:** 88–93.

- What does the author report about the desire of many women, especially in poor countries, to have fewer children? What about the willingness of their husbands to cooperate towards that goal?
- Why is the author correct in saying that large families are a recent, and temporary, anomaly?
- Is the expectation that the availability of modern contraceptives reduces TFRs born out by data? If so, which data?
- What does Potts say about the financial assistance from developed countries for family planning in developing countries?

Feedback

On a scale from 1 to 10, the being the best, please rank the above reading for

1. Interest and Relevance to the topic (10 being most interesting/relevant) _____
2. Readability (10 being most clearly written, easy to understand) _____

In the remaining space, enter any comments that you may have on this reading.

CHAPTER TWENTY-THREE

EARTH'S CARRYING CAPACITY

The world population currently stands at 6.9 billion and is expected to increase beyond 9 billion in the year 2050. The most obvious problem of the growing human population is a shortage of food. Already, 720 children *per hour* die of malnutrition. Of the survivors, many will never grow to their full potential. They will have little resistance to disease, learn less in school, and earn less as adults. Two factors that are limiting food production in many parts of the world are arable land and water. Does the growing human population exceed the carrying capacity of planet Earth?

A. Easter Island as a Warning

The history of Easter Island has been reconstructed from logbooks of European explorers, pollen analysis, and archaeological evidence. It shows that humans have in the past exceeded the carrying capacity of their habitat and have crashed in warfare and cannibalism.

Easter Island is 166 square kilometers in size and lies very isolated in the Pacific Ocean, about 2,300 miles west of Chile. The island received its name from the Dutch sea captain Jacob Roggeveen, who became the first European visitor on Easter Sunday, April 5, 1722. The hallmark of Easter Island are *moai*, large monolithic human figures carved from rock between the years 1300 and 1500. Standing up to 30 feet tall and weighing up to 86 tons, they are impressive testimonials to the artistic and engineering skills of their makers. Their overly large faces seem to represent deified ancestors, gazing over the landscape and showing off the power of their clans (Figure 23.1).

FIGURE 23.1:
Stone statues (*Moai*) on Easter Island, which was completely deforested by a rapidly growing human population.
Image © Massimo Ripani/Grand Tour/Corbis

The island was first settled by Polynesian seafarers. Radiocarbon datings indicate the earliest settlements around AD 1200. The population grew to more than 10,000 people in 1680 and then plummeted to 2,000 within a hundred years. The growth of the population was accompanied by the complete destruction of forest and by soil erosion. Remarkably, deforestation was not halted even as it meant the total loss of wood, which must have been needed for fuel, for building boats, and for erecting the moai.

Stone tools and other archaeological findings indicate that the periods of final population growth and subsequent steep decline were marked by famine, gang wars, lawlessness, and cannibalism. Today Easter Island is a barren grassland without any trees, supporting only a quarter of its former population.

Easter Island is not a singular case. Around the Mediterranean, there are ruins of large Roman cities, which must have relied on surrounding farms and forests. Today, the landscape around many of these ruins is barren, supporting just a few flock of sheep.

B. Land Availability

Most food today is produced by **agriculture**, that is, growing domesticated plants on **cropland** and by herding domesticated animals on **pastureland**. Significant amounts of food are also provided by harvesting wild fish and by **aquaculture**, meaning the growing and harvesting of freshwater and saltwater organisms under controlled conditions. In **industrial meat production**, chicken, pigs, cattle, and other animals are kept in small enclosures and fed corn or other food items transported to the enclosures. These practices raise concerns about animal cruelty, waste management, and adverse health effects of the antibiotics, growth hormones, and other chemicals used in the process (see Chapters 8, 24, and 25).

Directly or indirectly, most food production depends on the availability of land. Following are some numbers about worldwide human land use.

- total land on Earth (including ice and tundra)   14.7 billion hectares
- pastureland   3.8 billion hectares
- currently used cropland   1.4 billion hectares
- cropland per person in 2005   0.22 hectares = 2,200 m^2

The last line above means that *worldwide* the average person today depends on 0.22 hectares of cropland, which is about half of a football field.

Some of the current pastureland and forest are being converted to cropland, a practice that is problematic because it reduces species diversity and long-term viability of ecosystems (see Chapter 27). Even so, the *overall* area of currently available cropland is decreasing in *quantity* due to urbanization and industrialization. Because of the increasing human world population, the picture looks even worse on a *per capita* basis. In the poorest and most populous countries, the per capita area of cropland has decreased steeply between 1960 and 2000 (Figure 23.2). The only exception is Brazil, which has been holding steady on a per capita basis.

Available cropland is also decreasing in *quality* because of **top soil erosion**. (Top soil, the upper 2–8 inches of soil, has the highest concentration of organic matter and microorganisms.) According to UN-sponsored studies, 17 percent of the Earth's vegetated land has been degraded by human overuse or inadequate agricultural practices. "Vertical farms" (tall glass buildings for growing plants hydroponically) have been built in large cities but are struggling to become cost effective.

The human use of land for the production of food competes with other demands, including the generation of energy and the sequestration of waste. A study by Mathis Wackernagel and coworkers has translated all human demands on the biosphere into **land requirements**, with weighting factors for the production of energy, food, and other goods and services (Figure 23.3; see the assigned reading at the end of this chapter). The biggest item in the Wackernagel et al. budget is the land required for plants to sequester enough carbon dioxide (CO_2) from fossil fuel burning, which causes global warming (see Chapter 24).

The calculations of Wackernagel et al. indicate that the human demands on the biosphere in 1999 were exceeding its sustainable supply capacity by more than 20 percent. Factoring in 12 percent for bioconservation, and the increase of the human population between 1999 and today, the current ecological overshoot is about 40 percent. Much of the overshoot is because we are burning fossil fuels as a convenient way of producing energy without sequestering the CO_2 generated in the process (Chapter 26). By analogy, we behave like a family that is living beyond their means, making ends meet in the short run by not getting trash hauled off, postponing critical repairs, and running up debts on credit cards.

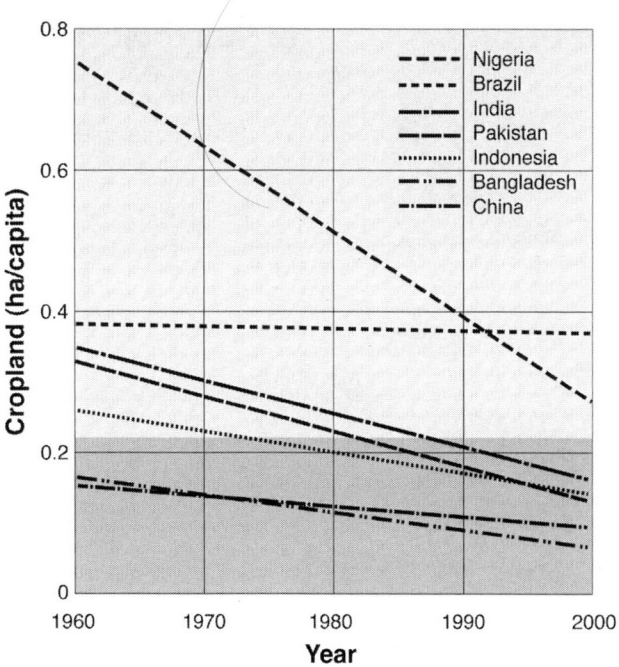

FIGURE 23.2:
Decreasing areas of cropland per person in developing countries. The level of the shaded area indicates the worldwide average of cropland per capita in 2005. Data source: Smil V. (2003) *Energy at the Crossroads*. Cambridge, MA: The MIT Press, Fig. 5.12

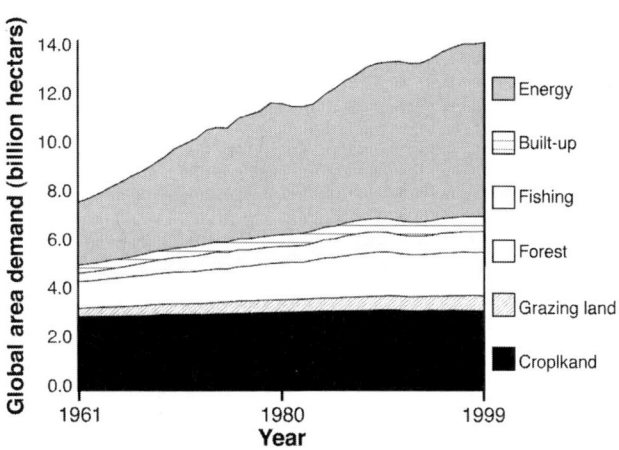

FIGURE 23.3:
Humanity's global area demands in six different categories. Global hectares are biologically productive areas with an average bioproductivity in that year. The total area demand of over 13 billion global hectares in 1999 exceeded the existing biocapacity by about 20%. Data source: Wackernagel M. et. al. (2002) *Proc. Natl. Acad. Sci. U.S.A.* **99**: 9266–9271, Fig. 2

C. Can Crop Yields Be Increased?

Given the need of a growing world population for food, and the limited availability of land, humans have looked for ways to increase the **crop yield**, measured in tons of grain equivalent produced per hectare of land.

Modern agronomical techniques have more than doubled food production between 1965 and 1990, a remarkable feat that has become known as the **green revolution**. It has been realized mostly by raising the crop yield to its current worldwide average of 2.1 through scientific breeding of high-yielding strains of wheat, rice, and corn. However, since the human population expanded rapidly as well, per capita food production has increased only modestly (Figure 23.4). In Africa it has actually declined. According to data from the UN Food and Agricultural Organization (FAO), recent increases in global cereal harvests have been only about 1.3 percent per year, just keeping pace with population growth. This *rate* of increase is half of what it was in the 1970s, suggesting that the easy gains in crop yield have already been made.

In fact, these gains may be difficult to sustain because they have been realized through large crop monocultures. Such monocultures reduce the variety of insects, earthworms, nematodes, and plant-symbiotic fungi, which are critical to the availability and recycling of soil nutrients. The planting of high-yielding crop strains over large areas also makes them vulnerable to pathogenic insects, fungi, bacteria, viruses, and occasional extremes in climate. Relying on insecticides and fungicides to control pests comes with severe health risks for humans and with the evolution of resistant pests in the long run (see Chapters 8, 24, and 25).

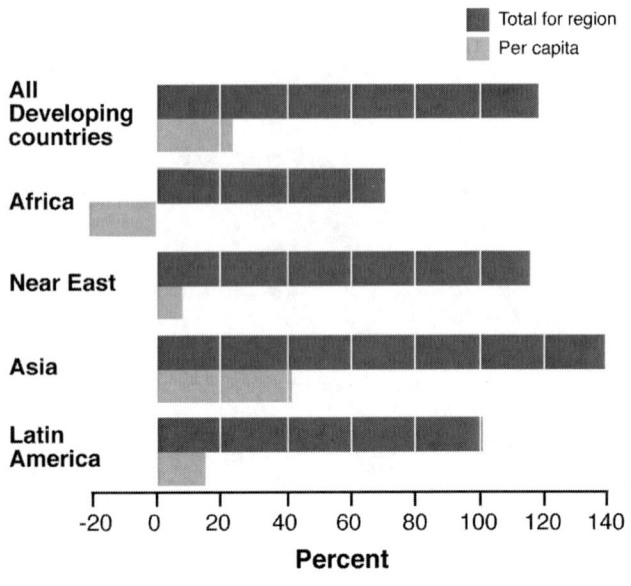

FIGURE 23.4:

Green revolution. Columns indicate the change in food production between 1965 and 1990. Note that improvements were more modest per capita than overall. Data source: Bongaarts J. (1994) Can the growing human population feed itself? *Scient. Amer.* **Mar. 1994:** 36–42, p. 40

In summary, even if crop yields could be increased somewhat by improved agricultural methods, they are likely to be cancelled out by the growing world population, so that the prospects for overcoming hunger are not good.

D. Genetically Modified Foods to the Rescue?

Hunger, unbalanced calorie intake, and vitamin/mineral deficiencies cause much of the world's disease burden. With successes of the first green revolution leveling off, hopes are now being staked on a second green revolution; this one based on *transgenic plants*.

Techniques for making *transgenic organisms* have greatly improved our ability to engineer plants and animals for human benefit (see Chapter 11). Crop plants modified by transgenes include plants that

- grow faster, allowing multiple harvests per season
- produce natural "antifreezes" that allow them to grow in colder climates

- have more favorable growth patterns, which make them less susceptible to damage from wind and rain, or produce more fruit and less stem tissue
- have a delayed rotting process, reducing losses on the way to consumers
- are more resistant to insect and fungal pests
- are more resistant to insecticides, fungicides, or herbicides that kill weeds (These improvements, of course, rely on the application of these chemicals, with possibly deleterious effects on human health, see Chapters 24 and 25.)
- yield diets with more protein and cis-unsaturated fat, which would make vegetarian diets more nutritious
- produce vitamins or other dietary components that are critically missing from the food supplies of malnourished people

An example of potentially significant improvement has been the engineering of a new rice variant that synthesizes ß-carotene, which is readily metabolized to vitamin A by humans. This variant has been cross-bred with another transgenic rice that accumulates iron in a form that is promptly absorbed by humans (Figure S23.a). Rice is a staple food in large parts of the world. However, the grain lacks essential nutrients including ß-carotene (pro-vitamin A) and iron in a form that can be utilized by humans. These deficiencies afflict more than 3 billion people, resulting in blindness, anemia, and susceptibility to infections. By adding daffodil and bacterial genes to a laboratory strain of rice, scientists at the Swiss Federal Institute of Technology of Zürich have generated a rice variant that synthesizes ß-carotene in the edible part of the grain (endosperm). This variant has become known as "golden rice" due to its color and anticipated dietary benefits.

The same group of scientists has generated another transgenic rice strain that enhances the storage of iron in a form that is readily absorbed by humans. By cross-breeding their ß-carotene- and iron-rich strains, the researchers have generated hybrids that combine both improvements. Unfortunately, severely malnourished people will be unable to benefit from the transgenic rice because their diet does not contain enough fat to dissolve pro-vitamin A. Further research may be required to improve the lipid content and other properties of the rice.

An ambitious project for the future will be to engineer rice, a C3 plant (producing a three-carbon compound as the first step in carbon fixation), into a C4 plant, which could be much more efficient at converting solar energy into biomass.

It may be possible to combine some of the advantages of **genetically modified (GM)** crops with the benefits of organic farming, that is, farming that minimizes the use of synthetic chemicals. Unfortunately, the organic farming community seems to be ideologically opposed to GM technology. There has been much concern on the part of the general public and many scientists about the dangers of genetically GM food. These dangers can be classified as adverse health effects on the consumer or as potential ecological problems of growing GM plants in the open.

GM foods can be unhealthy for certain consumers. For example, people allergic to Brazil nuts are also allergic to soybeans that have been nutritionally enhanced by a transgene for a Brazil nut protein. Such problems should be manageable by appropriate design and testing of GM food, and by warning labels.

The risk of ecological damage from growing GM plants in the open environment is much more serious, and unfortunately, very difficult to assess. The greatest concern is that genes conferring resistance to herbicides or insect pests may be outcrossed from

GM plants to their natural relatives, creating "superweeds" that might wreak havoc on ecosystems.

The risks of GM food must be weighed against the urgent need to feed a growing world population. This situation pits ecologists against engineers, small farmers against big business, and developing countries against developed countries. Resolving these conflicts will take time and international cooperation.

It needs to be kept in mind that faster-growing GM foods require more fertilizer, which may create environmental problems (Chapter 24), and more water, which is in short supply (below).

E. Water Scarcity and the Hydrological Cycle

In addition to land, water is emerging as a limiting factor to the effort of feeding a growing human world population.

Most of the Earth's water is saltwater; only about 2.6 percent are freshwater. Most freshwater is locked up in glaciers and ice caps, only about one quarter is present as fluid or vapor. Only a small portion of the fluid water is present as **surface water** in the atmosphere, vegetation, soil, and superficial rock, rivers, and lakes. Below a certain level known as the **water table**, any spaces between soil and rock particles are completely filled with water. This water is called **groundwater**, and the geological formations containing groundwater are called **aquifers**. Some groundwater travels quickly and close to the surface, discharges into springs and streambeds, and is replenished by rain. Deeper layers of groundwater are covered by impermeable rock and move only very slowly. As groundwater approaches sea level it becomes salty like seawater.

People can dig wells and pump out groundwater. If the water is pumped rapidly, the water table sinks unless it is restored by heavy rainfall. If in response people drill deeper through impermeable rock, they tap into very slowly moving groundwater that is replenished only in millennia if at all. This "fossil water" is a nonrenewable, one-time reserve.

A *renewable but finite supply* of freshwater for human use comes from the **water cycle**, or *hydrological cycle*, which is water moved around by solar energy and gravity. Being cyclic, the water cycle has no starting point. We begin our short description with the oceans, where most water exists (Figure 23.5). As the sun warms the ocean water, some of it evaporates. Rising air currents take the water vapor into the atmosphere, where cooler temperatures cause it to condense into tiny droplets visible as clouds. Winds move the clouds around the globe, some more from over ocean to over land than the other way. Under certain conditions, cloud moisture forms drops and crystals that are large enough to fall down as **precipitation** in the forms of rain, snow, or hail. The latter forms may accumulate in ice caps and glaciers before they melt. Some of the melted water and rain flow over the ground as **surface runoff**, forming creeks and rivers returning water to the oceans. The remaining precipitation infiltrates soil and rock, replenishing the groundwater in the aquifers. Groundwater flow and surface runoff add up to the **return flow** back into the ocean. Other components of the water cycle are evaporation from lakes and other freshwater surfaces, and transpiration from plants; these two components together are called **evapotranspiration**. And of course, much precipitation falls right back into the oceans. A simplified version of the water cycle, along with quantitative estimates of its major components, is shown in Figure 23.6.

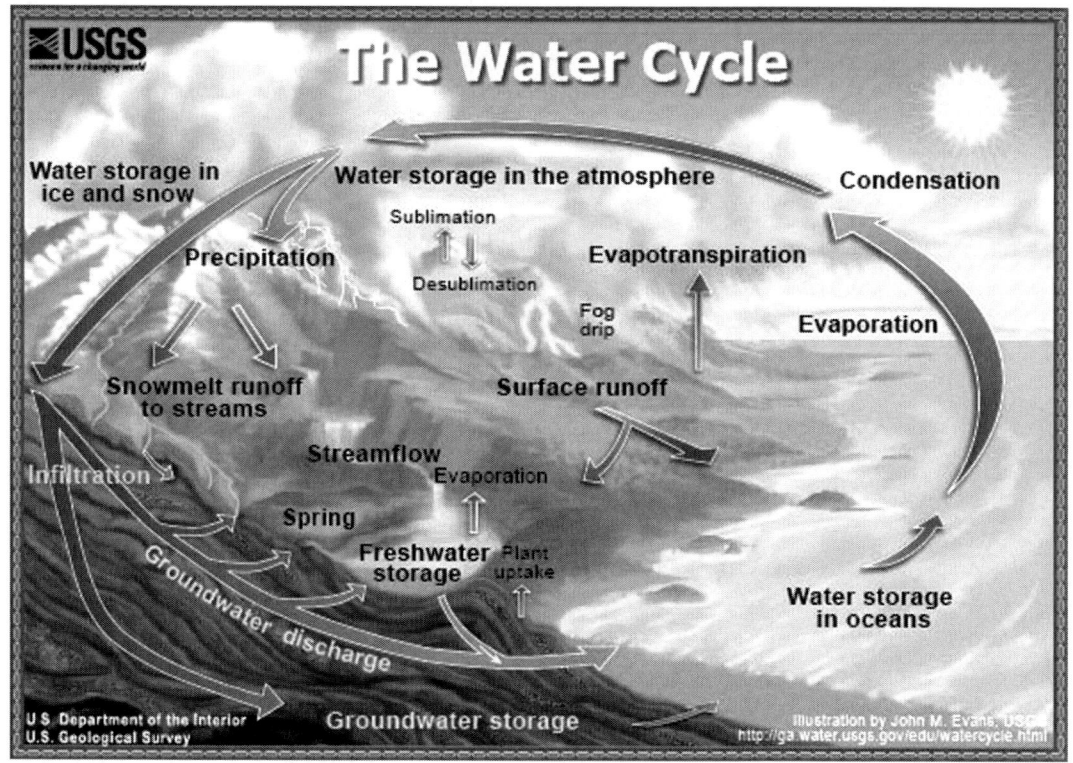

FIGURE 23.5:
The water cycle is a finite amount of freshwater moved around by solar radiation and gravity. A simplified version of the water cycle is shown in Figure 23.6. Found at http://ga.water.usgs.gov/edu/watercycle.html

Some of the surface runoff occurs in remote areas or as floodwater. The remaining portion, from which humans satisfy their needs, is called the **accessible runoff (AR)**. The AR is estimated at 12,500 km³ per year. Of the AR, about 2,350 km³ per year, or 19 percent, are *used instream* for fishing, recreation, navigation, dilution of pollutants, and generation of hydroelectric power. An additional 4,430 km³ per year, or 35 percent of the AR, are *withdrawn* for irrigated agriculture, industry, and municipalities. Between instream uses and withdrawal, the total **human appropriation** of the AR was already at 54 percent in 1995.

F. Major Types of Human Water Use

Of the total water withdrawn, more than 70 percent is used for **irrigation** in agriculture (Figure 23.7). Smaller amounts of water are used in industry and for municipalities, or are lost as evaporation.

Even though humans withdraw water from the hydrological cycle mostly for agricultural use, this

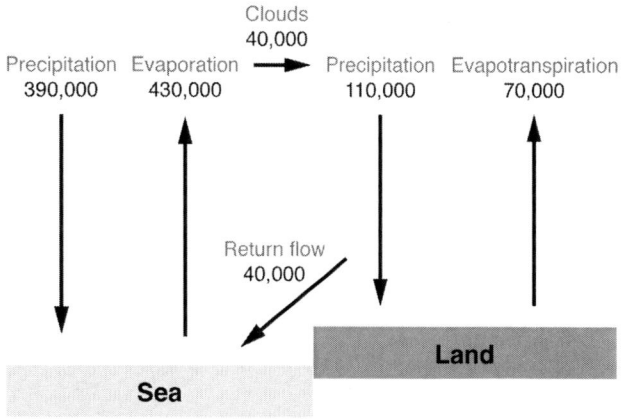

FIGURE 23.6:
Simplified depiction of the water cycle. Numbers indicate cubic kilometers per year. The arrow drawn at an angle, representing the return flow from land to sea, includes the accessible runoff of about 12,500 cubic kilometers per year. Data source: Postel S.L. et al. (1996) Human appropriation of renewable fresh water. *Science* **271**: 785–788, p.785

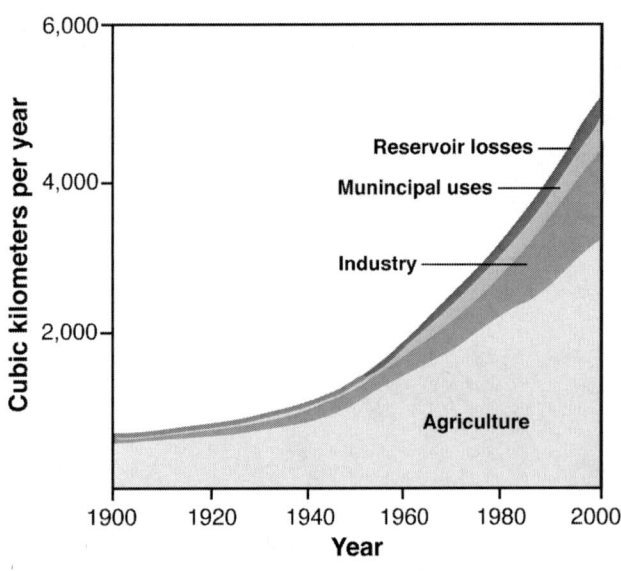

FIGURE 23.7:
Principal human uses of water withdrawn from accessible runoff worldwide. Data source: Postel S. (1993) Facing water scarcity. In: Brown L.R. State of the *World* 1993. pp. 22-41. New York: W.W. Norton & Company.

water does not drive *all* agriculture. Currently, about 60 percent of the agricultural production worldwide receives its water supply from rain, and only 40 percent from irrigation. Thus, doubling the amount of water used for irrigation would increase total food production only by 40 percent. However, doubling the withdrawal for irrigation would use up most of the accessible runoff freshwater, at enormous ecological costs.

Major efforts are therefore being made to use water more sparingly in agriculture. Breeders select for plants that can tough out droughts. For example, wheat varieties with long coleoptiles (seed sprouts) can be sown deeper into the soil, where moisture is retained longer. Losses to evaporation during irrigation are estimated to range between 20 percent and 50 percent, depending on climate, crops grown, and irrigation technique. Evaporation losses during irrigation also exacerbate the problem of soil salinization (salt deposition), which eventually renders soil useless. Highly developed regions with scarce water, such as Israel and northwest Texas, spend much effort on techniques for precision irrigation ("spoonfeeding") combined with occasional "flooding" to counteract salinization.

The amount of water used in **industry** is less than used for agriculture, and it can often be returned to the water cycle with minimal expense for cleaning. Worldwide, industries account for less than a fourth of human water withdrawal. As governments have imposed standards for release of industrial water into the environment, water quality in many rivers has improved tremendously.

Municipal use is estimated at 300 km^3 per year worldwide, less than one tenth of all human withdrawal. More than 80 percent of the municipal water withdrawn is quickly returned to the environment. However, there are special problems. The water needs of large cities can be difficult to meet because they are focused on small areas, forcing engineers to reach out to ever more distant sources. Municipal water and wastewater systems are expensive to build and maintain, and keeping water safe for humans to drink is an ongoing concern.

About half of the world population lives in large cities. The percentage is increasing, especially in developing countries, which often do not have the quality infrastructure needed for safe water supply. Already 1.2 billion people have no access to safe drinking water, and 2.6 billion have little or no sanitation. Each day, 3,900 children die from diseases transmitted by unsafe water and human excrements.

G. Uneven Water Distribution and Use Patterns

If the accessible runoff were distributed proportional to the density of the human population, it would be enough to sustain the current world population. However, because water is expensive to transport, *locally uneven supply and demand* make water a limiting factor in many areas. Water withdrawal per capita per year averages around 800 m^3 but ranges from 2,200 m^3 in the United States to 35 m^3 in Egypt and Ghana.

Many areas are already relying on fossil water, a one-time supply of deep groundwater that is not recharged from precipitation. This use is not sustainable. In Saudi Arabia, three-quarters of all water supplies comes from fossil water, which is expected to be depleted in 50 years or less. Other countries in the Middle East, North Africa, Thailand, Mexico, and some southwestern U.S. states, including Texas, also rely on fossil water. In large countries depending on irrigated agriculture, including India and China, the exhaustion of water wells is already causing social unrest.

H. Can Accessible Runoff Be Increased by Building Dams?

Accessible runoff (AR) is being increased mainly by building **dams** on large rivers. The human-made water reservoirs typically serve other purposes as well, such as recreation and generating electricity. However, the dams also come with long-term problems.

The number of large dams (storing > 3 million cubic meters of water) has increased from 5,700 in 1950 to about 50,000 in 2004. The peak years of dam building were the 1970s, but construction continues, especially in developing countries. For example, the Chinese government has embarked on two gigantic water management projects on the Yellow River and the Yangtze River.

However, building more dams will not be able to satisfy the needs of a much greater world population. Assuming 350 new dams per year for the next 30 years, with similar reservoir capacities as built in the past, and assuming further that water in each reservoir would turn over once a year, then accessible runoff could be increased to about 13,500 km^3 per year). If the entire increase were used for agriculture, food supplies from irrigated agriculture could increase by about 25 percent, but since only 40 percent of all food comes from irrigated (as opposed to rain-fed) agriculture, total food production would increase only by about 10 percent. While this increase would be significant, it would not nearly offset the expected increase in world population.

One of the problems with dams is that they interfere with traditional farming along rivers, which relies on seasonal mud deposition by floods. Also, dams need to be decommissioned when structural damage accumulates to the point of making them unsafe. The high costs of decommissioning dams are usually ignored when the dams are built. As an alternative to building dams, some communities have begun to rely more on "rainwater harvesting" and measures to slow down the runoff of rainwater from the ground.

I. The Ecological Costs of Increased Water Withdrawal

Further increases in water withdrawals will compromise both the *instream* functions of water and the *ecosystems* that depend on natural flow of water in rivers and aquifers.

One of the instream uses of water is the dilution of pollutants that escape the organized methods of disposal. However, many major rivers, including the Mississippi, Ganges, and Nile, are already a health risk to human bathing. Further deterioration is likely to result in major losses to freshwater fisheries and recreational uses of water.

Damming and diverting water from rivers, and polluting it in the process, wreaks havoc on downstream wetlands, deltas, and riverine habitats. In North America, an estimated one-third of the continent's fish, two-thirds of its crayfish, and nearly three-fourths of its mussels are now rare or imperiled. The Okavango delta of Botswana is world-renowned for its antelopes, elephants, and other wildlife migrating there during the dry season (Figure 23.8). Plans to siphon off water from the delta for agricultural, industrial, and municipal use have been defeated for now, but the struggle is likely to continue.

J. Alternate Sources of Water

Alternate sources of water include fossil water (water not recharged from precipitation) and desalinized seawater. Since the former is obviously unsustainable, we will focus on the latter.

Because of the polar nature of the water molecule, **seawater desalination** requires a lot of energy. Therefore, this process currently provides only for 0.2 percent of the world's water use. However, it is an option for water-scarce but energy-rich locations, such as Saudi Arabia.

Modern reverse osmosis technology (pressing seawater through filters that pass water while retaining salts) has made desalination more efficient. Plants of this type in Florida and Israel already produce 0.035 km^3 and 0.1 km^3 per year, respectively. Texas plans to build similar desalination plants in Freeport, Corpus Christi, and Brownsville.

Worldwide, to increase irrigated food production by about 10 percent, one would need an additional 1,000 km^3 of water per year. This would require 30,000 desalination plants, each needing energy at a rate of 10 MW (a big power plant produces 500 MW). Nuclear power may be the only way of generating so much additional energy without driving global warming to unacceptable levels. Unfortunately, in an age of terrorism, nuclear energy comes not only with safety concerns but also with great security risks (see Chapter 26).

FIGURE 23.8:
The Okavango delta in Botswana is world-renowned for its wildlife, which migrates there during the dry season.
Image © Gavriel Jecan/CORBIS

K. Summary

The prospects for feeding a growing human population are indeed limited by the availability of land and water. While there is room for improvement, each improvement seems to come with additional costs and/or risks.

- The human demands on land use are already unsustainable if carbon dioxide from fossil fuel combustion is to be sequestered by green plants.
- Significant increases in crop yield require large amounts of fertilizer, which is environmentally damaging, and large amounts of water for irrigation, which is not available in many parts of the world.
- GM foods may bring much-needed relief, but their possibly devastating ecological effects make it necessary to proceed with great caution.
- Water for sustainable use must be obtained from the hydrological cycle or by seawater desalination. The accessible runoff may be increased by about 10 percent over the next 30 years by building more dams. Improvements in irrigation techniques may eventually conserve about 20 percent of the water now withdrawn.
- Water desalination is feasible but very expensive in terms of energy.
- The current level of human water appropriation is already damaging many ecosystems that depend on natural flow of water in rivers and aquifers. Any increase in withdrawal is likely to cause severe ecological damage.

It is therefore imperative to curb fertility rates in order to slow down and eventually reverse the growth of the human population.

EXERCISE PAGE FOR CHAPTER 23

Student Last Name _____

Student First Name _____

Discussion Section _____

On the course web site, use the link on the **syllabus** to find a pdf file of your **assigned reading** for this chapter. The same reading is referenced below. Use the bulleted list of questions to test your knowledge of this reading.

Wackernagel M., et al. (2002). Tracking the ecological overshoot of the human economy. *Proc. Natl. Acad. Sci. USA* **99**: 9266–9271.
- What are the six types of land use defined by the authors?
- How did they adjust for the different bioproductivities of these land categories?
- Which human demand generates the greatest land requirement in the authors' budget? What data did they use to compute this land demand?
- How did they compute the total demand and the total existing biocapacity figures in their Table 1?
- How did they compute the "ecological overshoot" of 20% in 1999?
- The *worldwide average* human "footprint" (area demand per capita), according to Wackernagel et al., is 2.33 global hectares. What are the average human footprints in the U.S., England, and Germany?
- If we would get serious about sequestering the CO_2 that we release by burning fossil fuel, with which other uses of the Earth would the sequestration compete?

Feedback

On a scale from 1 to 10, the being the best, please rank the above reading for

1. Interest and Relevance to the topic (10 being most interesting/relevant) _____
2. Readability (10 being most clearly written, easy to understand) _____

In the remaining space, enter any comments that you may have on this reading.

CHAPTER TWENTY-FOUR

ENVIRONMENTAL DEGRADATION

"We have become, by the power of a glorious evolutionary accident called intelligence, the stewards of life's continuity on earth. We did not ask for this role, but we cannot abjure it. We may not be suited for it, but here we are." (Stephen Jay Gould, 1941–2002, evolutionary biologist at Harvard University)

A. Overpopulation Breeds Poverty, Environmental Degradation, and Lawlessness

The human impact on the environment has grown as a result of both population growth and the "energy burning lifestyle" of developed countries.

Human population growth began with the arrival of agriculture, which can sustain many more people than hunting and gathering could. Agriculture has replaced much natural vegetation with cropland. This reduces biodiversity, makes ecosystems more vulnerable to climate extremes and pests, and depends on fertilizer, which pollutes our freshwater. Growing cities have destroyed forests in order to obtain fuel and timber for construction. The ensuing erosion of soil has devastated large areas until today.

Scientific progress and modern medicine have initiated the demographic transition discussed in Chapter 23. Death rates fell while birth rates stayed high, leading to frightening growth spurts of human populations. In many countries, the demographic transition was accompanied by industrialization, meaning the mechanization of farming and the mass production of machines and consumer goods. Industrial development has raised the standard of living but has also polluted air, water, and soil to the point of serious health hazards for humans. It has also stoked the **energy burning lifestyle**, in which material possessions that consume energy to make and to run are used carelessly. Indeed, these possessions are marketed as status symbols and mate attractants, and their nonchalant use enhances these functions.

Overpopulation is linked in vicious circles with poverty, environmental degradation, and anarchy.

Overpopulation breeds poverty as family estates are progressively subdivided. Today, nearly 1 billion humans live in poverty: Many are malnourished to the point of being unable to carry out normal activities and to fend off diseases. They have no clean drinking water and no decent sanitation facilities. Conversely, *poverty exacerbates overpopulation* instead of correcting it. Aside from "the attraction between the sexes," as Malthus put it, destitute parents also regard children as cheap labor and as caregivers for their old age. Not surprisingly, the poorest nations tend to have the highest TFRs.

Poverty also promotes environmental degradation. People desperate for firewood simply collect it without replanting trees. The denuded soil erodes quickly. Likewise, overgrazing ruins the vegetation and causes soil erosion. Water that previously infiltrated the soil slowly and replenished the groundwater is now running off fast. In turn, *environmental degradation exacerbates poverty* because water scarcity and soil erosion reduce crop yields and grazing opportunities for livestock.

Poverty may easily become linked in another vicious cycle with lawlessness, terrorism, and warfare. In humans as well as in other animal species, crowding and hunger lower the threshold for aggression. People who cannot buy food or grow it themselves will start looting and fomenting unrest. *Conversely, as law and order break down, poverty, mass migration, and epidemic diseases are bound to increase.* In several countries, these problems are coming to a head in the form of failed governments. This is already happening in Somalia, and may happen also in Sudan, Zimbabwe, Chad, and the Democratic Republic of Congo.

In dealing with these problems we cannot rely on instinct or tradition. Although we have excellent nervous and hormonal responses to emergencies, we tend to ignore disasters that creep upon us slowly. Being genetically programmed to maximize our inclusive fitness, we take care of ourselves, our families, and our friends. In a process of gene-culture coevolution, we have learned to cooperate with our neighbors and countrymen. We are naturally inclined to protect our private land, county, state, and nation. However, we have not been selected for caring about the *world* or *humankind*.

As the human population and its impact on the environment have grown, environmental degradation no longer stops at regional boundaries. Solutions will require global research efforts and international cooperation. The following section on global warming will illustrate the magnitude and difficulty of this task.

B. Global Warming

The most pervasive kind of environmental degradation today is a global climate change. During the twentieth century, the **global average temperature** of Earth's near-surface air and the oceans has risen by 0.7°C, or 1.3°F[(Figure 24.1). This increase, called **global warming**, seems to be an accelerating trend, which is likely to have devastating consequences for the future.

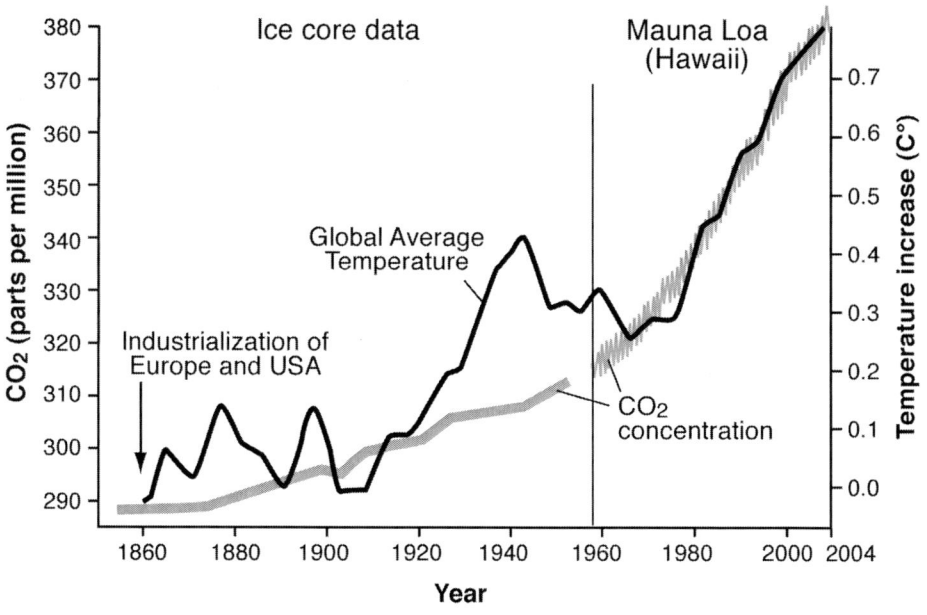

FIGURE 24.1:

Atmospheric concentration of carbon dioxide (CO_2, left ordinate) and global warming (right ordinate). Data source: IPCC Report 2007

As Earth is hit by sunlight, it warms up and emits infrared radiation. The infrared is absorbed and scattered in all directions by atmospheric gases, which increase their kinetic energy (temperature) in the process. Thus, in a way somewhat similar to a greenhouse, the atmospheric gases trap solar energy before eventually releasing it back into space (Figure 24.2). This is called the **greenhouse effect**, and the atmospheric gases causing it are known as **greenhouse gases**. The greenhouse effect has been going on for hundreds of millions of years, keeping Earth warm and hospitable for life as we know it. The concern is about the *recent increase* in the amount of energy trapped temporarily on its way back into space, with the result of global warming.

The principal greenhouse gases are water vapor, carbon dioxide, methane, ozone, and nitrous oxide. Most of global warming is ascribed to increases in atmospheric CO_2.

Water vapor (H_2O) is a potent greenhouse gas, but its warming action is mitigated by the shading effect of the cloud cover. Water vapor is produced naturally by evaporation from water surfaces and transpiration from plants, and by human activities including the burning of fossil fuel. The effects of human activities on water vapor concentration and global warming are just beginning to be investigated.

The atmospheric **carbon dioxide (CO_2)** concentration increased from 290 ppm (parts per million) in 1860 to currently above 380 ppm, an increase of

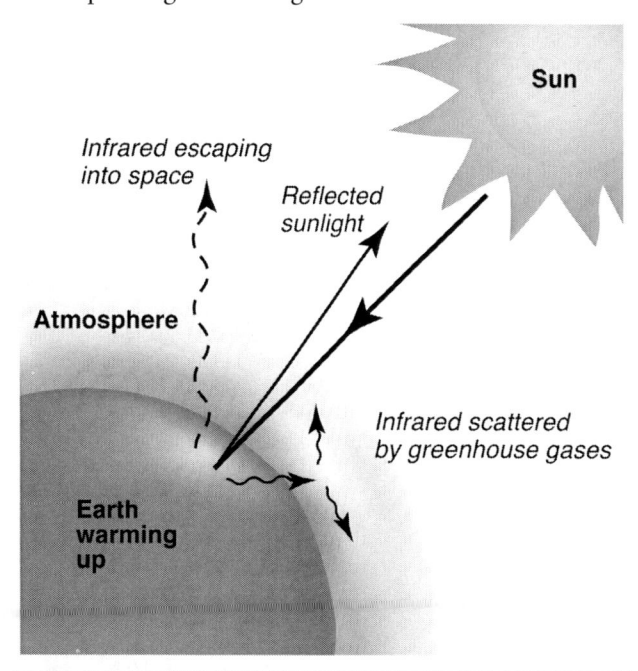

FIGURE 24.2:

The greenhouse effect. After a drawing originally prepared by Dr. Nelson Guda.

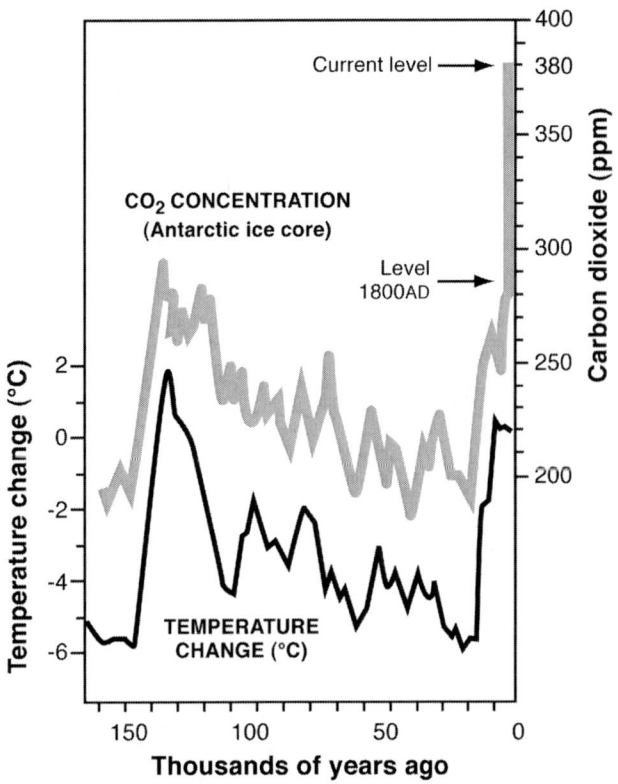

FIGURE 24.3:
Atmospheric carbon dioxide concentration and global temperature change. The correlation between atmospheric CO_2 and temperature over the past 160,000 years suggests a causal relationship. Note that both the current CO_2 level and its rate of change are outside the bounds of previous variation. Data source: IPCC Report 2007

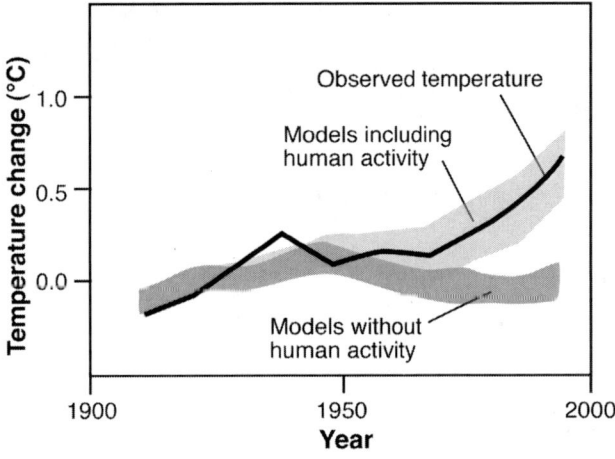

FIGURE 24.4:
Modeling temperature change. Models including human activities fit the observed change in global temperature better than models not including human activities. Redrawn from IPCC Report 2007

30 percent since the beginning of industrialization in Europe and the United States (see Figure 24.1). This is serious because CO_2 has a lifetime of more than 100 years in the atmosphere.

Analysis of air bubbles trapped in the ice of Antarctica and Greenland extends the correlation between global average temperature and greenhouse gas concentration much further back in time. Carbon dioxide and methane concentrations in air bubbles of the ice are correlated with local temperature at the time of ice formation measured by oxygen isotopes in the ice (Figure 24.3).

The explanation of global warming by greenhouse gases has been the subject of intense scientific and political debate. However, most scientists today agree that this explanation is probably (>90 percent) correct and that the recent increases in CO_2 are human-made. Evidence includes the following:

- exact correlation between temperature increase and CO_2/CH_4 concentrations
- current *level of* CO_2 concentration is outside the bounds of variation since 200 kya (see Figure 24.3)
- recent *rate of increase* in CO_2 concentration is also unprecedented
- CO_2 levels began to rise with the beginning of industrialization (see Figure 24.1)
- climate models including anthropogenic effects explain the available data much better than models accounting only for natural events (Figure 24.4)

Of the CO_2 produced by human activities, about 30 percent are sequestered by the oceans, 20 percent are taken up by the terrestrial biosphere, and 50 percent are staying in the atmosphere. Human activities producing CO_2 include the following:

- burning of fossil fuels in electric power plants, which is currently the principal way of generating commercial energy (see Chapter 26).
- combustion of fossil fuels in automobiles. Unfortunately, this is a major contribution to global warming (see Chapter 26).
- biomass burning to clear forests. This is the most destructive human activity producing CO_2 because it also causes species extinction (see Chapter 27), disturbs the hydrological cycle, and diminishes a much-needed sink for CO_2.

Other greenhouse gases, generated in part by human activities, include methane and nitrous oxide.

Methane (CH$_4$) is generated by decomposing organic matter. Related human activities include rice farming, garbage accumulation, and cattle ranching. CH$_4$ has increased to about 2.5 times its preindustrial level and persists in the atmosphere for nearly 10 years. Much CH$_4$ is stored in the permafrost of arctic lakes. As this ice is thawing due to global warming, CH$_4$ is bubbling up into the atmosphere. Estimates are that escaping CH$_4$ could drive up human-made global warming 20 to 40 percent by the year 2100. The only way to slow down this vicious cycle is to reduce CO$_2$ emissions.

Nitrous oxide (N$_2$O) is released from soil containing agricultural fertilizers, by biomass burning, and by fossil fuel combustion.

C. Major Consequences of Global Warming

The **Intergovernmental Panel on Climate Change (IPCC)**, consisting of more than 1,000 scientists and reviewers from over 100 countries, writes periodic reports on global warming. Based on peer-refereed scientific papers, the IPCC reports are published after a complex series of draft writing and reviewing by scientists and government officials. Former vice president Al Gore has produced and distributed a documentary video on global warming. Both Gore and the IPCC (including Dr. Camille Parmesan from UT Austin) were awarded the 2007 Nobel Peace Prize.

The IPCC has drawn attention to global warming since 1988, summarizing the evidence that global warming is occurring and is caused mostly by human activities. The IPCC report published in 2007 in the leading U.S. journal *Science* marks a *shift to predicting the further course of global warming and its dire consequences*. The report predicts a further global warming between 1.1 and 6.4°C (2.0 to 11.5°F) until the year 2100, depending on release of greenhouse gases.

In an **optimistic scenario**, the IPCC predicts that global warming will level off after a temperature increase of 2.5°C above the preindustrial level, at which point nearly 15 percent of all global ecosystems will be transformed and 20 percent of extant species committed to extinction (Figure 24.5). *This scenario requires a reduction of greenhouse gas emissions to 30 percent of their 1990 levels by the year 2050 and no net emission after 2080.* In a **pessimistic scenario**, the IPCC assumes that global warming will proceed beyond 4.5°C above the preindustrial level, at which point more than 40 percent of ecosystems will be transformed and a similar percentage of species committed to extinction.

The predicted global warming will have major effects on the viability of ecosystems, the ranges of species, land availability, weather patterns, and the prevalence of tropical human diseases. Some of these effects will cause severe economic strains, especially in developing countries. A few major effects are listed below:

- Melting of the polar ice sheets and glaciers now covering Greenland, Antarctica, and the Himalayas will deprive animal species, such as the polar bear, of their natural habitat (Figure S24.a).
- Melting ice and thermal expansion of ocean water will cause major coastal flooding and large-scale human migration.
- For statistical reasons, even a small increase in mean temperature translates into substantially longer and more intense heat waves, posing a major threat to the health of susceptible humans (Figure 24.6).

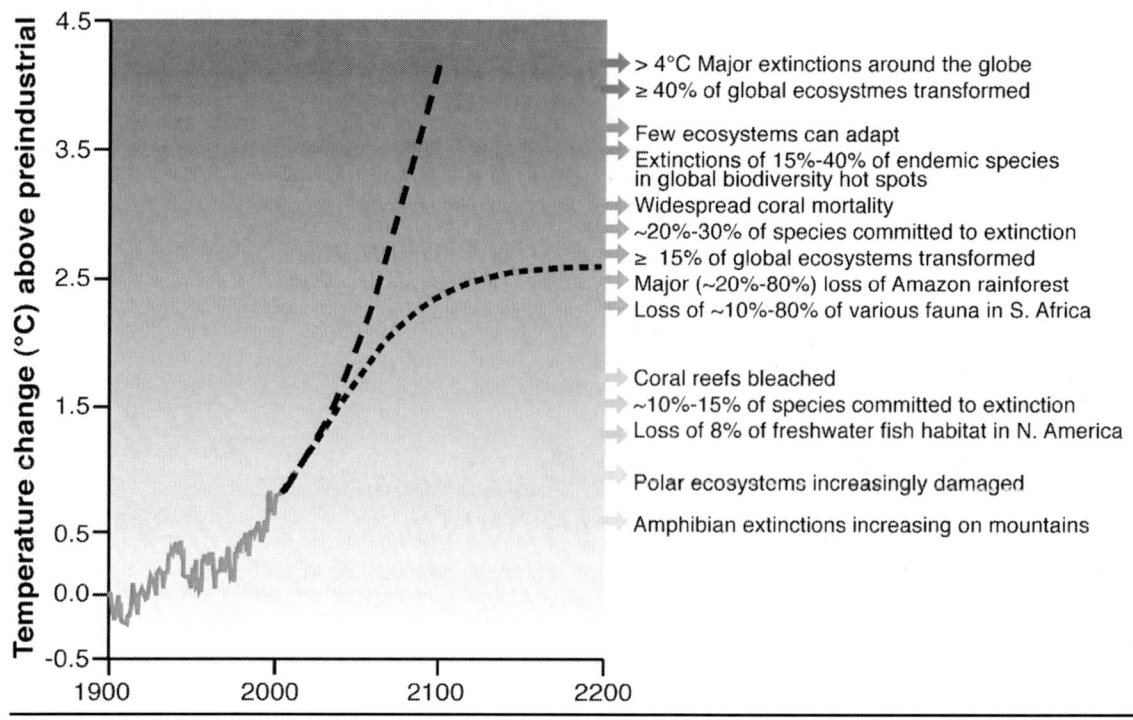

FIGURE 24.5:

Global Warming and Consequences. Temperature given in °C above preindustrial level. Data from IPCC report 2007, as redrawn in *Science,* 23 Nov. 2007, p. 1231

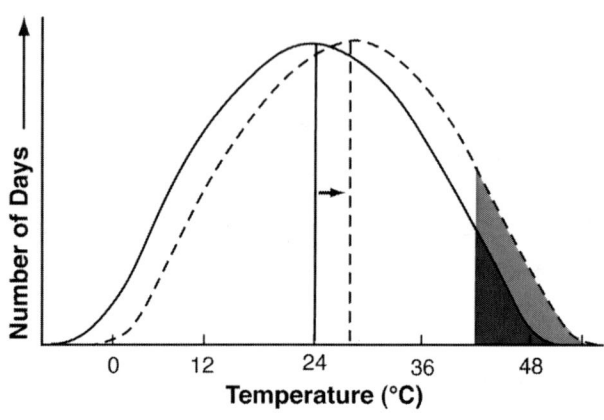

FIGURE 24.6:

More heat waves. Because of the bell-shaped distribution of average daily temperatures over the year, a small shift in mean temperature translates into a substantial increase in hot days. In this diagram, the hot triangle to the right has increased in size by a factor of 2.4.

- Animal ranges will shift poleward in latitude and upward in altitude. Evidence of these trends was first presented in 1996 by Camille Parmesan for *Euphydryas editha,* a Californian butterfly (Figures 24.7, S24.b, S24.c; see the assigned reading at the end of this chapter). Subsequently, this trend was found in 1,700 other species.
- The expansion of tropical and subtropical habitats would cause a spread of diseases like malaria, yellow fever, and viral encephalitis.
- Because of an acceleration of temperature-driven air circulation patterns, some geographical areas are predicted to experience more droughts, while others can expect heavier rains and flooding (Figures S24d,e). For instance, Bangladesh stands to lose much of its land to rising seawater and flooding rivers (Figure 24.8). At the same time, in Arizona and other western states of the United States, droughts are making the land volatile for persistent wildfires.
- Major "climate drivers," such as ocean currents, may be halted or "flipped" into a new stable state, causing major changes in global temperature and precipitation patterns.

FIGURE 24.7:
Edith's Checkerspot butterfly (*Euphydryas editha*) has already shifted its range to higher altitude as well as poleward in latitude in response to global warming. Photograph courtesy of Dr. Michael C. Singer.

FIGURE 24.8:
Rising waters in Bangladesh. The country's capital, Dhaka, has been hit by repeated flooding.
Image © Kapoor Baldev/Sygma/Corbis

CHAPTER TWENTY-FOUR: Environmental Degradation

- Many agricultural crops are now grown near their optimal temperature. Shifting temperatures will cause reduced yields or force new crops to be grown. The resulting changes will be disruptive in most cases.

It is therefore imperative to conserve energy, to develop ways of generating energy that avoid global warming, and to foster the international cooperation necessary to accomplish these changes (see Chapter 26).

D. Threats to Water Quality

Water is critical for human health and well-being as well as for fisheries and agriculture (see Chapter 23). However, during the hydrological cycle, freshwater may become contaminated with various pollutants. Coastal waters are the nurseries of many marine species. Yet here is where humans discharge sewage and toxic waste through myriads of poorly controlled pipes. From here, pollutants diffuse into the open oceans. Smaller seas, such as the Baltic or Mediterranean, are already heavily polluted.

In the atmosphere, carbon dioxide combines with water to form carbonic acid. Nitric oxide, nitrogen oxides, and sulfur dioxide also dissolve in atmospheric water vapor and come down as **acid precipitation**, consisting mainly of nitric acid and sulfuric acid. In soil containing aluminum, acid precipitation can "mobilize" aluminum ions, which are normally present in an insoluble nontoxic form of aluminum hydroxide. Both the acidity and the higher aluminum concentrations in surface water can cause damage to plant roots, fish, and other aquatic animals. As lakes and rivers become more acidic, biodiversity is reduced. In some areas of the United States, acid rain has eliminated insect life and fish species, including the brook trout.

Runoff water is contaminated with pollutants, of which some are biodegradable, while others are not. Synthetic fertilizer as well as animal and human waste are fully biodegradable but nevertheless harmful. Human waste contains pathogens including those for cholera, typhoid fever, and dysentery. In developing countries, millions of people die each year from waterborne diseases.

Large amounts of fertilizer cause algal overgrowth and oxygen depletion in lakes, rivers, and coastal waters. The magnitude of this problem became apparent from the discovery of **dead zones**, lifeless areas discovered by shrimpers and divers in coastal waters near the mouths of large rivers. According to a report released by the U.N. Environment Program in 2004, there are 146 dead zones—most in Europe and the U.S. East Coast—ranging from under a square mile to up to 45,000 square miles. One of the most infamous zones is where the Mississippi river empties into the Gulf of Mexico. This dead zone extended over 7,000 square miles, an area the size of New Jersey, in the summer of 1997.

The origin of the dead zones is explained as follows. Fertilizers from intensive agriculture and excess waste from farm animals held at high density are carried by runoff from rainwater. In rivers, and eventually in the coastal waters of the oceans, the fertilizers support excessive growth of algae, which eventually sink to the bottom of the sea, where they are degraded by bacteria. The breakdown of excessive biomass depletes oxygen from the water until *no* higher organisms can survive. The situation is not normally relieved by water circulation because the oxygen-rich freshwater is lighter than the oxygen-poorer saltwater so that the two form separate horizontal layers.

Another problem is the leakage of industrial and organic wastes into *groundwater tables*, where pollutants are not degraded because there is little oxygen to support the microorganisms that do the aerobic degrading. In the United States alone, the chemical industry produces some 70,000 different compounds, with about 1,000 new ones added each year. This pace is overwhelming the capacity of government laboratories to test each compound for health hazards and for environmental impact.

Some toxic chemicals are long-lived and not degraded by naturally occurring bacteria. For example, the half-life of the **insecticide DDT** in soil may range up to 30 years. Being extremely hydrophobic, it accumulates in the food web and reaches the highest levels in top predators. In animals and humans, it has long-lasting deleterious effects by interacting with DNA and by disrupting the endocrine system (see Chapter 25).

Another nonbiodegradable pollutant is **mercury**, which is released into the air from coal-burning power plants, small-scale gold mining operations, and many other sources. From the atmosphere, the mercury enters the hydrological cycle with precipitation and accumulates in the aquatic food chain. In many freshwater and saltwater fish, the accumulation of mercury has reached levels that have led to advisories against the consumption of fish.

Even the oceans are becoming sites of major, and potentially irreversible, pollution. Pollutants enter the ocean with surface runoff and groundwater flow from the hydrological cycle. Acid precipitation enters the oceans directly, diminishing the ability of corals and many other marine organisms to grow. Other contaminations come from offshore mining, drilling for oil, ship wreckages, waste dumped from cruise ships, and so forth. A circular pattern of ocean currents in the northern Pacific Ocean is collecting an area of floating plastic particles and other trash known as the **Great Pacific Garbage Patch**, which is estimated to be the size of Texas or larger.

Continued pollution of the oceans would be disastrous because cleanup is difficult, if not impossible, and healthy marine ecosystems provide services to humans on a gigantic scale. Marine photosynthesis curbs global warming by sequestering about 30 percent of all carbon dioxide generated by human activities. Marine fish supplies 19 percent of animal protein consumed by humans. This enormous resource is already approaching a worldwide crisis from overharvesting, a problem that does not need to be exacerbated by pollution.

E. Loss of Ecosystem Services to Humans

Human life depends on diverse ecosystems that churn our soil, purify our water, pollinate crops, and maintain the food chains to the animals we eat. The explosive growth of the human population, and the energy-burning lifestyle of the developed countries, are placing ever-growing burdens on the environment and endanger the ecosystems that sustain us. As a result, services to humans that used to be provided by natural ecosystems for free have to be manufactured at cost, and in ways that place more burdens on the environment. Two examples may illustrate this point.

Both freshwater and saltwater fish provide a substantial share of protein in the human diet, as mentioned earlier. This resource was traditionally available at low cost by fishing. Since overharvesting and pollution have exhausted natural fisheries, it has become economical to raise fish in fish farms. However, these fish have to be confined and fed at cost, and they are treated with chemicals to accelerate growth, combat infections, and make their meat look good in the store.

The second example illustrates a service provided to humans by an intact ecosystem that was endangered by carelessness but has been restored. New York City historically has enjoyed exceptionally clear water courtesy of the Catskill mountains, which capture rainwater and return it filtered and ion-exchanged (Figure S24.f). As Catskill forests were developed into farms, homes, and resorts, the water quality fell below EPA standards. The New York City administration then faced a choice. They could build treatment plants at 6 to 8 billion in capital investment and 300 million annual running costs. Or they could create laws and economic stimuli to upgrade septic tanks and finance other measures to protect the natural environment, which they did for 1 billion and minimal maintenance costs. And their water quality was restored.

The services to humans of ecosystems depend on their species diversity. Impoverished ecosystems provide reduced services and are more susceptible to climate extremes and destruction by pests. The recreational and aesthetic values of unspoiled wilderness are hard to quantify, but the popularity of most State and National Parks demonstrates a strong desire of city dwellers to get out into pristine areas. Thus, bioconservation serves not only the utilitarian interests of human but is also an ethical imperative. We will revisit this topic in Chapter 27.

F. The Tragedy of the Commons

The connection between human population growth and environmental degradation was epitomized in an article by ecologist Garrett Hardin, entitled "The Tragedy of the Commons" and published in *Science* **162:** 1243–1248 (1968).

Hardin uses the term **commons** for any resource used as if it belonged to all. When anyone can use a shared resource simply because he/she wants or needs it, then he/she is using a commons. A classical example are alpine meadows, which are still being used for free by multiple farmers to graze their cows and other livestock in the summer. Modern examples are public roads or restrooms.

The problem, or "tragedy," of the commons is that it can be destroyed by overuse. Hardin illustrates this point by the following scenario:

Picture a pasture open to all. It is to be expected that each herdsman will try to keep as many cattle as possible on the commons. Such an arrangement may work reasonably satisfactorily for centuries because tribal wars, poaching, and disease keep the numbers of man and beast well below the carrying capacity of the land. ...

As a rational being, each herdsman seeks to maximize his gain. Explicitly or implicitly, more or less consciously, he asks, "What is the utility to me of adding one more animal to my herd?" This utility has one negative and one positive component.

The positive component is a function of the increment of one animal. Since the herdsman receives all the proceeds from the sale of the additional animal, the positive utility is nearly +1.

The negative component is a function of the additional overgrazing created by one more animal. Since, however, the effects of overgrazing are shared by all the herdsmen, the negative utility for any particular decision-making herdsman is only a fraction of -1.

Adding together the component partial utilities, the rational herdsman concludes that the only sensible course for him to pursue is to add another animal to his herd. And another. ... But this is the conclusion reached by each and every rational herdsman sharing a commons. Therein is the tragedy. ...

Most human societies have solved the problem of the common pasture by individual ownership of land, which rewards sustainable use because this promotes the owner's inclusive fitness. Some systems of individual ownership may be flawed or unfair, but at least they avoid "ruin to all." However, major components of the environment, such as air and water, are still used as commons. Anyone may use them for breathing and drinking, and with few restrictions, for dumping waste and pollutants into them. Problems have arisen again as users have become too many, and new solutions have to be found.

Payments related to the quantities used, as well as caps and trade allowances for waste generated, appear to be reasonable principles. For instance, if the rights to harvest fish stocks were given to individual fishermen as "fish shares," the collapse of fisheries would become less likely. More detailed analyses addressing the governance of commons have been carried out by economists. Elinor Ostrom and Oliver E. Williamson received the 2009 Nobel Memorial Prize in Economic Sciences for their "analysis of economic governance, especially the commons."

G. Conclusions

Ecologists warn that satisfying the needs and wants of a growing human population is setting us on a collision course with preserving the ecosystems that sustain us. Ecosystems are inherently complex and hard to predict. Given the magnitude of what is at stake, it is imperative to err on the side of caution. This means to

- reduce environmental pollution at all levels
- conserve energy and promote environmentally friendly ways of generating energy
- slow down and eventually reverse the human population growth

Biologists have an additional obligation to raise public consciousness about the connection between sexual selection, material wealth, and environmental degradation. Traditionally women have preferred mates who are good financial prospects (see Chapter 19). Men flaunting ostentatious houses, big cars, gas-guzzling power boats, and so on often enjoy not only these products but also the companionship of attractive women. As long as these energy-consuming possessions work as mate attractants, men will use them to their advantage. Women seem to have more cultural flexibility in their mate choice criteria than men (see Chapters 19 and 20). If women could gage the financial appeal of men using environmentally sensible criteria (energy-efficient homes, fuel-efficient cars, sailboats instead of power boats, etc.), this would go a long way towards protecting the environment.

EXERCISE PAGE FOR CHAPTER 24

Student Last Name _____

Student First Name _____

Discussion Section _____

On the course web site, use the link on the **syllabus** to find a pdf file of your **assigned reading** for this chapter. The same reading is referenced below. Use the bulleted list of questions to test your knowledge of this reading.

Parmesan, C. (1996). Climate and species' range. *Nature* **382**: 765–766.

- What hypotheses on the biological consequences of global warming did the author test?
- Which species did she study, and which life history trait of this species was especially useful for her work?
- What were the major results? How did they fit the prediction that global warming affects wildlife?

Feedback

On a scale from 1 to 10, the being the best, please rank the above reading for

1. Interest and Relevance to the topic (10 being most interesting/relevant) _____
2. Readability (10 being most clearly written, easy to understand) _____

In the remaining space, enter any comments that you may have on this reading.

CHAPTER TWENTY-FIVE

ENDOCRINE-DISRUPTING CHEMICALS

Some chemicals synthesized by humans and released into the environment are known as *endocrine-disrupting chemicals* because they interfere with hormonal signals in wildlife as well as humans. The threat posed by these chemicals has not been fully realized until recently because they act in extremely small concentrations and because their effects may not show up immediately but only in following generations. A good summary of what was known in the 1990s has been reviewed in a book by Colborn, T., Dumanoski, D., and Myers, J. P. (1996). *Our Stolen Future.* New York: Penguin Books.

A. The Endocrine System Involves Complex Hierarchies and Feedback Loops

Animals and humans use two communication systems to coordinate the actions that keep them alive.

- The *nervous system* for fast and selective communication, for example, between the pain receptors that sense a prick in your foot and your muscles that lift the foot away.
- The *endocrine system* relying on **hormones**, chemical signals produced in specific glands and distributed via blood. Hormones activate matching receptors and their signal transmitters in certain cells. For example, insulin causes cells in muscle, liver, and fat tissue to take up glucose. So indirectly, insulin regulates the glucose level in blood. We have already discussed the profound effects

of sex hormones on our development and behavior (see Chapters 16 and 17). For each hormone, there are some tissues in which the hormone's receptors and downstream signal transmitters are especially prevalent and active. These tissues are called the hormone's **target tissues**. Being a target tissue for one or more hormones is part of each cell's *differentiation* in the course of development.

The nervous and endocrine systems of vertebrates are integrated by the activities of the *hypothalamus* and the attached *pituitary gland*. The latter is composed of two different parts (Figure 25.1). The **posterior pituitary gland** is an extension of the hypothalamus. It stores and releases two **neurohormones**, meaning hormones produced and released into the bloodstream by cells that also act as neurons. The two posterior pituitary hormones are *oxytocin* and *vasopressin*, which we have discussed in Chapter 13. Other neurosecretory cells in the hypothalamus produce **releasing hormones**; they are carried by a system of **portal blood vessels** to the **anterior pituitary gland**, which is derived from epithelial tissue that covered the roof of the mouth in the early embryo.

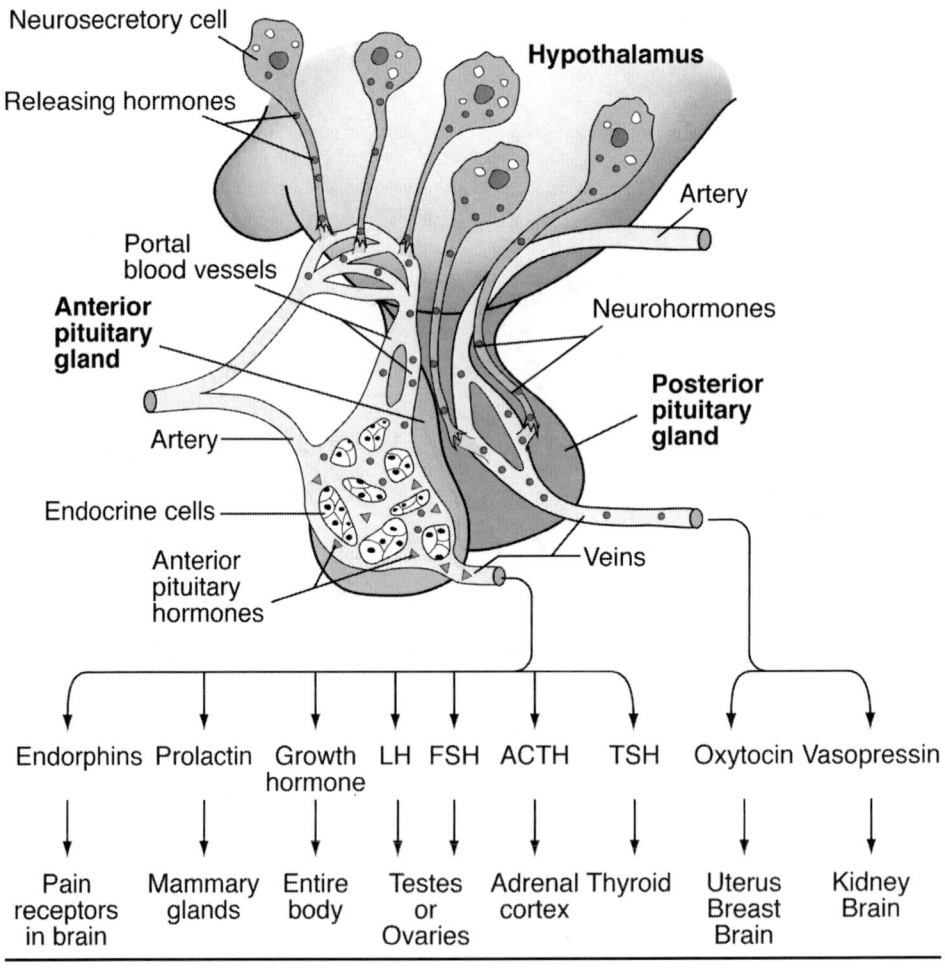

FIGURE 25.1:

Hypothalamus, pituitary, and subordinate glands in mammals.

Stimulated by the releasing hormones from the hypothalamus, cells of the grown anterior pituitary gland produce six different hormones, which in turn have other hormone glands as targets (see Figure 25.1). For example, thyroid-releasing hormone from the hypothalamus stimulates specific pituitary cells to secrete thyroid-stimulating hormone (TSH). TSH enters the general blood circulation and binds to matching receptors in its target organ, the thyroid gland. In addition, the anterior pituitary produces endorphins, which act on brain cells as "natural painkillers."

Of particular interest for the present chapter is the **hypothalamus-pituitary-gonadal axis** in vertebrates (Figure 25.2). It includes the *hypothalamus,* which produces—as one of its releasing hormones—**gonadotropin-releasing hormone (GnRH)**. GnRH stimulates the anterior pituitary gland to produce—in both sexes—**luteinizing hormone (LH)** and **follicle-stimulating hormone (FSH)**.

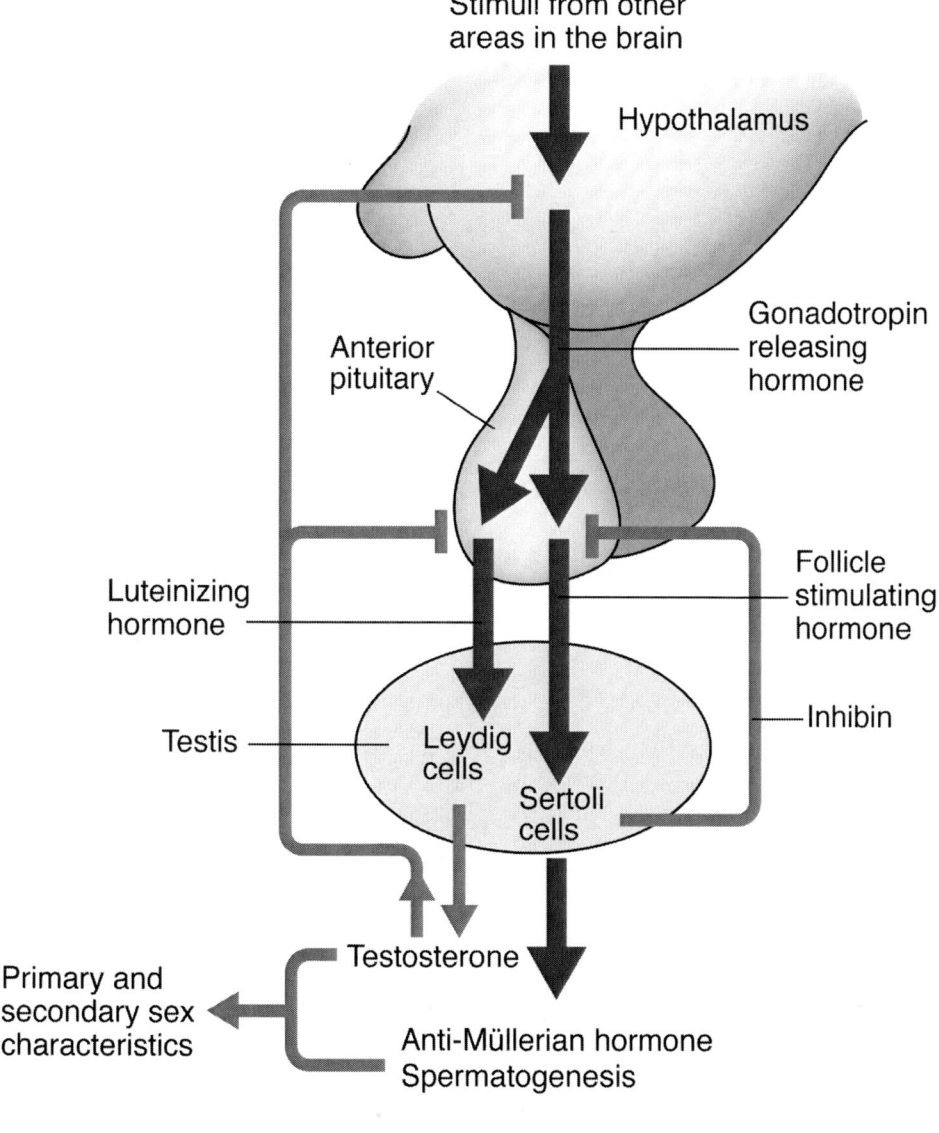

FIGURE 25.2:

Hypothalamus-pituitary-gonadal axis in mammalian males

In males, the primary targets of LH are the *Leydig cells* of the testes, which produce *testosterone (T)*. T and its metabolites are key hormones controlling male sexual development and behavior, as discussed in Chapters 16 and 17. T also affects sperm development, presumably via Sertoli cells. Also in males, the primary targets of FSH in the testes are the *Sertoli cells*, which make another male sex hormone, *anti-Müllerian hormone (AMH)*. AMH is known to be critical for the degeneration of the Müllerian ducts in males but is also thought to be involved in other steps of male sexual development, such as testicular descent.

In females, a surge of LH triggers ovulation and the subsequent conversion of the remaining ovarian follicle into the *corpus luteum*, the event for which LH has been so named. FSH has also been named after its function in the ovary, where it stimulates follicular growth, especially in the *granulosa cells*.

The normal function of the hypothalamus-pituitary-gonadal axis depends on negative feedback loops. In males, T dampens the production of GnRH and LH in the hypothalamus and anterior pituitary, respectively (see Figure 25.2). In parallel, the hormone **inhibin** produced by the Sertoli cells dampens the production of FSH by the pituitary gland.

B. Human-Made Chemicals Can Interfere With Hormonal Signals

The complex hierarchies and feedback loops of the endocrine system are subject to disturbance by human-made chemicals. Over the past 100 years, humans have synthesized and released into the environment tens of thousands of chemicals, and about another thousand are added each year. Each of these chemicals has been designed for a well-intentioned purpose, but there is growing evidence that some human-made chemicals have disruptive side effects on the endocrine system. Such chemicals are therefore called **endocrine-disrupting chemicals (EDCs)** or *endocrine disruptors* for short. The may act by

- mimicking a hormone (agonistic effect)
- blocking a hormone receptor (antagonistic effect)
- interfering with a hormone's synthesis, transport, or breakdown (metabolistic effect)

EDCs are taken up by humans and wildlife with food, drink, and air. Many EDCs are fat-soluble (rather than water-soluble). This means they accumulate in the food chain and reach unhealthy concentrations in top predators (Figure 25.3). From the long and growing list of EDCs, a few notorious examples follow:

- **DDT, for dichlorodiphenyltrichloroethane**, is an insecticide that is very effective in fighting malaria and other insect-borne diseases. The Swiss chemist Paul Hermann Müller was awarded the Nobel Prize in Physiology or Medicine in 1948 "for his discovery of the high efficiency of DDT as a contact poison against several arthropods." It has been used massively (80,000 tons per year) and indiscriminately in farms and households until it became apparent that DDT and its metabolites caused the thinning of egg shells in birds and were acting as estrogen mimics in humans. While DDT some of its metabolites, such as DDE and dicofol, are banned in the United States now, it is still used selectively in malaria-infested parts of Africa.

FIGURE 25.3:
Endocrine-disrupting chemicals include nonylphenol (bold formula at top), dichlorodiphenyltrichloroethane (DDT, second from top), polychlorinated biphenyls (PCBs, bottom left), and bisphenol A (BPA, near bottom right). Being fat-soluble, they accumulate in top predators. Eagle (c) 2010, Shutterstock, Inc.

- **Nonylphenol** is a breakdown product of a class of chemicals called alkylphenol polyethoxylates. It acts as a weak estrogen mimic, but hundreds of millions of tons of these chemicals are used each year in the United States alone as additives to pesticides, detergents, personal care products, and plastic materials.
- **PCBs, for polychlorinated biphenyls**, are a family of more than 200 chemicals designed for use as lubricants, hydraulic fluids, cutting oils, and liquid seals. Even though toxic effects of PCBs were discovered by companies producing them in the 1930s, they released millions of pounds of them into the Hudson river and into the soil elsewhere, prompting costly cleanup operations after the adverse health effects of PCBs on humans and wildlife had been publicized.
- **Bisphenol A (BPA)** is used as a monomer for polycarbonate plastic and epoxy resins. These are common materials for drinking bottles and inner linings of soda cans. Ongoing studies strongly suggest that BPA causes disorders in humans ranging from heart disease to diabetes to breast cancer.
- **Dioxin** is a member of a family of highly toxic by-products generated during the synthesis of various chlorine-containing chemicals used as pesticides and wood preservatives. Dioxin is also released by bleaching papers with chlorine, by incinerating plastics, and by burning fossil fuels.

Following are some of the observations EDCs collected over the past decades. Many of these observations have been illustrated by in a BBC documentary broadcast on September 4, 1994, on The Discovery Channel.

C. "Lesbian" Sea Gulls on Channel Islands in Southern California

Most mimics of steroid hormones, like these hormones themselves, are fat-soluble. They accumulate in fat tissues, from which they are mobilized again during fasting. Because they are passed up the food chain, the highest concentrations are found in birds of prey and in carnivorous mammals including humans. Female mammals also pass them on in their milk to offspring.

Around 1970, in **sea gull** colonies on San Nicolas Island and Santa Barbara Island, females were found to share nests with other females. Many chickens from these colonies did not hatch, and those who did were often grossly abnormal. Males in particular had feminized reproductive organs and had little interest in mating. The "lesbian sea gull" pairs apparently formed because functioning males were so rare that many females were only able to snatch an extra-pair copulation with a roaming male but could not form heterosexual pairs to raise a brood. The same set of symptoms could be reproduced by injecting eggs from healthy gulls with estrogen.

Similar female pairs were found in colonies of herring gulls at the Great Lakes and among other gulls in Puget Sound. In colonies of terns, relatives of sea gulls, living near toxic waste sites off the coast of Massachusetts contaminated with PCBs, many chicks did not hatch, and among those that did, most of the males had feminized sex organs. In various species of birds high up in the food chain, biologists observed nesting sites in areas of high chemical contamination where adult birds were not interested in nest building or mating, or did not defend their broods against predators.

In an attempt to find the cause of these problems, Michael Fry, a wildlife toxicologist at UC Davis, injected sea gull eggs from healthy colonies with DDT or DDE. He found that hatching males had feminized reproductive organs and females had duplicated and disrupted reproductive organs. Other scientists showed that male chickens or quail injected with estrogen during early development showed the same changes in behavior and anatomy.

The simplest explanation of all these observations is that DDT, DDE, PCBs, and possibly other man-made chemicals, mimic the effects of estrogens on the brains and reproductive organs of birds during early development.

D. Alligators in Lake Apopka, Florida

A well-documented case of animals harmed by EDCs are **alligators** of Lake Apopka in Florida. Located on the lake is a now defunct plant of Tower Chemical Company, which made the pesticide dicofol, which contained up to 15 percent DDT or DDE. In 1980, there was a major spill of dicofol into the lake. Subsequently, biologists observed serious problems that were specific to Lake Apopka and were not observed in other Florida lakes.

Between 80 and 95 percent of the alligator eggs in Lake Apopka failed to hatch. Of those that did, roughly half died within the first two weeks. Penises of developing males in Lake Apopka were small to nonexistent. Adult alligators had highly abnormal ratios of testosterone and estrogen, with males having ratios like normal females, and females having estrogen levels much higher than normal. Similar observations were made on turtles living in Lake Apopka. The maladies of alligators and turtles were more severe near the Tower plant than in other parts of Lake Apopka.

A team led by Louis J. Guilette, Jr., reproductive endocrinologist at the University of Florida at Gainesville, tested the hypothesis that the abnormal hormone levels in the adult alligators were caused by exposure of the developing eggs to DDE or dicofol. They collected alligator eggs from Lake Woodruff and painted them with dicofol and DDE at concentrations that generated the tissue contaminations typical of hatchlings from Lake Apopka. The hatchlings from these experimental eggs also showed the same aberrant concentrations of testosterone and estrogen observed in Lake Apopka hatchlings, as predicted by their hypothesis.

The results of Guilette and coworkers show that DDE and/or dicofol affects the development of reptiles in the egg shell so as to change their balance of sex hormones, raising the estrogen levels and depressing the testosterone levels that these animals show later in life, and disrupting their ability to reproduce.

E. Estrogen Mimics from Plastics, Detergents, and Toiletries

Endocrine disruptors may leach from **plastic** containers in which we keep our water, soft drinks, and canned foods. The same uncanny scenario may unfold for soda cans and other metal cans with inner linings of polycarbonate and other plastics. This concern is illustrated by the following episode.

Drs. Ana Soto and Carlos Sonnenschein at Tufts Medical School in Boston worked on the control of cell proliferation using cultured cells that divided only in the presence of estrogen. During their investigation, previously estrogen-dependent cells began to divide wildly in the absence of estrogen.

After several months of trying to track down the apparent estrogen contamination in their laboratory, the researcher found the cause of their misery: A new formulation of the plastic of their test tubes, used to store serum for their cell cultures, turned out to be the culprit. The manufacturer of the tubes, Corning, admitted that they had changed the formula for these tubes without changing the catalog number or otherwise alerting their customers. However, they refused to share the nature of the change they had made, claiming protection of trade secrets under the law.

Soto and Sonnenschein teamed up with chemists at the Massachusetts Institute of Technology, who identified the new component in the test tubes as p-nonylphenol. Many manufacturers of plastic ware add nonylphenols to common plastics, including polystyrene and polyvinylchloride, to make them less breakable. Nonylphenols also originate as breakdown products of chemicals called alkylphenol polyethoxylates, which are produced in million ton quantity each year for use in pesticides, industrial detergents, and personal care products.

Coincidentally, a research team at Stanford Medical School discovered another estrogen mimic, bisphenol-A, leaching from another type of plastic, polycarbonate.

F. Reproductive Deficiencies in Human Males

Over the past 50 years, reproductive problems of human males have increased dramatically. These problems can be attributed, at least in part, to exposure of male fetuses to EDCs.

The incidence of **testicular cancer** in Denmark has more than tripled over the past 50 years and continues to grow. Similar increases have been recorded for other

Scandinavian countries, Scotland, and the United States. The U.S. National Cancer Institute reports a 126 percent increase in **prostate cancer** between 1973 and 1991. After correction for an increasingly older population and improved detection methods, these numbers still reflect an alarming increase. In the United States, prostate cancer is now the most common cancer in older men.

European data covering 14,947 men show that **sperm counts** have fallen from 113 to 66 million per ml semen between 1938 and 1990. Similar data for the United States show a decrease from 134 to 56 million per ml. During the same time, the **volume of semen** per ejaculation has fallen an average of 19 percent, so that the absolute number of sperm per ejaculate is now between a half and a third of what it was 50 years ago. Moreover, the percentage of abnormal sperm has increased dramatically.

The incidence of **cryptorchidism** (undescended testicles) in newborn boys has increased from 1.6 percent in the 1950s to 2.9 percent in the late 1970s.

The incidence of **hypospadia** (imperfect closure of the urogenital sinus into the penile urethra) in boys has more than doubled between 1964 and 1983.

All of these problems are overrepresented among men born by mothers who during their pregnancy received DES, a potent estrogen mimic prescribed liberally in the 1950s and 1960s for women who had a history of miscarriages. It is therefore a reasonable hypothesis that the reproductive deficiencies of men born by non-DES mothers may be caused by other estrogen mimics during pregnancy.

A possible causative chain between environmental estrogens and male reproductive deficiencies has been suggested by Richard M. Sharpe in the BBC documentary quoted earlier: The secretion of follicle-stimulating hormone (FSH) in the pituitary gland of young males is very sensitive to inhibition by exogenous estrogens. FSH is required for the formation of Sertoli cells in developing testes. The Sertoli cells in turn synthesize hormones and growth factors promoting male sexual development and sperm production.

G. Reproductive Problems in Human Females

The reproductive problems caused by endocrine disrupting chemicals in human females are no less severe than those of males.

In **ectopic pregnancies**, the fertilized egg implants itself in an oviduct instead of in the uterus. This is a dangerous condition, which may cause the loss of the woman's oviduct or even her life. The rate of ectopic pregnancies in Wisconsin increased 400 percent between 1970 and 1987. Daughters born to mothers treated with diethylstilbestrol (DES) suffer three to five times more tubal pregnancies than unexposed women, suggesting that estrogen mimics may be a major contributing factor in the general population as well.

Endometriosis, a painful inflammation of the inner lining of the uterus, is a major cause of infertility. The National Institute for Child Health and Human Development estimates that endometriosis afflicts 10 to 20 percent of women of childbearing age in the United States, including an increasing fraction of very young women. Prior to 1921, there were only 20 reports of this disease worldwide. German researchers report that women with endometriosis have higher levels of PCB in their blood than women without the disease. Rhesus monkeys exposed to dioxin developed endometriosis ten years later.

Women who suffered **miscarriages** were found to have higher PCB levels in their blood than women who have normal pregnancies. Studies on rats and mice indicate that PCBs accelerate the breakdown of progesterone, a hormone that is necessary to sustain pregnancy.

The rising frequency of **breast cancer** is by far the most alarming trend for women. In the United States today, one in eight women will be diagnosed with breast cancer in her life time, whereas 50 years ago the same risk was only one in twenty. Among U.S. women between 40 and 50 years of age, breast cancer is the leading cause of death. A particular increase is noted for estrogen-responsive breast cancers. A likely cause is the increased overall exposure of women to estrogens caused by earlier puberty, later menopause, a reduced number of pregnancies, the use of estrogen-based oral contraceptives, and various environmental estrogen mimics. It is also possible that modern diets have changed the bacterial flora of the intestine so that estrogen, which would otherwise be excreted via bile and feces, is recycled into the bloodstream.

H. Disruptions of Human Mental Development

In addition to causing reproductive problems in men and women, some endocrine disruptors also act as **neurotoxins**, causing serious disturbances in the mental development of infants and children.

The physical and mental development of human babies seems to be adversely affected by PCBs in their mothers' blood and breast milk. Water and fish in the Great Lakes are notorious for having PCB content much greater than what is considered safe for human health. The problem is not easily remedied because PCBs are very long-lived.

In related studies, babies of mothers who ate fish from these lakes were born sooner, weighed less, and had smaller heads than babies of mothers who did not eat these fish. The babies' deficiencies increased with the amount of fish their mothers had eaten. The same babies showed weak reflexes and more jerky movements than their counterparts from mothers who had not eaten Great Lakes fish. At seven months of age, the affected children showed a reduced ability to distinguish faces. The children's test scores were negatively correlated with the PCB levels in their mothers' breast milk. When retested at age four, seventeen children, all of them born to mothers with the highest PCB levels, were hyperactive and easily frustrated. An independent study on 866 infants found the same negative correlation between their motor skills and PCB levels in maternal breast milk.

In another study, humans and rats were tested in parallel for the effects of eating fish contaminated with PCBs. Children born by women eating different amounts of Lake Ontario fish were compared as described above. In parallel, experimental rats were fed different amounts of contaminated Lake Ontario salmon, while control rats received equivalent amounts of uncontaminated salmon from the Pacific coast. The parallels in the effects of these diets on humans and rats were astounding. The offspring from rats fed on Lake Ontario salmon were described as "hyper-reactive to even mildly negative situations." Babies whose mothers had eaten Lake Ontario fish adapted less readily to disturbances in sleep or feeding regimens.

A study on the children of women born to mothers in Taiwan who consumed cooking oil accidentally contaminated with PCB and furans revealed a similar picture. The children showed persistent problems with motor skills, scored lower on IQ tests, were hyperactive, and had attention deficit disorder. Boys born by these mothers also had shorter penises than boys born by uncontaminated mothers.

I. Search for Molecular Mechanisms of Endocrine Disruption

The molecular mechanisms of endocrine disruption are diverse and complex, and much more research is needed to gain a better understanding of their actions.

Steroid hormones bind to receptor proteins that act as transcriptional regulators (see Figure 17.1). Estrogen binds to two different receptor proteins known as ER alpha and ER beta, which have different tissue distributions and target specificities. Loaded ERs bind to DNA sequences known as estrogen response elements in 50 to 100 vertebrate genes, the "target genes" of estrogen. When bound to response elements, the loaded ERs interact with other gene-regulatory proteins to activate or inhibit their target genes.

Some biological effects of estrogen are caused not by the principal estrogen, 17ß estradiol, but by its physiological breakdown products. One of these breakdown products, 17-epiestriol, helps to maintain the strength of bone. Lack of 17-epiestriol causes osteoporosis, a demineralization that makes a person prone to bone fractures. To prevent osteoporosis, many postmenopausal women are taking estrogen pills. A problem with this replacement therapy is that it may increase the risk of breast cancer.

So 17-epiestriol as well as a drug known as raloxifene seem to protect against osteoporosis without increasing the risk of breast cancer. Both molecules interact with ER in a way that does not require the ERs DNA-binding domain and depends on the presence of "adapter" proteins. The 17-epiestriol/raloxifene-receptor-adapter complex also binds to a gene regulatory element other than the typical estrogen response element. While these results are of great medical interest, they also illustrate the uncanny ability of ERs loaded with different estrogen metabolites or mimics to **act cooperatively** with other proteins.

Similarly, research into the mechanism of dioxin action revealed an unexpected way of acting as an estrogen mimic. Dioxin binds to an aryl hydrocarbon receptor (AhR), which then recruits another protein called aryl hydrocarbon nuclear translocator (Arnt). The AhR/Arnt complex in turn "hijacks" *unliganded* ER, which then binds to estrogen response elements, causing the transcriptional activation of estrogen target genes without estrogen.

Estrogen mimics may also tap into the *negative feedback loops* of sex steroid hormones that down-regulate the production of gonadotropin-releasing hormone in the hypothalamus and of LH/FSH in the pituitary gland (see Figure 25.2). In particular, down-regulation of Sertoli cells could explain the multitude of disorders observed in males exposed to estrogen mimics. However, these loops are known only from adults and may be different in the fetus.

Some pesticides, including DDT and its metabolites, act by antagonizing the androgen receptor. Other endocrine disruptors may interfere with processes involved in hormone synthesis or breakdown, for instance, by overloading the liver enzymes that break down steroids.

Moreover, some chemicals act as endocrine disruptors and independently as **neurotoxins**. For instance, PCBs affect psychomotor development, cognitive functions, and other behavioral traits as described in section H.

Hormones act at exceedingly small concentrations, often measured in parts per trillion (ppt; one ppt corresponds to one drop of gin in 700 tank cars of tonic!).

Even at these low concentrations, hormone levels are still regulated by binding to blood serum proteins that render the hormones temporarily unavailable.

Endocrine disruptors are typically active at concentrations 1000-fold or more above the concentrations of the hormones they mimic. The mimics thus compensate their suboptimal fit to the hormone receptors with higher concentrations. Endocrine disruptors are not known to bind to serum proteins, and are therefore much less regulated than hormones in this regard.

Transcriptional regulators including ERs act in **synergistic ways**, potentially making combinations of two or more different regulators much more effective than each one by itself. The reason for this synergism is that the gene-transcribing enzyme (RNA polymerase II, or Pol II) is positioned at the transcription start point by a three-dimensional complex of several regulatory proteins (Figure 25.4). Different proteins, for example, ERs loaded with different estrogen mimics, may contribute more effectively to the optimal landscape for Pol II than two ERs loaded with the same estrogen mimic (Figure S25.a).

It is currently unclear whether estrogen mimics do indeed act synergistically *in vivo*. If this were the case, then adequate testing of human-made chemicals for endocrine disruption would be technically and legally very difficult.

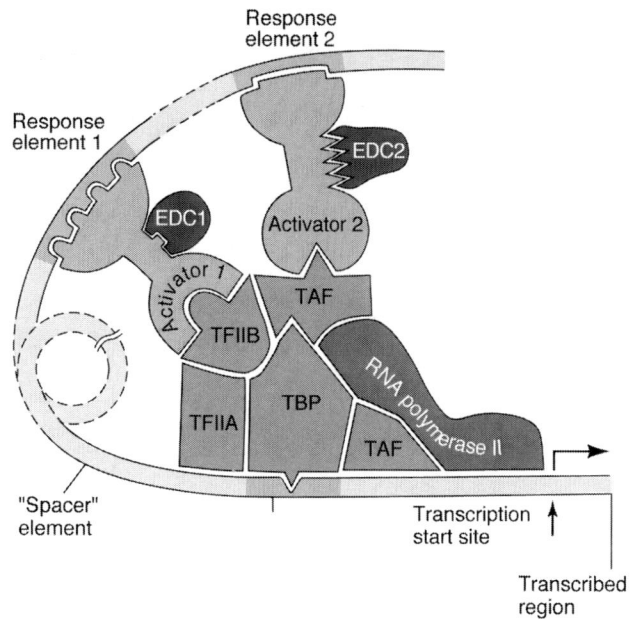

FIGURE 25.4:

Transcriptional activators act cooperatively by stabilizing a transcription initiation complex of multiple proteins at the promoter region of a target gene. The transcription initiation complex directs the critical enzyme, RNA polymerase, to the transcription start point. Transcriptional activators bind to specific DNA sequences, called DNA response elements. Such elements may be positioned at some distance from one another and from the promoter region.
Modified from Kalthoff, K. (2nd ed. 2001) *Analysis of Biological Development*; New York: McGraw-Hill, p. 409, Fig. 16.4

J. Summary and Conclusions

1. In human males, there has been an unprecedented decrease in the quantity and quality of sperm, accompanied by significant increases testicular and prostate cancer. Other male reproductive abnormalities including hypospadia and cryptorchidism have also increased. In women, the frequency of breast cancer has risen to an alarming level, while endometriosis and ectopic pregnancies have also become more common.

2. Many wildlife populations show similar symptoms including abnormal reproductive organs, infertility, abnormal behavior, and death.

3. These reproductive problems in humans and animals are caused, at least in part, by disruption of the endocrine system, especially during fetal development.

4. Human-made chemicals released into the environment can disrupt the hormonal controls systems of animals including humans. Many of these chemicals mimic the functions of estrogen, while others seem to antagonize androgens or disturb the regulation of sex hormone production.

5. Many endocrine disruptors accumulate in fat tissues and are passed up the food chain. Through blood and milk, mammals including humans hand down these chemicals to their young.

6. The conclusions drawn from observations on wild animals living in contaminated environments are confirmed by laboratory experiments, by accidental contaminations, and by unexpected side effects of medications given to humans.
7. Some endocrine disruptors may act synergistically, meaning that in combination they act at much lower concentrations than each one does by itself.
8. Much more research is needed to better understand the molecular mechanisms of endocrine disruption by human-made chemicals.
9. Strong efforts are warranted to reduce the environmental load of human-made chemicals.

EXERCISE PAGE FOR CHAPTER 25

Student Last Name _____

Student First Name _____

Discussion Section _____

On the course web site, use the link on the **syllabus** to find a pdf file of your **assigned reading** for this chapter. The same reading is referenced below. Use the bulleted list of questions to test your knowledge of this reading.

Hayes, T. B.; Collins, A.; Lee, M.; Mendoza, M.; Noriega, N.; Stuart, A. A.; and Vonk, A. (2000). Hermaphroditic, demasculinized frogs after exposure to the herbicide atrazine at low ecologically relevant doses. *Proc. Natl. Acad. Sci. U.S.A.* **99:** 5476–5480.

- How do the authors justify their focus on atrazine and *Xenopus laevis*?
- Which specific hypothesis do they set out to test?
- At which stages of development were the frogs exposed to atrazine, and at which concentrations? How do these atrazine concentrations compare to the allowed concentration in drinking water, and to concentrations found in riverbeds in Midwestern agricultural areas?
- What are the results shown in Figures 1 through 4? At which atrazine concentrations were the effects observed?
- Which indicators of frog health were NOT affected by atrazine at the concentrations tested in this study and in an earlier study by Allran and Karasov (2001), as quoted by the authors?
- How do the authors interpret their results? Which observations from other studies do they quote to support their interpretation?
- What are the general lessons to learn from this study about testing the environmental impact of human-made chemicals?

Feedback

On a scale from 1 to 10, the being the best, please rank the above reading for

1. Interest and Relevance to the topic (10 being most interesting/relevant) _____
2. Readability (10 being most clearly written, easy to understand) _____

In the remaining space, enter any comments that you may have on this reading.

CHAPTER TWENTY-SIX

ENERGY PRODUCTION AND CONSERVATION

In developed countries, the twentieth century has brought better health care, less hard labor, and more cultural diversity. These improvements in the quality of life were tied to an unprecedented rise in the consumption of energy, mostly in the form of fossil fuels. However, the use of these fuels generates greenhouse gases and other pollutants that threaten the ecosystems on which we rely (see Chapter 24). Satisfying our growing energy needs while preserving the environment is one of the greatest challenges we currently face, as discussed in a book by Vaclav Smil. (2003). *Energy at the Crossroads.* Cambridge, MA: The MIT Press.

A. History of Energy Supply

The oldest method for generating energy, still widely used in developing countries, is the burning of biomass, such as wood or animal dung. The energy obtained was, and is, used mostly for heating and cooking.

Coal was first obtained in the form of charcoal, by removing water and volatile constituents from wood and other organic matter. This was done by heating under the exclusion of oxygen. The mass production of charcoal has often led to deforestation. As energy demands increased, charcoal was replaced with "fossil coal" or "stone coal," dug first from beds exposed at the surface and later from coalmines. Thus, coal became the first **fossil fuel**, that is, fuels formed by anaerobic decomposition of dead organisms. The age of fossil fuels is typically millions of years, sometimes exceeding 650 million years.

In 1769, James Watt patented his steam engine, which was heated with coal. Mobile steam engines ("locomotives") powered by coal were used to pull trains, for human passengers and all kinds of loads, including more coal from coalfields. The modern era had begun. Today in developed countries, coal is used mainly to generate electricity. Currently, the share of commercial energy supplied by coal is 23 percent worldwide.

Petroleum (Greek for "rock oil") is another naturally occurring fossil fuel. It is a mixture of hydrocarbons of various molecular weights found along with natural gas in geologic formations beneath the Earth's surface. We will refer to it as **oil** for short. Oil is fractionated into a wide range of flammable liquids used for many applications, including heating oil and gasoline. Gasoline in particular became the primary fuel for transportation in cars and airplanes. Due to their superior energy density and ease of transportation, oil products have surpassed coal as the most important fossil fuel in 1966. Oil products have also become the raw material for a myriad of synthetic products.

Natural gas is the lightest fossil fuel and consists mostly of methane. It has become the most widely used energy source next to oil. Although not as easy to transport as oil, natural gas has become the preferred fuel for space heating, and its use in electric power plants is also increasing.

Waterwheels as sources of mechanical energy were first used by ancient Greeks for grinding grains. Later on they served in many applications including small-scale industries. Today, **hydroelectric plants** convert the gravitational energy of water into electricity. Having been in use since the 1880s in the United States and Canada, they are stable and renewable source of commercial energy. They currently generate nearly 20 percent of all electricity worldwide.

Solar power focused by lenses and mirrors was used by classical Greeks and Romans to light torches, burn ants, and set fire to enemy ships. The same principle is still used in **solar concentrator** power plants, which use large systems of adjustable mirrors to focus sunlight on a boiler. A different use of solar power is based on the **photovoltaic effect**, first described by the French physicist Antoine-César Becquerel. When he experimented with a solid electrode in an electrolyte solution, he saw a voltage develop when light fell upon the electrode. Since the 1930s, selenium cells were used as light measuring devices in the emerging field of photography. These cells converted less than 1 percent of solar energy to electricity, whereas modern solar panels boast efficiencies near 20 percent. Photovoltaic electricity accounted for less than 0.1 percent of all electricity sold through large grids in 1999 but is growing fast, especially in remote locations.

The first **windmills** were developed in Persia for pumping water and grinding grain. Like waterwheels, wind energy has been harnessed for centuries to provide mechanical power for various applications. Today, large modern wind turbines are generating electricity that is sold through the grid. They currently account for about 1 percent of all installed electricity-generating capacity worldwide, but for more in Europe. For instance, Denmark meets roughly 20 percent of its energy needs from wind.

Nuclear power plants have been in use since the 1950s. They are based on the *induced fission* of uranium-235. If a free neutron hits a U-235 nucleus, the nucleus will split in two smaller nuclei, while releasing several neutrons and heat. By setting up a controlled chain reaction, induced fission can be used to generate large amounts of heat. This is used to generate water vapor, which then drives turbines. Nuclear power plants can generate large amounts of energy without greenhouse gases but are problematic with regard to safety and security. They also generate a lot of heat, adding to global warming. Worldwide, about 16 percent of all energy is currently generated by nuclear power.

Our major problems with energy today are the following:

- The world population is still growing (see Chapter 22).
- The energy consumption per capita is increasing steeply with economic development.
- The prevailing method for generating energy today is to burn fossil fuels, which produces CO_2 and other greenhouse gases (see Chapter 24).

The United States in particular produce about 85 percent of their energy by burning fossil fuels (Figure 26.1). Because of their chemical nature, the fossil fuels differ in the amount of carbon dioxide released per unit of energy generated. The degrees of carbonization of the atmosphere compare as 25 : 19 : 14 for coal : refined oil : natural gas. Thus, natural gas is the "cleanest" fossil fuel, but leaks from pipelines are a concern because CH_4 is a potent greenhouse gas itself. Coal, on the other side, is the "dirtiest" fossil fuel. Unfortunately in this regard, coal is also the cheapest fossil fuel and is present in large supplies. Thus, our growing use of energy, the chemical nature of fossil fuels, and the laws of supply and demand are setting us on a collision course with the need to keep global warming from reaching catastrophic levels (see Figure 24.5).

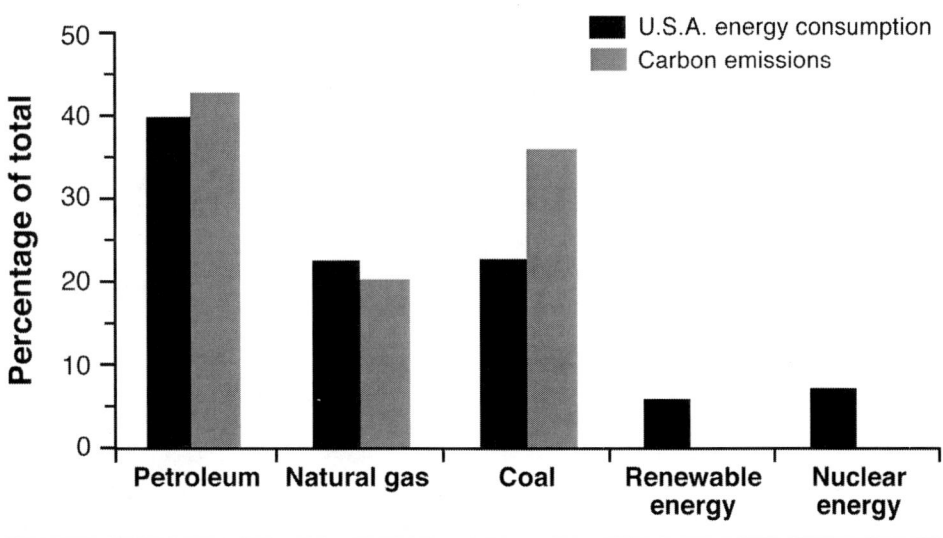

FIGURE 26.1:

Energy consumption and carbon emission in the U.S.A. for the year 2003. Total energy consumption was about 100 quadrillion Btu, and total carbon emission was about 6 billion tons. Data source: Shinner R. and Citro F. (2006) A road map to U.S. decarbonization. *Science* **313**: 1243-1244.

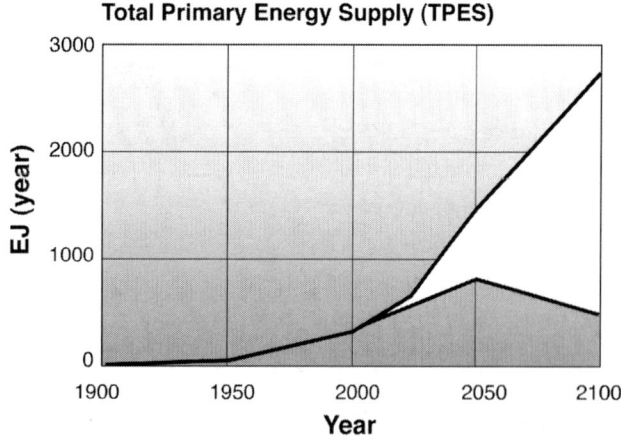

FIGURE 26.2:
Global total energy supply. Data source: Smil V. (2003) *Energy at the Crossroads,* Cambridge, MA: The MIT Press. 1 EJ =1 Exajoule =10^{18} Joules.

B. Patterns of Energy Use

Between years 1900 and 2000, worldwide annual consumption of commercial energy (fossil fuels plus nuclear and hydroelectric energy) increased by a factor of 16.6, from 22EJ/yr to 365 EJ/yr (Figure 26.2; 1 EJ = 1 Exajoule = 10^{18} Joules). During this time interval, the human population nearly quadrupled from 1.6 to 6.1 billion. At the same time, the annual *per capita* consumption of energy more than quadrupled, from 14 GJ to 60 GJ (1 GJ = 1 Gigajoule = 10^9 Joules) worldwide. The latter increase happened despite major improvements in the efficiency of both power plants and our appliances. Thus, unless further improvements in efficiency can be made, the per capita consumption stands to increase even more steeply in the future.

There are large **disparities in energy consumption** between rich and poor countries. In the year 2000, the United States accounted for about 5 percent of the world population but consumed 27 percent of the world's energy. At the same time, the G7 nations (United States, Japan, Germany, France, United Kingdom, Italy, and Canada) accounted for 10 percent of the world population and consumed 45 percent of all energy. In contrast, the poorest quarter of mankind (including much of sub-Saharan Africa, Nepal, Bangladesh, rural India, and Indochina) consumed merely 2.5 percent of the world's energy.

Large developing countries, such as China and India, are emulating the rich countries' economy and lifestyle, and are rapidly catching up in energy consumption. If the world population would grow according to the U.N. medium variant (see Chapter 22), and if all countries would raise their energy consumption to the level of the G7, then the global energy consumption would increase almost by a factor of eight (see Figure 26.2). Conversely, if the human population growth could be reversed, and if economic development worldwide were slow and energy-efficient, then the energy consumption (and hence, energy supply) would decrease after the year 2050. IPCC estimates range from 550 to 2700 EJ/yr in 2100. It is mostly due to this **uncertainty about future energy levels** that the global warming predicted in the IPCC report of 2007 covers such a wide interval: between 1.1 and 6.4°C (2.0 to 11.5°F) until the year 2100 (see Chapter 24).

C. Energy and Quality of Life

Energy is generally used to improve the quality of life. For the world's 57 most populous nations, Smil (2003) plotted several different indicators for **quality of life** over energy consumption (expressed in kilograms of oil equivalent per capita per year). The following measures for quality of life were used.

- average female life expectancy (Figure 26.3)
- lack of infant mortality (Figure S26.a)
- average per capita food availability (Figure S26.b)

- human development, a composite index based on life expectancy, adult literacy, educational enrollment, and per capita gross domestic product (Figure S26.c)

Invariably, these plots indicate significant improvements in the quality of life up to 4,000 kgoe, but then the curves enveloping the plots level off, showing little if any improvement between 4,000 kgoe (Japan, European countries) and 8,000 kgoe (United States, Canada). These data indicate energy consumption promotes quality of life up to a point, beyond which energy expenditures increase while quality of life remains nearly the same. In other words, the richest countries could cut their energy consumption in half without loosing quality of life. This is a very important conclusion, given that developing countries are striving to emulate the lifestyle of developed countries. It means that, if the developed countries could manage to cut their energy consumption in half, then the future energy needs of the entire world would be halved without significant losses in the quality of life to which all countries aspire.

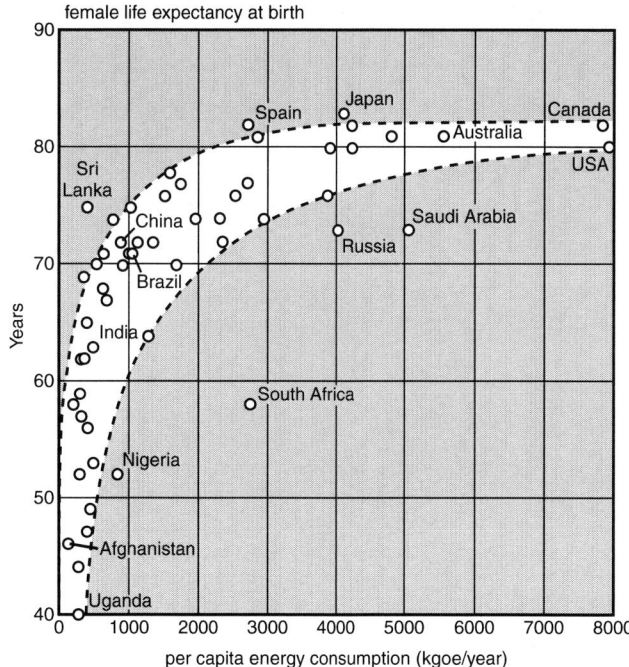

FIGURE 26.3:

Life expectancy and energy consumption per capita in different countries.
From Smil V. (2003) *Energy at the Crossroads*. Cambridge, MA: The MIT Press. Page 100, Fig. 2.11
1 kgoe = 1 kilogram of oil equivalent

D. The Future of Fossil Fuels

Nearly all previous attempts at long-range forecasting future energy sources and consumption levels have failed. The overall picture is that the relatively clean fossil fuels (oil and natural gas) are coming to an end, and that the limited supplies left must be used wisely to develop new methods of generating energy in environmentally acceptable ways.

Estimates of **future oil availability** vary widely. Because oil is found in different qualities and under varying geological conditions, its recovery costs depend on location and technological progress. Most of the known oil fields are located in regions that are politically unstable (Figure 26.4). Exploitation of other fields is expensive, or risky, or both. Thus, oil will probably not run out but will gradually be replaced by new energy sources that are cheaper, politically more stable, and/or environmentally more acceptable. A likely scenario seems to be that conventional oil production will peak around 2030 and then decline to about 10 percent of its current level in 2100. At that time, most energy would have to come from other sources.

Natural gas will probably be the fastest-growing component of the world's fossil fuel supply in decades ahead. However, estimates of **natural gas availability** seem as difficult as those for oil. According to Peter Odell, Professor Emeritus of International Energy Studies, Erasmus University of Rotterdam, conventional gas production will peak around the year 2050, with a gradual decline toward 2140 (Figure 26.5). "Nonconventional" gas production from coal and methane hydrates (methane molecules surrounded by a cage of water) may last much longer but awaits technical advances. One of the challenges in mining these hydrates is to prevent the escape of methane, a potent greenhouse gas.

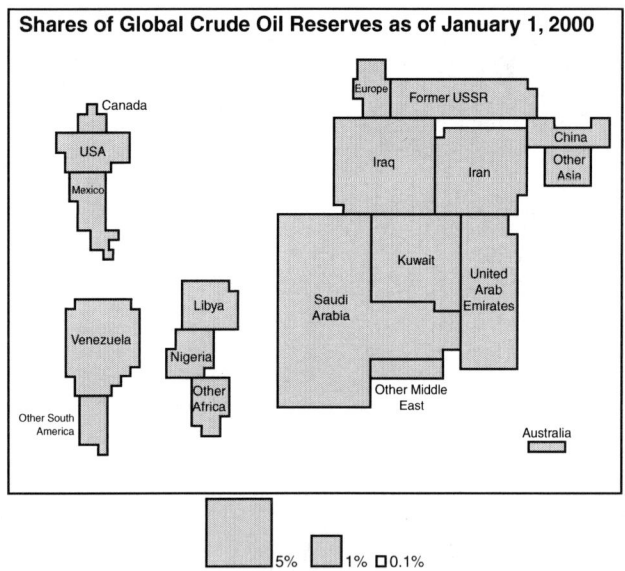

FIGURE 26.4:
Shares of crude oil reserves. Each country is drawn as an area representing its share of oil supplies.
From Smil V. (2003) *Energy at the Crossroads.* Cambridge, MA: The MIT Press. Page 209, Fig-4.14

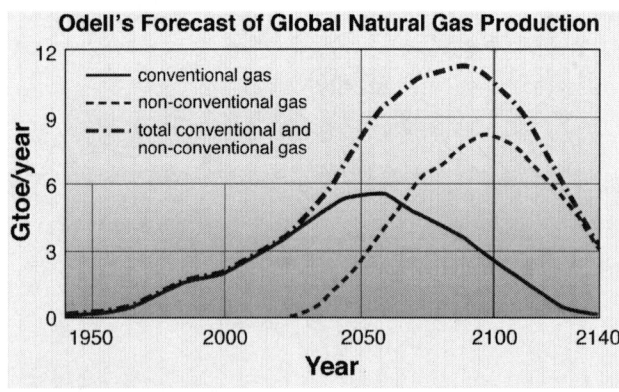

FIGURE 26.5:
Odell's forecast of natural gas production. 1 Gtoe = 1 Gigaton of oil equivalent. From Smil V. (2003) *Energy at the Crossroads.* Cambridge, MA: Fig-4.19

Natural gas is used increasingly for generating electricity, with conversion efficiencies surpassing 50 percent as compared to just over 30 percent for coal and oil. Natural gas can also be used to generate methanol (CH_3OH) and hydrogen, which can be used as fuels for automobiles. Price-wise, neither fuel can currently compete with gasoline, but this may change with further technical innovation and rising oil prices. Of course, running cars on methanol still produces CO_2.

Coal is in abundant supply. The known **coal reserves** are immense; no shortages are expected for the next 200 years. However, the big problem with burning coal for generating electricity is the large amount of CO_2 released per unit of energy generated. In addition, burning coal causes emissions of particulate matter, sulfur dioxide, and nitrogen oxides. Traditional coal-fired plants are also inefficient. All of these problems have affordable technical solutions, but there is much inertia in implementing them.

The release of CO_2 by stationary power plants fired with coal and other fossil fuels can be reduced dramatically by various ways of **capture and storage**. Technologies in advanced stages of testing include:

- an **integrated gasification combined cycle (IGCC)**, which makes CO_2 easier to capture than the traditional coal steam cycle
- **oxyfuel combustion**, a process that separates oxygen from other air components by established cryogenic methods and then burns coal or gas in oxygen, which makes the resulting CO_2 easier to capture
- **underground injection** into geological formations containing saltwater capped with impermeable rock
- **above-ground neutralization** of carbonic acid with milled rock rich in magnesium silicates to form water-insoluble carbonates

Each of these technologies would make a major contribution towards reducing global warming but would add some cost to the electric energy produced. The safety of the storage needs to be monitored and warranted for hundreds of years, well beyond the time horizon of typical commercial enterprises. Introducing CO_2 capture and storage technologies is therefore not in the economic interest of private power companies or in the short-term interest of political parties governing communally regulated power companies.

There is currently a disconnect between the consensus that carbon emission should be reduced sharply (to 30 percent of its 1990 level by the year 2020!) and the slow pace at which CO_2 capture and storage technologies are adopted by power plants. It takes true political leaders—and educated followers—to bring about technological transitions that are required for the common good but do not happen as a result of market forces.

E. Biofuels Are of Limited Utility

Because of the environmental problems associated with burning fossil fuels, there have been many plans to replace them with **biofuels**, that is, fuels derived from living, or recently killed, organisms. The simplest use of biofuels is **biomass burning**, such as the burning of wood, other plant materials, and dung. Biomass burning generates about 10 percent of all energy produced worldwide, and about 80 percent of all energy in the poorest countries. However, biomass burning is very damaging to the environment because it creates greenhouse gases and smoke. Also, collecting biomass often leads to soil erosion with the result of environmental degradation and further losses in bioproductivity.

More refined biofuels are flammable fluids such as ethanol and biodiesel. **Ethanol** (CH_3-CH_2-OH) is the most common biofuel worldwide. Alcohol fuels are produced by fermentation of wheat, corn, sugar cane, and any sugar or starch that alcoholic beverages can be made from. The methods involved include enzymatic digestion to release sugars, fermentation, and distillation. The latter process requires significant input of heat, which leaves the problem of generating energy without degrading the environment.

Making biofuels from edible crops competes directly with providing food for a growing world population in a market that is already pricing poor people out of food (see Chapter 23). Indeed, running all vehicles in the United States on ethanol made from corn would use up the country's entire cropland. Making biofuel from algae or nonedible plant materials (which could be fermented by genetically modified bacteria) could avoid the competing uses of land. Related research and development are underway at UT Austin and elsewhere.

Biodiesel consists of alkyl esters of fatty acids and can be made by reacting waste oils and fats with alcohol. While this procedure avoids the direct competition between fuel production and growing crops, the amount of biodiesel that can be produced this way is very limited. And, of course, using ethanol or biodiesel as fuel still generates CO_2.

An important concept in judging the utility of any method for generating energy is the **power density**, that is, the amount of power produced per time and per area. For example, the power density of photovoltaics is about 20 to 50 Watts per square meter, which is sufficient to provide the energy for a single-story home by covering its roof with solar panels (Figure 26.6). In contrast, to power the same house by biofuel, which has a power density between 0.2 and 1 W/m^2, one would need to plant an area of 100 times the footprint of the house.

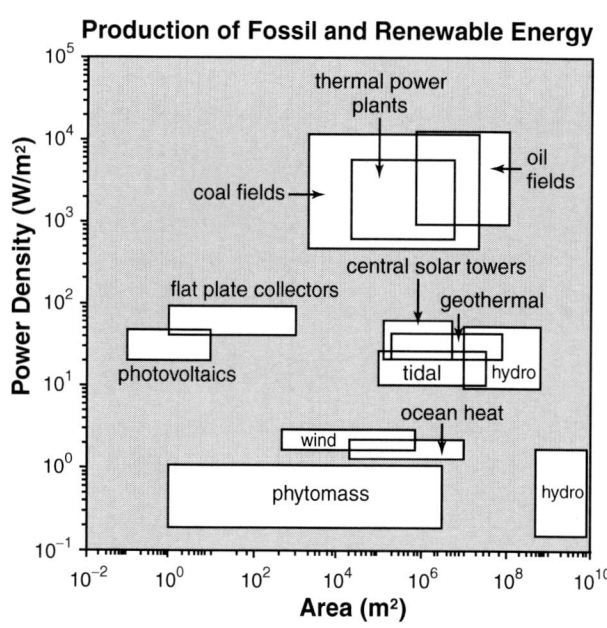

FIGURE 26.6:

Production of fossil and renewable energy. The footprints and power densities of typical power plants is plotted on axes with logarithmic scales.
From Smil V. (2003) *Energy at the Crossroads*. Cambridge, MA: The MIT Press. Page 242, Fig. 5.2

F. Hydroelectric Power Is Losing Its Green Appeal

Hydroelectric plants currently provide nearly 20 percent of the world's electricity, and their water reservoirs often serve additional purposes, such as irrigation, recreation, and protection against floods. For many developing countries, hydroelectric plants are the major source of energy. However, these plants also have disadvantages. The associated water reservoirs cover large areas, reducing the power density of hydroelectric plants (see Figure 26.6). Depending on climate, the water reservoirs also become breeding grounds for vectors of malaria and schistomoniasis.

Moreover, the green appeal of hydropower has been undermined by recent measurements of the amounts of CO_2 and CH_4 produced by the decay of organic matter in water. The resulting greenhouse effect seems to be in the same order of magnitude as that of a fossil fuel plant producing the same energy.

Finally, deterioration of dams makes it necessary to decommission them after 50 to 100 years, an expensive task that is rarely considered during the design phase. These drawbacks have persuaded many governments in developed countries to scale back or even abolish the building of new hydroelectric plants.

G. Electric Cars Need Better Batteries

About two-thirds of the 20 million barrels of oil consumed in the United States each day are used in motor vehicles. The amount of carbon released into the atmosphere is enormous. A single average family sedan driving at 65 mph with a fuel efficiency of 26 miles per gallon is releasing the equivalent of three charcoal briquettes per minute! Carbon dioxide from the exhausts of automobiles and airplanes cannot be captured and stored because the equipment to do this would be too bulky. *This is a major unresolved problem in urgent need of attention.*

One possible solution is the use of **electric cars** with batteries that are recharged with electricity from stationary power plants that allow sequestration of CO_2 (Figure 26.7). Currently, the use of electric cars is limited by the heavy weight and small energy storage capacity of available batteries. The recharging of batteries also takes significant amounts of time. Thus, the development of better batteries seems to be the key to a wider usage of electric cars. And it needs to be kept in mind that the savings of greenhouse gases by electric cars depend on the way the electricity is generated in power plants.

H. Hydrogen Presents Engineering Challenges

Instead of electricity, cars could also use hydrogen as an intermediate energy carrier (see Figure 26.7). Indeed, internal combustion engines that now use gasoline can be modified fairly easily for using hydrogen as fuel. The tailpipe emission of CO_2 would be zero. There are also **hydrogen fuel cells**, which convert hydrogen and oxygen into water, and in the process produce electricity, which can be used to drive a car.

Hydrogen as a pure gas does not occur naturally but must be manufactured at considerable cost in energy and money. For example, a hydroelectric power plant converts the gravitational energy of water via a turbine into electric energy, which can be used for the electrolysis of water into oxygen and hydrogen (Figure 26.8).

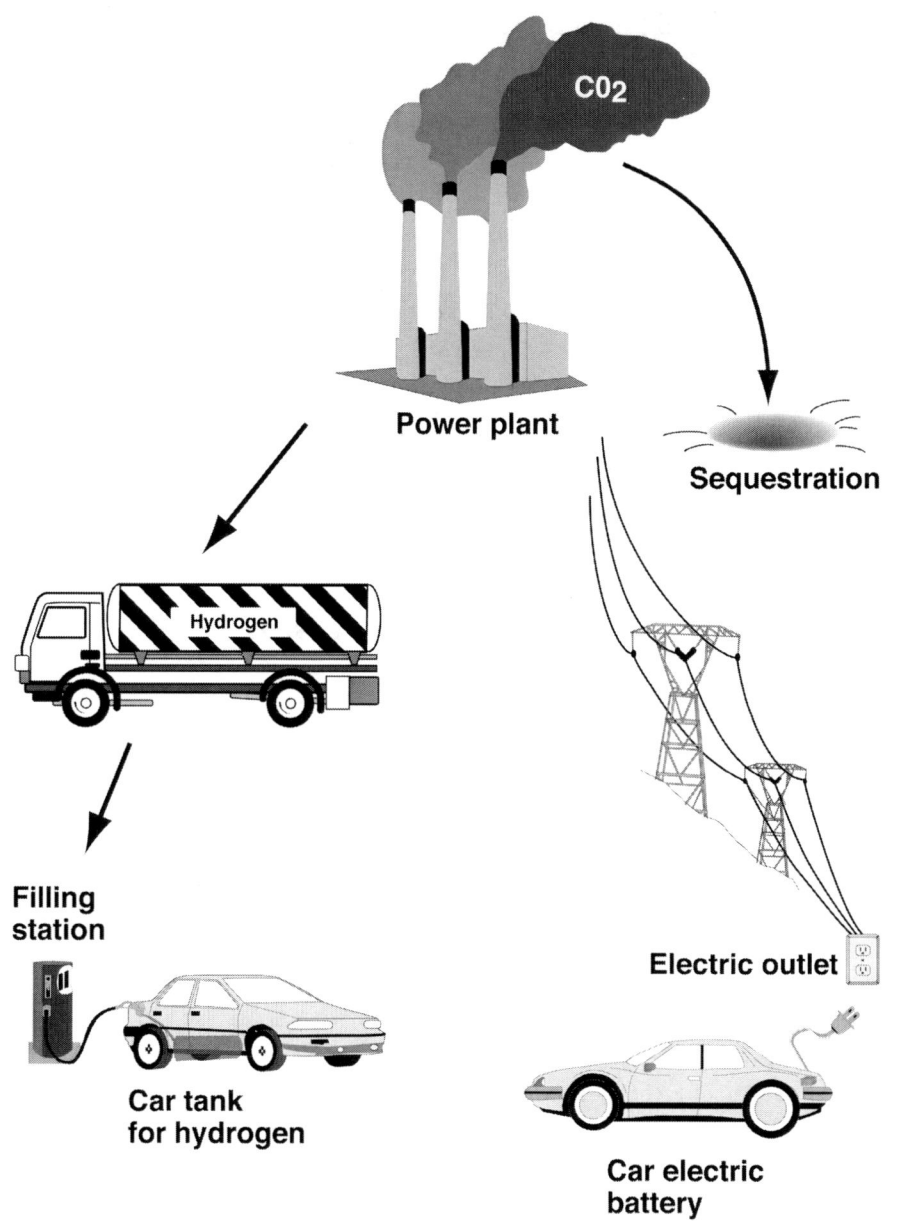

FIGURE 26.7:
Hydrogen or electricity can be used as intermediate energy carriers to power automobiles. Using these intermediates reduces carbonization of the atmosphere only if carbon dioxide is sequestered at the power plants that produce the intermediates.

Hydrogen can be compressed, stored, and transported in tanks. Fuel cells revert the electrolysis by generating electricity from hydrogen and oxygen. The electricity can then be converted into mechanical energy by an electric motor, which reverses the action of the turbine.

Unfortunately, generating hydrogen by electrolysis uses up large amounts of energy. Alternative ways of producing hydrogen include steam reforming of methane with water vapor to produce hydrogen and carbon dioxide, which can be (and should be!) sequestered in stationary power plants. Another option is the thermochemical production of hydrogen from water, which does not produce greenhouse gases directly but needs

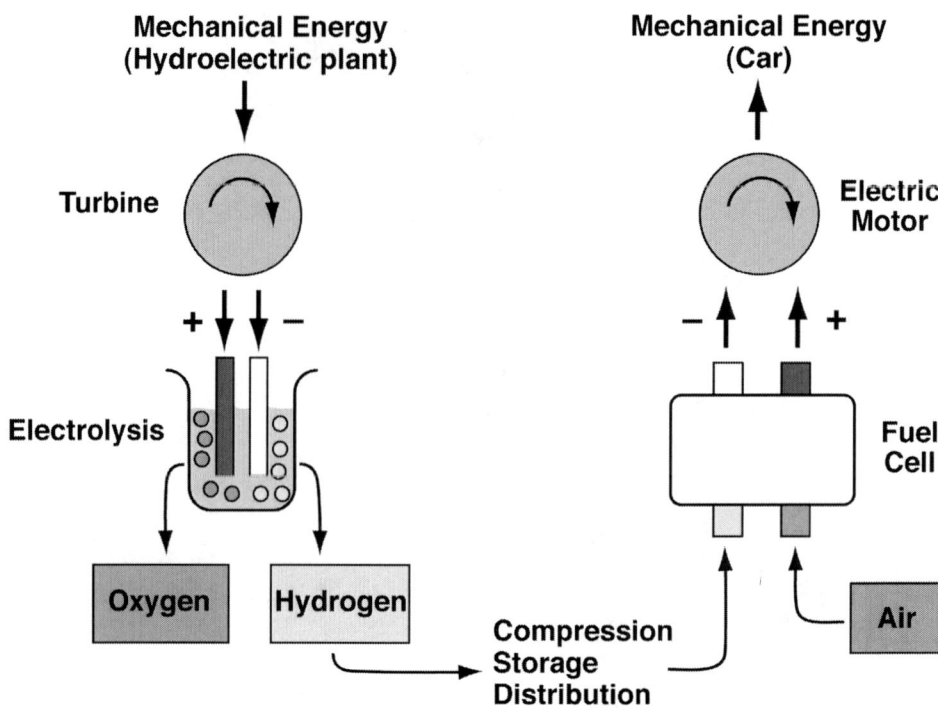

FIGURE 26.8:

Hydrogen as an energy carrier. Hydrogen generated by electrolysis can be compressed, stored and distributed as a fuel. Fuel cells using hydrogen (and oxygen from air) reverse the process of electrolysis.

high temperature (near 1000°C). The latter may be generated by nuclear power plants. However, it would take about 400 power plants of 1 GW each to satisfy the current demand of automobile fuel in the United States by hydrogen. As we will discuss below, nuclear power plants are expensive and come with great risks in safety and security.

The storage and transportation of hydrogen also presents major problems. Hydrogen being the lightest gas, it takes extreme cooling or very high pressure to make it fluid. Fuel tanks for hydrogen-powered vehicles will therefore have to be much bulkier than gasoline tanks today (Figure S26.d). The engineering challenges posed by these requirements are compounded by the fact that hydrogen is odorless and very explosive. Establishing a network of filling and supply stations for hydrogen will therefore need major investments and time for adaptation.

I. Nuclear Power Comes with Safety and Security Risks

Nuclear power plants currently generate 16 percent of commercial electricity worldwide, with major differences among nations (France 76 percent, United States 20 percent, Russia 15 percent, China 1 percent). Nuclear power can provide abundant energy without carbonization of the atmosphere. Due to their high-power density, nuclear plants demand only minimal land. Unfortunately, nuclear technology raises major concerns.

- Many people perceive nuclear power plants as inherently unsafe, even though accidents like those at Three Mile Island and Chernobyl could probably be avoided by better designs in the future.

- Spent nuclear fuel is highly radioactive and needs to be stored safely for very long time periods. The amount of energy recovered from uranium 235 can be doubled by recycling. However, current recycling technology also generates plutonium, which may be used for nuclear weapons. Spent nuclear fuel in any form can be used to make "dirty bombs."
- The events of 9/11/2001 have made it clear that nuclear plants are prime targets for terrorist attacks. While the reactors can probably be secured, protecting the entire delivery and storage system seems very difficult.
- The conversion from nuclear to electric energy entails major losses of heat, which will drive up global warming.

Because of these concerns, nuclear power currently depends on large subsidies, and construction of new plants has stagnated over the past 20 years. Global warming, rising energy consumption, and political dynamics have prompted at least nine countries, including the United States, to join in the research and development of "Generation IV" nuclear energy systems, which may be safe, but will not be as secure as they should be. Living with nuclear power in times of terrorism means living with great risk.

J. Wind Power Is Soaring

Wind energy is plentiful, renewable, and does not produce greenhouse gases or other pollutants. Windmills have a long history as local providers of mechanical energy. Modern **wind turbines** have rotors with diameters greater than 100 m and generate commercial electricity at 1 MW or more per machine. They are often clustered in "wind farms" connected to the electric power grid (Figure 26.9). Wind power densities

FIGURE 26.9:
Wind farm. Wind turbines are a clean, renewable source of electricity at a competitive price.
Image © Terrance Emerson, 2010, Shutterstock, Inc.

are favorable because most of an area occupied by wind facilities can still be used for other purposes. Large wind turbines can also be installed offshore, which reduces their visual impact and makes them aesthetically more acceptable.

By the end of 2009, wind-powered generators produced about 2 percent of all electric energy used worldwide. In several countries, where governments have subsidized the development of wind energy, its share of commercial energy is much higher: 19 percent in Denmark, 13 percent in Spain and Portugal, and 7 percent in Germany and Ireland. In windy areas of the United States, including Texas, unsubsidized wind-generated electricity is already competitive with fossil fuel–generated electricity. In some countries, utility companies have begun to buy back surplus electricity generated by small domestic turbines.

In contrast to the power generation techniques discussed so far, wind power is only sometimes available. This **intermittency** is seldom a problem when wind power supplies only a small portion of the total demand. As the proportion of wind power rises, so does the need for electric transmission lines over long distances and other load management techniques.

J. Solar Power May Shine

Solar power is defined here as the generation of electricity from sunlight. This can be done with **photovoltaic (PV) materials**, in which light energy is used to mobilize electrons. Solar energy can also be harnessed by **sunlight concentration** with sun-tracking lenses or mirrors to boil water and drive a turbine with the water vapor. Either way, solar power is plentiful, renewable, and environmentally sound. In terms of power density, PV energy is about 100x better than phytoenergy and hydroelectric energy (see Figure 26.6).

As in the case of wind power, the development of solar energy initially depended on governmental subsidies but is now attracting enough venture capitalists to support an annual growth rate of 25 percent. The current trend seems to be towards large PV power stations with capacities of over 100 MW, close to those of coal-fired plants. On a smaller scale, and again similar to wind power, some utility companies have begun to buy back surplus electricity generated by domestic solar panels.

Further extending the similarity to wind power, sunlight is available intermittently, although in a more predictable pattern. The challenges of intermittency may be avoided by orbital solar power collection, which requires satellite launching and beaming the collected power to receiving antennas on Earth. This technology is in its earliest beginnings, and its eventual usefulness remains to be seen. The important point now is that generating energy in environmentally acceptable ways will be promoted by technological innovation, and that it will be wise to use the current window of cheap fossil energy wisely for the development of technologies that can sustain future generations.

K. Energy Conservation Is the Key

In regard to the availability of cheap, safe, and convenient energy, the twentieth century has been a blessed one. We have been able to tap into a rich supply of fossil fuels that have brought unprecedented improvements in quality of life to people in developed nations, and a chance for similar improvements in developing nations.

Fossil fuels are convenient and safe, but burning them generates CO_2 and other pollutants. This connection has become a big problem due to the dramatic increase of the worldwide human population, which is still going on (see Chapter 22). The problem is compounded by a strong aversion to disadvantageous inequity, which humans share with other primates and which seems to have genetic roots (see Chapter 21). This keen sense of fairness will have people in developing countries aspire to the same standard of living now enjoyed in developed countries. If the required energy is produced by burning fossil fuels without CO_2 sequestration, global warming will become disastrous.

Sequestration of CO_2 seems feasible and affordable in stationary power plants, but developing the technologies and international cooperation to fairly share the associated costs will take time. Because CO_2 sequestration does not seem practical in automobiles and airplanes, technologies for driving these vehicles by intermediate energy carriers, such as electricity or hydrogen, need to be developed.

In the meantime, because the lifetime of CO_2 in the atmosphere is long, it is critical to curb CO_2 emissions. This can be done effectively by reducing the level of energy consumption in the most developed countries, which can be cut in half without much loss in quality of life. The savings will apply worldwide because all other countries will then aspire to responsible, rather than wasteful, levels of energy consumption.

Your reading assigned to this topic describes ways to hold carbon emissions in check using currently available technologies.

EXERCISE PAGE FOR CHAPTER 26

Student Last Name _____

Student First Name _____

Discussion Section _____

On the course web site, use the link on the **syllabus** to find a pdf file of your **assigned reading** for this chapter. The same reading is referenced below. Use the bulleted list of questions to test your knowledge of this reading.

Socolow, R. H., and Pacala, S. W. (2006). A plan to keep carbon in check. *Scient. Amer.* **Sep. 2006: 50–57.**

- Where do the authors draw the line between a "truly dangerous" and a "merely unwise" *carbon amount* (in billion tons) in the atmosphere? How did they arrive at this limit, and what CO_2 *concentration* (in ppm) does it translate into?
- In order to stay under the critical limit of atmospheric CO_2 *concentration*, what needs to happen with yearly *carbon emission* (in billion tons per year), over the next 50 years and beyond? [*Scient.Amer.* **Sep. 2006: 52 b, a**]
- Explain the concept of the "stabilization triangle" and its "wedges."
- On page 54, the authors show 15 different measures grouped in five (color-coded) categories. Each measure is to realize one wedge of emission reduction using currently available technologies. Name the five categories and one measure in each category.
- Which economic tool do the authors propose to make emission reductions happen?
- What are the authors' suggested contributions of developed and developing countries to the goal of worldwide emission reduction?

Feedback

On a scale from 1 to 10, the being the best, please rank the above reading for

1. Interest and Relevance to the topic (10 being most interesting/relevant) _____
2. Readability (10 being most clearly written, easy to understand) _____

In the remaining space, enter any comments that you may have on this reading.

- There's been an increase in extinctions ever since modern H. sapiens.
  - increments have been seen in Americas & Australia

---

* Δ: What can we do?

① Promote birth control
② Capture & store CO_2// * Unless H is prod
③ Use electricity & H was intermediate energy carriers for cars // w/ water/...
                                                                   // were not doing any
④ SOLAR & WIND ENERGY
⑤ TREAD VERY CAREFULLY W/ NUCLEAR ENERGY
⑥ CONSERVE ENERGY (SOFAR ARTICLE W/ 7-WEDGES)
⑦ AVOID OVER~~~ slide! CONSUMPTION AS STATUS SYMBOL
⑧ JOIN AN ENVIRONMENTAL ORGANIZATION (EDUCATE YOURSELF)

SLIDE: "Can we avoid the worst in global warming?"
  - "I have showed this to 3 times"
  - "This is a glaring point"

CHAPTER TWENTY-SEVEN

SPECIES EXTINCTION AND SYNOPSIS

The disappearance of plant and animal species from Earth has been part of the evolutionary process all along. However, pollution, overharvesting, habitat destruction, and other human activities have accelerated species extinction to a disastrous pace. This is a major ethical and economical problem, which is ultimately caused by human population growth and overconsumption in the developed countries. A recent summary of the subject has been given by Wilson, E. O. (2002). *The Future of Life*. New York: Alfred A. Knopf.

A. Described and Estimated Number of Species

Only a fraction of the *extant* (currently living) species are known. Some 1.8 million have been described by taxonomists and given a scientific name. (No one has counted the exact total yet from the literature.) Expert opinions on the actual number of species range widely, but most estimates are around 10 million. Intensified zoological explorations, driven by a sense of urgency over vanishing habitats, have revealed surprising numbers of *new* species including vertebrates.

- The amphibian species count grew from 4,003 to 5,743 between 1985 and 2004.
- The mammalian species count grew from about 4,000 to about 5,000 over the last two decades. The new species

include four large mammals (hare- to cow-sized) found in the Annamite Mountains between Vietnam and Laos.
- Even among the primates, the most intensely sought animals in the wild, 9 new species were recently added to the 275 known in 1990.

B. Current and Previous Rates of Species Extinction

Species formation and species extinction have always been part of evolution. Only recently has the rate of extinction increased by a factor of about 1,000 over the pre-human levels in taxonomically diverse groups.

The rate of species extinction is best measured for organisms that are large and/or have hard shells or skeletons, which are easy to observe and leave a clear fossil record.

The **edenic rate** (predating modern *Homo sapiens*) of extinction, based on the fossil record, is estimated at an *average* of one species per million species each year. Extinction rates have varied between different parts of the world and between different groups of organisms, for example, from 2 per million species per year in mammals to 0.2 in echinoderms.

Periods of relative tranquility have been punctuated by occasional periods of mass extinction. Such natural catastrophes were typically associated with major changes in climate. They were followed by episodes of rapid evolution, when new species arose to fill the niches vacated by the extinct species. The edenic birth rate of new species formation was somewhat higher than the extinction rate, as species diversity doubled over the past 450 million years.

The appearance of modern *Homo sapiens* was followed regularly by the disappearance of large animals, which were most likely hunted to extinction.

- Pre-human Australia was home to many exceptionally large animals including giant lizards (up to 23 feet in length) and large flightless birds (up to 220 pounds). This megafauna came to an abrupt end when the first humans arrived about 50 kya.
- Similarly, most of America's large mammals became extinct as the continent was overrun by modern *Homo sapiens*.
- Large European mammals including mastodons and saber-toothed tigers vanished when agricultural *Homo sapiens* spread from the fertile crescent to England.
- Pre-human Madagaskar was an untouched zoological garden with tortoises measuring 4 feet across, a mongoose the size of a lynx, an "aardvark" so unusual in its anatomy that zoologists placed it into an order of its own, and giant eagle-like birds that could seize and carry away large mammals. About 400 years after the oldest known human settlements on the island, all animals larger than 20 pounds had disappeared.
- In each single case it may be argued that species extinctions could have resulted from nonhuman causes, such as unknown viruses. But the regular sequence of human arrival and species extinction in different times and places strongly suggests that human hunting was the major factor.

The current rate of species extinction is estimated most directly by tracking species through the **Red List** of the International Union for the Conservation of Nature

(IUCN). Many species pass from "least concern" to "near threatened" to "vulnerable" to "endangered" to "critically endangered" to "extinct in the wild" to "extinct." Very few species reverse to a safer category.

The 2008 Red List of the IUCN encompasses 44,838 species, up from 41,415 in 2007. Of these 16,928 are threatened with extinction, up from 16,306 in 2007. By conservative estimates, 21 percent of all mammals are threatened with extinction. Based on Red List tracking and two independent, although indirect, methods the current rate of species extinction is estimated to be near 1,000 species per year, three orders of magnitude above the edenic rate. This rate is likely to accelerate. Even under the optimistic scenario in the IPCC (2007) report, 20 to 30 percent of all known species will be committed to extinction by the year 2100 (see Figure 24.5).

C. Causes of Species Extinction

Most paleontologists and ecologists agree that the current acceleration in the rate of species extinction is not based on natural causes, as were earlier waves of extinction. Rather, the current increase in species extinction seems to be human-made.

An endangered species is typically *not* like a dying human patient whose care is futile to prolong. To the contrary, the majority of declining species are composed of healthy individuals, who will thrive if their normal living conditions are restored. Many animals, including the mountain gorilla (Figure 27.1), and the Sumatran

FIGURE 27.1:
Not sick. Critically endangered species typically consist of healthy individuals, such as this gorilla female with her baby.
Image © Ronald van der Beek, 2010, Shutterstock, Inc.

orangutan, as well as hundreds of other species, are nearly extinct but could probably be brought back with appropriate effort.

- The Californian condor, *Gymnogyps californianus*, had been reduced to a few dozen individuals by habitat destruction, hunting, and food poisoning. When offered protection and uncontaminated food in a breeding colony near San Diego, the birds flourished, and some individuals could be re-released.
- The Mauritian kestrel, a small hawk, was down to a single breeding pair in 1974. Owing to the heroic efforts of a bird breeder, Carl Jones, the population is back now to about 200 breeding pairs, with no apparent genetic defects.
- The tallest bird in North America, the whooping crane, was down to two small breeding populations in 1937 and is now back to more than 700 individuals due to habitat protection in the Aransas National Wildlife Refuge and elsewhere (See Figure 27.2. Boat tours provide excellent views of the whooping cranes during spring time and leave from Fulton, Rockport, and Port Aransas.)

Ecologists and conservation biologists have identified four major causes of species extinction. All of them are related to human activities and therefore tend to become worse as the human population grows.

- **Habitat destruction**, such as the clearing of forest to plant crops
- **Overharvesting** to the point where a natural population cannot recover

FIGURE 27.2:

Whooping Crane (*Grus americana*). Standing nearly 5 feet, these are the tallest birds of America. Adults are white with a red and black crown. Chicks are light brown. Image © Al Mueller, 2010, Shutterstock, Inc.

- **Environmental degradation** by polluting air, water, and soil with human waste, industrial products, greenhouse gases, and so on; also by animal waste and fertilizer runoff from intensive agriculture
- **Invasive species**, that is, species introduced by humans that displace or prey excessively on native species

Typically, these causes of species extinction act in combination, or they are compounded by a natural event, such as the spread of a parasite. For example, chimpanzees and gorillas in West Africa are currently decimated by human poaching and by the Ebola virus. The final phase of species extinction is often inbreeding. Under these conditions, a deleterious allele in one animal may no longer be covered by the wild-type allele of a mate.

D. Examples of Species Extinction

The Hawaiian Islands illustrate all major causes of species extinction. Once they were home to more than 10,000 endemic (genetically and morphologically unique) plant and animal species. They had had made an unassisted landfall many mya and then, due to Hawaii's isolated location, evolved in distinct ways.

There have been about 135 endemic Hawaiian bird species including eagles, long-legged owls, painted honeycreepers, and flightless birds. Most of these are now extinct. Of the 35 original bird species that still exist, 24 are endangered.

Species extinction on Hawaii began when the first human settlers, Polynesian seafarers, arrived around AD 400. They found the flightless birds easy prey, and evidently hunted them to extinction. Next the Polynesians cleared forests and grasslands to plant bananas, breadfruit, and other plants they had brought with them on their boats, along with pigs and other domesticated animals. European and American settlers later cleared almost all of the primary vegetation in order to make room for cash crops such as pineapple and sugarcane. These activities destroyed almost all of the native vegetation, leaving only the steepest mountain slopes untouched.

When Hawaii became the commercial and military transportation hub of the Pacific in the twentieth century, humans and their ships carried in alien plants, animals, and microbes from all over the planet. Many invaders, having evolved in more diverse communities, would displace the native species. While most of the new arrivals were innocuous, some became real pests. In their original habitats, they had been hemmed in by predators and competing species. When freed from these constraints, they crowded out native species that had not been selected to resist. Notorious invasive species include the African big-headed ant (*Pheidole megacephala*), which destroyed most other insects in its path. Easier visible to the human eye, feral strains of domestic pig, cat, mongoose, and rat have wreaked havoc on native plants and birds.

Of the 1,935 free-living flowering plant species on Hawaii, 902 are alien (human-introduced). The leis placed around the necks of arriving visitors are made mostly from flowers of the *Plumeria* tree, an alien species.

The Hawaiian situation is not unique. Notorious examples of alien species that were introduced to the United States mainland by human activity and have caused enormous damage include the following.

- The Asian chestnut fungus (*Cryphonectria parasitica*) was accidentally introduced to North America around 1900 by chestnut lumber or trees. The

fungus has virtually wiped out the American chestnut, which used to be the dominant tree in many parts of Eastern America.
- The red imported fire ant (*Solenopsis invicta*) was introduced by ship from South America to Mobile, Alabama, and has spread throughout the Southern United States.
- The zebra mussel (*Dreissena polymorpha*), introduced from the Black Sea or Caspian Sea to the Great Lakes in the 1980s, forms contiguous beds on the floors of lakes and streams, crowding out native mollusks.

One of the massive losses of species in recent decades has been the die-off of amphibians. Almost a third of the world's amphibians are threatened with extinction.

- Australia's unique northern gastric breeding frog (*Rheobatrachus vitellinus*) vanished shortly after it was discovered in a National Park in Queensland.
- The golden toad of Costa Rica, *Bufo periglenes,* a zoological celebrity because of its spectacular color, plummeted from huge numbers to zero within two years.
- The leopard frog, *Rana pipiens*, has disappeared from 60 percent of its range in Canada, including a complete disappearance from British Columbia.

Some herpetologists think that environmental degradation is the primary cause for the amphibian decline. Amphibians use their moist skin for gas exchange, thus making it into an adsorbent pad for poisons and parasites. An intensive study (see the assigned reading for Chapter 25) has shown that atrazine, the most commonly used herbicide in the United States, demasculinizes frogs at concentrations found naturally in agricultural areas. Other herpetologists ascribe amphibian decline to a parasitic fungus, *Batrachochytrium dendrobatidis*, which may have been carried around the world by commercial trade in *Xenopus laevis*, a species used in many laboratory studies.

The most devastating form of habitat destruction is the clearing of forests. The maximum extent of Earth's forest cover was reached some 6 to 8 kya, after the last Ice Age and before the spread of agriculture. During the twentieth century, about half of the original forest had been replaced with farm and timber land, and the remaining old forest was being cut at an accelerating pace. This loss of forest is one of the most profound and rapid environmental changes in the history of our planet.

Forest clearing involves biomass burning, which releases large amounts of CO_2 and soot. The crops and timber replacing the forest will sequester less CO_2 in the future. As the size of forests is shrinking, so is their ability to sustain a diverse community of species. It is therefore critical to prevent the fragmentation of forests (and other wildlife habitats) into smaller and smaller pieces.

The most valuable forests in terms of biodiversity are the **rainforests** (Figure 27.3), defined by annual rainfall of more than 1900 mm (76 inches). Although they cover only 6 percent of the land surface, they support more than half of all known species. Tropical rainforests are destroyed rapidly by logging companies, which sell hardwoods like teak, mahogany, rosewood, and other timber. What they leave behind is burned by squatters who use the land for grazing cattle or planting cash crops.

The worldwide rate of rainforest clear-cutting has been close to 1 percent per year since the 1980s, according to estimates of the Food and Agricultural Organization

(FAO) of the United Nations. Ironically, the clear-cutting rarely attains its ostensible goal of generating farmland. With most of the biomass stored above ground, the soil of rainforests is generally poor and erodes fast, prompting disappointed would-be farmers to move on after a few years and repeat the cycle of destruction.

Land ownership seems to be poorly regulated in much of the Amazon, a situation that invites use for short-term profit rather than long-term cultivation. The Brazilian government, which oversees two-thirds of the Amazon, has established the goal of protecting 10 percent of its rainforest. However, this would still mean losing more than half of its current biodiversity, including in particular large animals, which do not survive in reduced habitats.

The clear-cutting of large forests also reduces atmospheric water and changes rain patterns. About half of the rain falling in the Amazon basin arises from evapotranspiration in the forest itself, as opposed to clouds blown over from the Atlantic Ocean. Many ecologists fear that the combined effects of clear-cutting and global warming will lead to a "savannaization" of the Amazon, and measures to avoid this fate have been proposed.

Coastal waters are also sites of highest biological diversity and productivity, supporting marine life from plankton to fish, turtles, and whales. These waters are especially hard hit by pollution from river discharges, excessive human waste from coastal urban areas, and offshore mining and drilling.

FIGURE 27.3:
Tropical rainforest in South Costa Rica
Photograph by Alanna Aspinall. File down-loaded with permission from http://www.tiskitalodge.co.cr/rainforest/images/ra inforest2.jpg

E. Worth of Species Diversity to Humans

Species extinction is an **ethical problem**. There are two common ways of discouraging unethical behavior. First, one can appeal to societal values or to a sense of fairness. For instance, speeding is unethical because it endangers innocent people. Second, one can point out penalties or unwelcome consequences, such as expensive citations by police.

Being outdoors and observing wildlife refreshes our bodies and minds. Extinct species are gone forever; we cannot replace them. The extinction of a charismatic species that elicits our admiration or compassion makes us feel guilty and sad: We have failed in our unwanted but inevitable role as caretakers of the planet. Future generations will have time and documentation to reflect on each species lost, and they will not thank us for the mismanaged and impoverished world we are leaving to them.

Species extinction is also bad business, because human life depends on ecosystems that churn our soil, purify our water, and provide wild fruit, game, fish, and so on. In the context of environmental degradation (Chapter 24), we have discussed two examples of ecosystem services to humans: fisheries and the Catskill mountains of New York. The robustness of animal and plant communities increases with their biodiversity. Many studies have demonstrated a positive correlation between the species diversity of ecosystems and their stability, bioproductivity and resistance to pests and climate extremes. Diminished ecosystems return diminished services.

There are many animal and plant species that can provide valuable resources to those who go out and study them, an activity now called **bioprospecting**. The basic idea is that the extant species are the winners of an evolutionary contest that has been going on for hundreds of millions of years. It is therefore reasonable to expect some highly effective adaptations in all kinds of organisms. Bioprospecting and other sustainable uses of rainforests may be the best way to save them from clear-cutting. It would also be a way of involving the indigenous people who have lived there for a long time.

For instance, bioprospecting may lead to the breeding of new crop plants. Most of the crop plants that we eat today have been bred from a few natural species that happened to be around where agricultures started. Other cultivable plant species may still be found from which more nutritious, tasty, or drought-resistant crop varieties can be derived (Figure S27.a).

Bioprospecting may also lead to the discovery of powerful medicines. Native tribes know how to obtain effective drugs from local plants and animals. For instance, natives of the South American rainforest rub the tips of their blowgun darts over the skin of certain frogs that secrete a deadly poison from their backs (Figure 27.4). A modified component of the toxin from one of these frog species, *Epipedobates tricolor*, turns out to be a powerful painkiller that promotes alertness instead of drowsiness and is said to be nonaddictive. Another valuable drug is an alkaloid from the Madagaskar

FIGURE 27.4:

Poison dart frog. This species, *Dendrobates tinctorius,* and related ones are brightly colored and move slowly. They can afford this behavior because their predators learn to avoid them. South American Indians turn their blow darts into deadly projectiles by rubbing their tips on the frogs' skin.
Image © Jeff Grabert, 2010, Shutterstock, Inc.

periwinkle, *Catharanthus roseus* (Figure 27.5). Known as vincristine, it has been used for chemotherapy of certain tumors.

Teams of renowned ecologists and economists have compared the economic benefits of destroying natural ecosystems by conventional logging, intensive farming, or destructive fishing, with the benefits of reduced-impact logging, small-scale farming and fishing, or leaving the ecosystems intact (Figure S27.b; see the assigned reading at the end of this chapter). Such studies show the need for interdisciplinary research to place political decisions about land use on a rational footing. Columbia, MIT, and other prestigious institutions already have such interdisciplinary graduate programs.

F. Biodiversity Hotspots

→ RICHLY - DIVERSE

Most species at risk of extinction live in biodiversity **hotspots**, that is, areas that cover only 1.4 percent of Earth's land surface but are the exclusive homes of 44 percent of all known plant species and more than a third of all land vertebrate species. Wilson (2002) names 25 hotspots, of which 11 are located in tropical rainforests while the others are covered with savanna and coastal sage brush. For example, more than 400 different trees have been found within one hectar (2.5 acres) of Brazil's Atlantic Forest. More than 1,300 butterfly species have been recorded in one corner of Peru's Manu National Park.

FIGURE 27.5:
Medicinal plants of the rain forest, such as this *Catharanthus roseus*, have been used widely in folk medicine. Pharmaceutical companies have isolated certain components for modern therapies.
Image © B. Holmes, 2010, Shutterstock, Inc.

By protecting this tiny fraction of our planet's land area, millions of species can be saved for our descendants. Hotspot preservation will not solve the problem of species extinction elsewhere, but it creates a reserve from which other areas can be repopulated if and when they recover. Many current conservation efforts are therefore focusing on biodiversity hot spots.

Institutions like Conservation International and the Nature Conservancy are buying up pieces of wilderness where it shows the greatest species diversity. Some enlightened land-rich individuals around the world are joining the effort by voluntarily agreeing to not develop their land. However, these alliances will have to deal with communal and state authorities when it comes to linking these enclaves into larger regions and to finding forms of managing them.

G. Species Extinction Is a Sign of Deeply Ingrained Problems

Beyond the urgent task of conserving biodiversity hotspots, it is important to realize that the sad reality of species extinction has deep roots in human nature and nurture. Human evolution has been precarious, and we have needed every bit of toughness and ingenuity to survive as a species. The same qualities have started the growth of the human population, which is now driving us to the brink of disaster. The poachers who kill gorillas and bonobos for "bush meat", and the squatters who occupy the clear-cuts of former rainforests, are people who need to eat. The problem is exacerbated by deeply engrained patterns of behavior, including a strong aversion to disadvantage, which is easily turned into consumerism ("keeping up with the Joneses"). Add to this the use of wealth as a mate choice criterion, and you understand the lure of overconsumption.

However, it is becoming clear that population growth and overconsumption have set us on a collision course, threatening a large fraction of extant species, and eventually ourselves, with extinction. Having too many people compete over limited resources leads to pollution, poverty, intensified agriculture, and the conversion of more wilderness into cropland (Figure 27.6). Together, these processes cause overharvesting, environmental degradation, habitat destruction, and the introduction of alien species. The combined effect of these human activities is an alarming acceleration of species extinction.

The long-term solution to these problems needs to come from reversing human population growth, curbing overconsumption, and protecting natural ecosystems. E. O. Wilson (2002) describes the current situation as a "bottleneck." In this view, our challenge is to pull most humans through the straining environment without loosing too many humans or too much of our current biodiversity. My purpose in writing this book was helping others to understand, and to appreciate, the magnitude of this challenge.

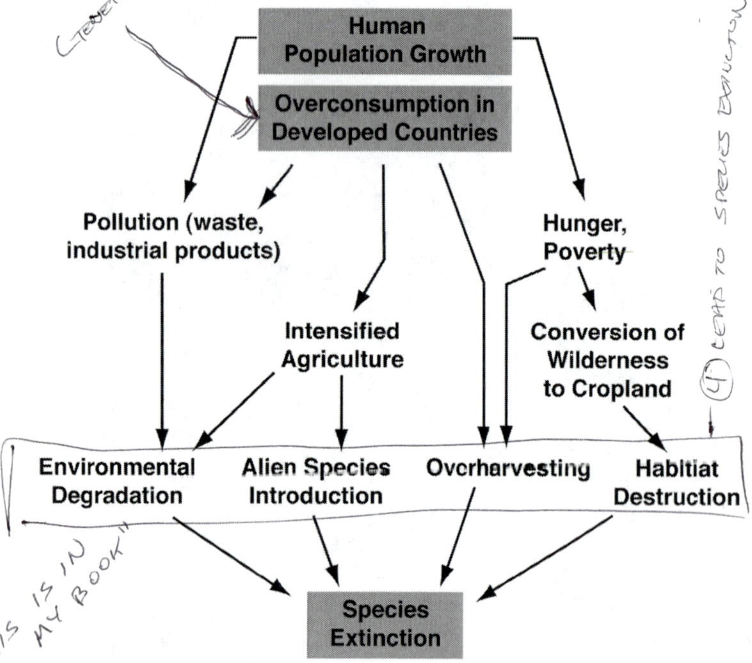

FIGURE 27.6:

Anthropogenic Causes of Species Extinction can be traced back to human population growth and overconsumption in developed countries.